An Introduction to Hydrogen Bonding

TOPICS IN PHYSICAL CHEMISTRY
A Series of Advanced Textbooks and Monographs

Series Editor, Donald G. Truhlar

F. Iachello and R.D. Levine, *Algebraic Theory of Molecules*
P. Bernath, *Spectra of Atoms and Molecules*
J. Cioslowski, *Electronic Structure Calculations on Fullerenes and Their Derivatives*
E.R. Bernstein, *Chemical Reactions in Clusters*
J. Simons and J. Nichols, *Quantum Mechanics in Chemistry*
G.A. Jeffrey, *An Introduction to Hydrogen Bonding*
S. Scheiner, *Hydrogen Bonding: A Quantum Chemical Perspective*

An Introduction to Hydrogen Bonding

George A. Jeffrey
University of Pittsburgh

New York   Oxford
Oxford University Press
1997

Oxford University Press

Oxford   New York
Athens   Auckland   Bangkok   Bogota   Bombay   Buenos Aires
Calcutta   Cape Town   Dar es Salaam   Delhi   Florence   Hong Kong
Istanbul   Karachi   Kuala Lumpur   Madras   Madrid   Melbourne
Mexico City   Nairobi   Paris   Singapore   Taipei   Tokyo   Toronto

and associated companies in

Berlin   Ibadan

Library of Congress Cataloging-in-Publication Data

Jeffrey, George A., 1915–
      An introduction to hydrogen bonding / George A. Jeffrey.
         p.   cm.
      Includes bibliographical references (p. 261–298) and index.
      ISBN 0-19-509548-0 (cloth). — ISBN 0-19-509549-9 (pbk.)
      1. Hydrogen bonding.   I. Title.
      QD461.J44   1997   96–26792
      541.2926—dc20   CIP

9 8 7 6 5 4 3 2 1

Contents

Preface

This introduction to hydrogen bonding is directed primarily to undergraduates who may be required to pay more attention to hydrogen bonding in the future than was necessary in the past. This is especially true for those students who become interested in supermolecular chemistry and molecular recognition in molecular biology. Graduate and post-doctoral students, whose knowledge of hydrogen bonding is based on the brief statements on the subject in standard physical and organic chemistry texts, will find that I have provided an introduction to the rapidly expanding literature using all the methods enhanced by modern technology.

Readers will note a strong emphasis on structure and structural correlations. This is because the author is a crystallographer, with a strong belief in a remark attributed to Charles Coulson, the theoretician, "No one really understands the behaviour of a molecule until he knows its structure."

The author is grateful to an old friend and colleague, Dick McMullan, for reading the first version of the manuscript and providing helpful suggestions. Once again I express my gratitude to Joan Klinger, who periodically came out of retirement to exercise her extraordinary ability for converting untidy handwritten pages into a polished manuscript. As usual, I thank my wife, Maureen, for her patience and assistance during the writing process. I dedicate this book to the memory of the late Linus Pauling.

I wish to acknowledge the assistance of the staff of Oxford University Press in producing this book. I also wish to thank the authors who provided me with the originals of figures in their publications and the following publishers for permission to reproduce figures: American Chemical Society, American Institute of Physics, Annual Reviews Inc., Cambridge University Press, Elsevier Science, Kluwer Publishing Company, North Holland Publishing Company, Royal Society of Chemistry, Plenum Press, Springer-Verlag Publishing Company, and VCH Publishing Company.

George A. Jeffrey
University of Pittsburgh

An Introduction to Hydrogen Bonding

C H A P T E R
Brief History

1

1.1 | INTRODUCTION

Science is a pyramidal endeavor, each layer of discovery arising out of previous ideas and research. It is also a very forward-looking endeavor. Scientists are generally more interested in the latest letter to *Nature* or *Science* than they are in the history of the development of an idea, concept, or methodology. This is understandable since the advances in chemical and physical sciences are closely linked to those in technology. In science, as in everyday life, the dramatic development of computers in the last 40 years has been the most influential technical advance. The majority of practicing scientists have to be specialists and be up-to-date in at least one methodology, with a general understanding of the capability of many others. However, the need for this specialization is diminishing as computer software increasingly provides the necessary specialized knowledge. This is particularly true of X-ray crystal structure analysis and nuclear magnetic resonance (NMR) spectroscopy which have become the most powerful methods for determining molecular structure. In consequence, the methods of analysis become *routine* and consequently less exciting to crystallographers or spectroscopists. This was predicted by Lipson (1970) for X-ray crystallography, who referred to it as *"The Penalty of Success."* One benefit is that the specialist can become a generalist able to apply a wide variety of methods to obtain answers to particular questions.

In the evolving history, some quite sophisticated methods have disappeared, such as microanalysis, which was painfully taught to students of chemistry in the 1930s. The configurational analysis of natural products by organic chemistry was superseded by crystal structure analysis. In turn, the precise and elegant subject known as crystallometry, which predated the discovery of X-ray diffraction, is rarely practiced today. Synthesis without the sharp melting point and chemical analysis of a crystalline product was unacceptable until NMR spectroscopy was developed. Gas electron diffraction as a means of studying molecular structure has been replaced by microwave spectroscopy. Theoretical chemistry is particularly frustrating since the extremely rapid expansion of computing power makes most calculations obsolete before they can be published. Will man catch up with nature one day? That is a good philosophical question to which I think the answer is "No."

When the concept of the hydrogen bond was being developed in the 1930s, two of today's three most powerful methods of study were not available. The principles of structure analysis by X-ray diffraction and structure prediction by quantum mechanics were known, but until the digital computers came along there were few prospects for implementing methods based on them. Neutron fluxes for diffraction and the methods of nuclear magnetic spectroscopy were unknown.

So the adage so aptly applied to world affairs, "those who don't read history are doomed to repeat it," barely applies to science. Nevertheless, some experiments are repeated and some concepts are rediscovered. Acronyms are very popular these days and old ideas are sometimes reinvented and made to sound original with the aid of an acronym. There used to be a saying, "If you think you have a new idea, it is wise to be sure that Linus Pauling did not publish it twenty years ago."

Books and review articles that include references with titles of papers or comments on their contents are particularly valuable in this respect. Browsing through the annotated references in *The Hydrogen Bond* by Pimental and McClellan (1960), who can resist searching out a 1938 paper on the infrared evidence for hydrogen bonds in proteins, hydrogen bonds important in chemotherapy in 1950; hydrogen bonds and blood clotting in 1949; and what Pauling had to say about antibody formation in 1948?

C—H hydrogen bonds have been rediscovered and are currently in fashion, yet they were reviewed more than 50 years ago by Hunter (1947). Aromatic rings are being considered as hydrogen-bond acceptors, as they were by Bamford (1954).

The bibliography of over 2000 annotated references in the first book devoted entirely to hydrogen bonding suggests that the concept of the hydrogen bond should have appeared much earlier than it did. It was certainly very ripe for discovery by a number of investigators when it did come. It is not surprising therefore that there was no Nobel Prize awarded specifically for the discovery of the hydrogen bond, now known to be one of the most important concepts, both in supramolecular chemistry and molecular biology.

1.2 WHO DISCOVERED THE HYDROGEN BOND AND WHEN?

The hydrogen bond has such an ubiquitous influence in gaseous, liquid, and solid-state chemistry that its consequences were observed long before it was identified and given a name. Any survey of late 19th and early 20th century literature shows references to many observations which, in retrospect, could be perceived as evidence of hydrogen bonding. The terms *nebenvalenz* (near valence) and *innere komplexsalzbildung* were used by the German chemists Werner (1902), Hantzsch (1910), and Pfeiffer (1914) to describe both intra- and intermolecular hydrogen bonds. Germany could claim that these were the discoverers of the hydrogen bond. Moore and Winmill (1912) used the term *weak union* in describing the properties of amines in aqueous solutions. Many early papers reference this article and use the same system as an example of hydrogen bonding. The British could claim these authors as the inventors of the hydrogen bond. Intermolecular hydrogen bonding effects were generally described as *associations* and intramolecular hydrogen bonding as *chelations*.[1]

The gas hydrates, which were extensively studied in the 19th century starting with Michael Faraday's (1823) discovery of chlorine gas hydrate, are hydrate inclusion compounds which depend on the hydrogen-bonding properties of water. This was not realized until 125 years later through a series of crystal structure analyses.

According to Linus Pauling, the concept of the hydrogen bond is to be attributed to M. L. Huggins and independently to W. M. Latimer and W. H. Rodebush. However, Huggins (1971) claimed that he was first.[2] "The hydrogen bond was proposed by me in 1919 and shortly after by Latimer and Rodebush. . . ." The 1919 reference is to a thesis in an advanced inorganic chemistry course at the University of California.[3] Latimer and Rodebush (1920) published a paper which contained the statement that "*The hydrogen nucleus held by two octets constitutes a weak bond.*" In a paper entitled *The Electronic Structure of Atoms*, Huggins (1922) stated that "a positively charge kernel containing no electrons in its valence shell (i.e., H^+) reacting with an atom containing a lone valence pair can form a weak bond" (emphasis added). Interestingly, both papers quote the example of $H:O:H \text{---} \overset{\text{H}}{\underset{\text{H}}{\text{·· N:H}}} \rightarrow H:\overset{-}{O} \text{----} \overset{+}{\underset{\text{H}}{\text{H:N:H}}}$ of Moore and Winmill.

Crystal structure analyses were beginning to appear, some of which are now known to involve hydrogen bonding. There was no mention of hydrogen bonding in the crystal structure analyses of $NaHF_2$ (1926), NH_4F_2 (1930), urea (1928), acetamide (1940), oxalic acid, and some oxalates (1935). In the oxalic acid paper it was suggested that the hydrogen H^+ ion was midway between two oxygen atoms. None of these authors used the term hydrogen bond in their original publications. *The descriptor hydrogen bond appeared after 1930.* Pauling (1931) wrote a general paper on the nature of the chemical bond, which was a precursor to his famous book. There he discussed the $[H:F:H]^-$ ion, using the term *hydrogen bond*, possibly for the first time. He remarked that such bonds are formed to some extent by oxygen and in some cases by nitrogen atoms. Huggins (1931) discussed the role of hydrogen in the conduction of hydrogen and hydroxyl ions in water, and used the term hydrogen bond.[4]

Four definitive papers on hydrogen bonding were published in 1935–1936 from the U.K. and the United States. These papers were by Pauling (1935) on hydrogen bonds in water and ice and Bernal and Megaw (1935) on *hydroxyl bonds* in metallic

---

[1]A good account of the earlier chemistry which could be, in retrospect, associated with hydrogen bonding is given by Huggins (1936b).

[2]This paper was presented in 1969 to the Royal Swedish Academy of Sciences, possibly to lay claim to being the inventor of the hydrogen bond.

[3]Attempts to retrieve this thesis have been unsuccessful.

[4]Huggins must have had misgivings about the word *bond*, since he later substituted *hydrogen-bridge*, which became the wonderful German word *Wasserstoffenbrückenbildung*.

hydroxides, minerals, and water. The concept of a hydroxyl bond O—H---O—H was developed to distinguish it from a hydrogen bond O—H---O=C. Although there is a clear distinction, the nomenclature did not persist. A year later, two important papers were published by Huggins (1936a, 1936b). One was on *hydrogen-bridges* in ice and water, the other was on hydrogen-bridges in organic compounds. In the ice and water paper, Huggins proposed synchronous jumps of the hydrogens across the hydrogen bridges to account for the high dielectric constraints of ice and water. He also discussed the possibility of single and double minimum *low barrier* hydrogen bonds associated with the HO_3O^+ oxonium ion, as shown in Figure 1.1.

In the longer organic paper by Huggins (1936b), a wide variety of both inter- and intramolecular hydrogen bonds are described having O—H and N—H as donors and O and N as acceptor atoms. He discussed the hydrogen-bonding patterns in carboxylic acids and pointed out the role of hydrogen bonding in the Astbury and Street (1931) and Astbury and Woods (1933) models for the folding of keratin chains (see Figure 1.2). Finally, he predicted that the "hydrogen bridge theory will lead to a better un-

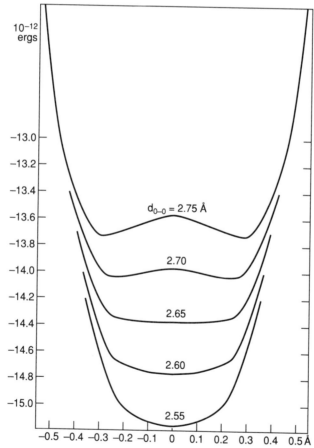

Figure 1.1.   An electrostatic calculation of the potential energy curve for O—H---O hydrogen bridges *vs* O---O distances from Huggins (1936a).

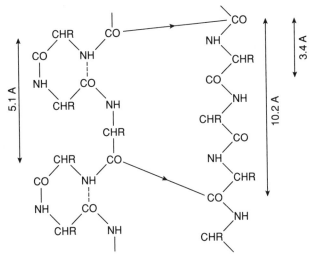

Figure 1.2. Astbury's (1933) models for the structures of keratin. Although the term hydrogen-bond was not used, the bridging concept was implied by the dotted lines.

derstanding of the nature and behavior of complicated organic structures, such as proteins, starch, cellulose, sugar and other carbohydrates, chlorophyll, haemoglobin and related substances." This was a remarkably prescient paper.

However, it was the chapter on hydrogen bonding in Pauling's (1939) *Nature of the Chemical Bond* that really introduced the concept of the hydrogen bond to the chemical world. Pauling expressed his view with two statements: "Under certain conditions an atom of hydrogen is attracted by rather strong forces to two atoms instead of only one, so that it may be considered to be acting as a bond between them. This is called a *hydrogen bond*" (p. 449, emphasis added). Pauling goes on further to say, "A hydrogen atom with only one stable orbital cannot form more than one pure covalent bond and the attraction of the two atoms observed in hydrogen bond formation must be due largely to ionic forces" (p. 449, 3rd ed.).

Before 1936, anomalous physical properties and thermodynamic measurements provided evidence of what is now known as hydrogen bonding. In 1936 it was realized that the relatively accessible method of infrared spectroscopy could provide a remarkably sensitive tool for identifying hydrogen bond formation through changes in the stretching frequency of the X—H bond. This was the starting point for infrared spectroscopy to become a primary method for studying hydrogen bonding in both liquid and solid phases. Even today, an average of 80 papers per year are published on hydrogen bonding using these methods.

There are a number of publications between 1922 and 1936 where a knowledge of the existence of the hydrogen bond would have been very relevant. The Astbury and Street (1931) and Astbury and Woods (1933) pioneer papers on the structure of hair, silk, and wool are examples. In their explanation of the changes in the X-ray diffraction patterns induced by the folding and unfolding of polypeptide chains, *bridge-atoms* are referred to and their figures, shown in Figure 1.2, clearly indicate hydrogen bonds, but that descriptor was never used.

In his paper on the structure of ice, Barnes (1929) did not mention hydrogen bonding but did suggest that the hydrogen atoms were midway between the oxygens. More surprising is that in the Bernal and Fowler (1933) classic paper on *The Theory of Water and Ionic Solutions*, the word hydrogen bond does not appear.

American chemists used the term *association* or *polymerization* for hydrogen-bonded complexes, while the British chemists, such as Sidgwick, preferred *chelation*, although in general, association is used for intermolecular bonding while chelation implies intramolecular hydrogen bonding. There was no mention of hydrogen bonding in several reviews on the effects of association on the infrared absorption bands of water published in 1931, or deviations from normal physical properties published in 1929. Even as late as 1937, in a Faraday Society discussion of "Structure and Molecular Forces in Pure Liquids and Solutions," only J. D. Hildebrand briefly mentioned hydrogen bonds. This was a period when Lewis in the United States and Sidgwick in the U.K. were having great success rationalizing constitution chemistry in terms of paired electrons and completed octets. That hydrogen should have a valence of two was probably heresy.

Huggins (1943) discussed the structure of fibrous proteins, as shown in Figure 1.3, and proposed the hydrogen-bonded helical and sheet structures for globular proteins shown in Figure 1.4. His helical structure is the 3.10 helix. If he had extended this model to one more peptide unit, he would have anticipated the a-helix by eight years. He also proposed models for hydrogen bonding in the aliphatic acids and an oxalate hydrate, shown in Figure 1.5.

Excellent reviews of knowledge concerning the physical and organic chemistry aspects of the hydrogen bond prior to 1947 are by Davies (1947) and Hunter (1947).

1.3 BOOKS ON HYDROGEN BONDING

The first international conference on hydrogen bonding was held in Ljubljana, Yugoslavia, and the proceedings were edited by Hadzi (1957). Among its contents was a discussion of hydrogen bonding in terms of electrostatic and exchange components by Coulson, and a clathrate hydrate model for liquid water by Pauling. The first text devoted entirely to hydrogen bonding was *The Hydrogen Bond* by Pimental and McClellan (1960). In this book, the definition of a hydrogen bond was made more general, as follows: "A hydrogen bond exists between the functional group, A—H, and an atom or a group of atoms, B, in the same or different molecules when (a) there is evidence of bond formation (association or chelation), (b) there is evidence that this new bond linking A—H and B specifically involves a hydrogen atom already bonded to A" (p. 6). This outstanding book describes all the phenomena associated with hydrogen bonding and is still useful today. It contains a table of nearly 300 entries of thermodynamic data for hydrogen bond formation in one-, two-, and three-component systems using a wide variety of methods.

The next book devoted entirely to hydrogen bonding was *Hydrogen Bonding in Solids* by Hamilton and Ibers (1968). By that time a significant number of X-ray and neutron diffraction crystal structure analysis studies of hydrogen bonded structures had

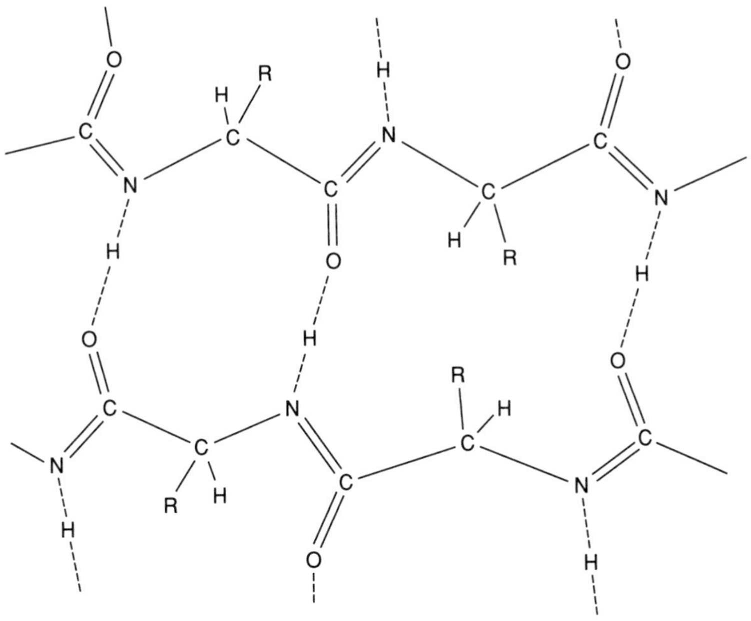

—Representing the structure of *beta* keratin

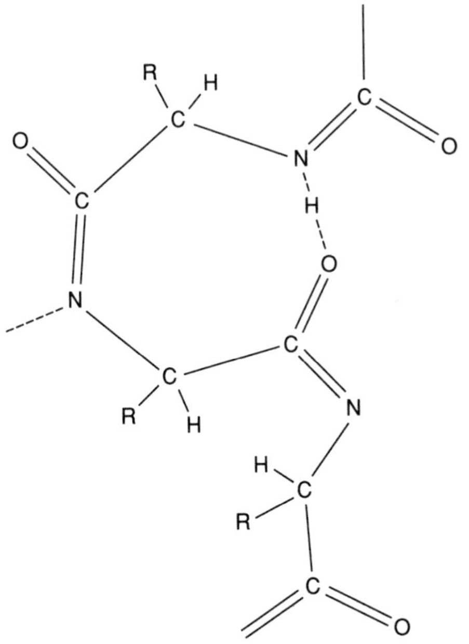

—Representing the structure of *alpha* keratin

Figure 1.3. Some postulated hydrogen-bonded structures from Huggins (1936b). Models for hydrogen-bonding in α and β keratin.

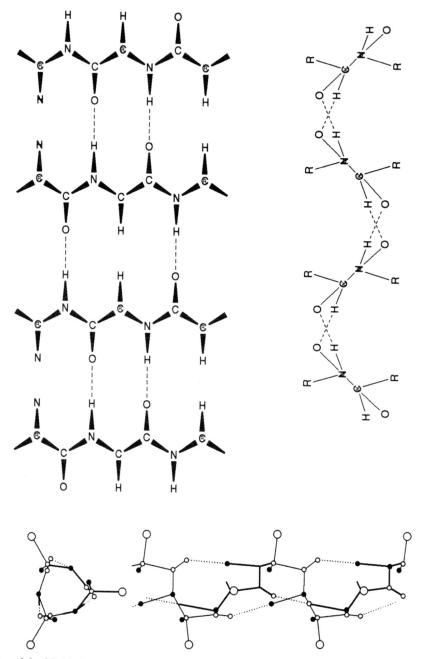

Figure 1.4. Models for hydrogen bond sheet and helical structures for polypeptides from Huggins (1943). The helix is a 3.10 helix. Had Huggins pursued this idea one peptide unit further, he would have discovered the α-helix eight years earlier.

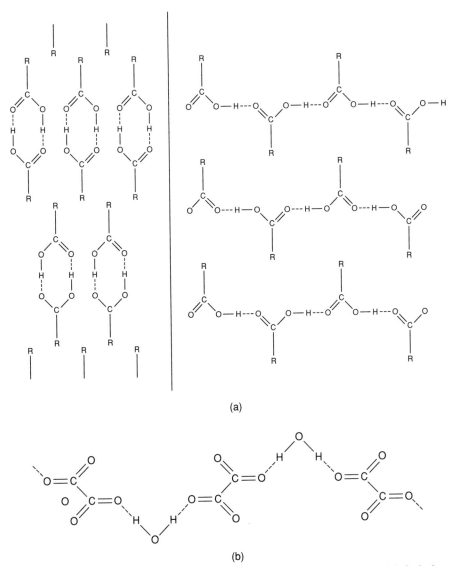

(a)

(b)

Figure 1.5. Some postulated hydrogen-bonded structures from Huggins (1936b). (a) models for hydrogen-bonding of aliphatic acids; (b) a model for the hydrogen-bonding in an oxalate hydrate.

been completed, prompting a book with emphasis on the solid-state. Prior to 1968 hydrogen atoms could not be seen by X-ray diffraction. Hamilton and Ibers therefore introduced a comparison with van der Waals radii that was subsequently given more prominence and application than the authors intended. They say guardedly, "A heavy atom distance less than the van der Waals distance is perhaps a sufficient, but not necessary, condition for the presence of hydrogen bonding" (p. 16).

A third book entitled *Hydrogen Bonding* by Vinogradov and Linnell (1971) gave a general review, updating that of Pimental and McClelland. The thermodynamics of hydrogen bonding was particularly well-discussed, with a useful appendix of 20 problems.

When advances in computer technology made possible theoretical quantum mechanical calculations on hydrogen-bonded complexes at the semi-empirical level, a fourth book on *hydrogen bonding* by Joesten and Schaad (1974) discussed the theory and gave a table with nearly 400 entries of theoretical studies of hydrogen-bonded systems from 1960 to 1973. These calculations are now outdated by the ab-initio calculations made possible by the more powerful modern computers, but comparisons are interesting. There is also an appendix of thermodynamic data and ν_s O—H frequency shifts between Lewis acids and bases from 1960 to 1973 containing nearly 2000 entries. The general annotated bibliography with nearly 3000 entries is particularly useful. General reviews relating specifically to the concept of hydrogen bonding were written by Kollman and Allen (1972), Allen (1975), Kollman (1977), and Morokuma (1977).

The most extensive coverage of particular aspects of hydrogen bonding came in a three-volume, multi-author publication entitled, *The Hydrogen-Bond. Recent Developments in Theory and Experiments*, edited by Schuster, Zundel, and Sandorfy (1976). This three-volume series has 29 chapters by specialists in their respective fields and covers all aspects of investigation and theory. Unfortunately, it has the disadvantage of many multi-author publications in that the various chapters have different styles and are uneven in the depth and length of discussion.

While publications relating to hydrogen bonding continued to accumulate at an estimated rate of one every fifteen minutes of every day, no further books were forthcoming until 1991. Instead there were periodical reviews of particular aspects of hydrogen bonding, which are referenced in the following chapters.

The importance of hydrogen bonding in the structure and function of biological molecules had been forecasted by the success of the Watson and Crick (1953) base-pairing in interpreting the structure of the nucleic acids in 1953 and by the a-helix and pleated sheet structures in proteins proposed by Pauling, Corey and Branson (1951). Subsequent research continued to emphasize this. It is rare these days that a paper on the structure or function of a biological molecule, however large, does not contain a reference to hydrogen bonding.

Hydrogen Bonding in Biological Molecules by Jeffrey and Saenger (1991) attempted to contain this field in less than 600 pages by focussing on some general principles and the experimental data available primarily from X-ray and neutron diffraction crystal structural studies.

The methods for structure determination that have made the greatest advances in the past decade are infrared, microwave and NMR spectroscopy, X-ray crystallography, and theoretical chemistry for molecular modelling. Just over 20 years ago, the complexity of hydrogen bonding patterns and the lack of reliable information concerning the positions of the hydrogen atoms led to the statement by Hopfinger (1973) that, "The one definite fact about hydrogen bonds is that there does not appear to be any definite rules which govern their geometry" (p. 99). This is certainly not true today, as I hope to illustrate in the following chapters.

Nature and Properties **2**

2.1 | A SIMPLE CRITERION AND SOME DEFINITIONS

Hydrogen bonding is a donor-acceptor interaction specifically involving hydrogen atoms. Following Pimental and McClellan (1960), A and B are used for the hydrogen bond donor and acceptor atoms because of the analogy with the Brönsted-Lewis acid which is a proton donor and the base which is a proton acceptor. There is, however, a fundamental distinction between proton donation and hydrogen bonding, as pointed out by Arnett and Mitchell (1971) and discussed in Chapter 7.

Hydrogen bonds are formed when the electronegativity, as defined by Pauling (1939), of A relative to H in an A—H covalent bond is such as to withdraw electrons and leave the proton partially unshielded. To interact with this donor A—H bond, the acceptor B must have lone-pair electrons or polarizable π electrons.

Despite this simple concept, understanding the electronic nature of the hydrogen bond appears to be more ellusive than for covalent and ionic bonds and van der Waals forces. This is because the term hydrogen bond applies to a wider range of interactions. Very strong hydrogen bonds resemble covalent bonds, while very weak hydrogen bonds are close to van der Waals forces. The majority of hydrogen bonds are distributed between these two extremes.

Pimental and McClellan (1960) in *The Hydrogen Bond* resolved this problem by the definition given in Chapter 1. This pragmatic definition satisfied all investigators at that time who, because of the different sensitivities of their methods and observational time-scales, may take different views as to what constitutes evidence of hydrogen bonding. Each experimentalist can answer the question *what is a hydrogen bond?* with reference to his particular method of investigation.

With this phenomenological definition, infrared, Raman and microwave spectroscopists define hydrogen bonds with respect to their effect on the vibrational motions

Table 2.1.
Properties of strong, moderate, and weak hydrogen bonds.

	Strong	Moderate	Weak
A—H····B interaction	mostly covalent	mostly electrostatic	electrostatic
Bond lengths	A—H ≈ H····B	A—H < H····B	A—H << H····B
H····B (Å)	~1.2–1.5	~1.5–2.2	2.2–3.2
A····B (Å)	2.2–2.5	2.5–3.2	3.2–4.0
Bond angles (°)	175–180	130–180	90–150
Bond energy (kcal mol^{-1})[a]	14–40	4–15	<4
Relative IR ν_s vibration shift (cm^{-1})[b]	25%	10–25%	<10%
H^1 chemical shift downfield (ppm)	14–22	<14	—
Examples	Gas-phase dimers with strong acids or strong bases Acid salts Proton sponges Pseudohydrates HF complexes	Acids Alcohols Phenols Hydrates All biological molecules	Gas phase dimers with weak acids or weak bases Minor components of 3-center bonds C—H····O/N bonds O/N—H····π bonds

[a]Suggested by Emsley (1980).
[b]Observed ν_s relative to ν_s for a nonhydrogen bonded X—H.

of the bonds immediately involved. NMR spectroscopists observe the chemical shift caused by the change in the electronic environment around the proton. The diffractionist observes characteristics of bond lengths and bond angles associated with hydrogen bonding. Thermodynamicists measure hydrogen bond energies and theoreticians calculate them and determine the configurations associated with the energy minima. These properties provide criteria which make it apparent that there are the different types of hydrogen bond shown in Table 2.1.

Prototypes of each category have distinctly different properties, although there is no clear demarcation in the boundaries between them. The failure to recognize these distinctions has led to some popular misconceptions in the past. For example, that hydrogen bonds are always almost linear and that the distance between the donor A and acceptor B atoms should be less than the sum of the van der Waals of A and B. While these are properties that are associated with strong hydrogen bonds, they do not necessarily apply to weaker hydrogen bonds.

2.2 | DIFFERENT CATEGORIES

Hydrogen bond energies which extend from about 15–40 kcal/mol^{-1} for strong bonds, to 4–15 kcal/mol^{-1} for moderate bonds and 1–4 kcal/mol^{-1} for weak bonds are evidence of a wider range of interatomic interactions than is observed for covalent or ionic bonds or van der Waals forces. Moderate and weak hydrogen bonds are very *soft in-*

teractions with stretching and bending force constants of an order of magnitude less than for covalent bonds. As a consequence, they have a wide spread of hydrogen bond lengths and angles when observed in the crystalline state where there is a compromise with other packing forces. Figure 2.1 shows some bond length distributions observed with different types of hydrogen bonds in a set of 32 amino acid crystal structures analyzed by neutron diffraction.

Strong hydrogen bonds are formed by groups in which there is a deficiency of electron density in the donor group, i.e., $-\overset{+}{O}$—H, $\overset{+}{>N}$—H, or an excess of electron density in the acceptor group, i.e., F^-, $\bar{O}$—H, $\bar{O}$—C, $\bar{O}$—P, $\bar{N}\langle$. This is to be expected since a deficiency of electrons of the donor group further deshields the proton thereby increasing the positive charge, while an excess of electrons on the acceptor group increases its negative charge and the interaction with the deshielded proton. For this reason, these are sometimes referred to as *ionic hydrogen bonds.*

Strong hydrogen bonds also occur when the configuration and conformation of a molecule is such as to force the neutral donor and acceptor groups into much closer than normal hydrogen bonding contact. They could be known as *forced strong H-bonds.*

Moderate hydrogen bonds are formed generally by neutral donor and acceptor groups, *i.e.*, —O—H, $>$N—H, ,—N(H)—H, and O$\langle$, O=C, N$\langle$, in which the

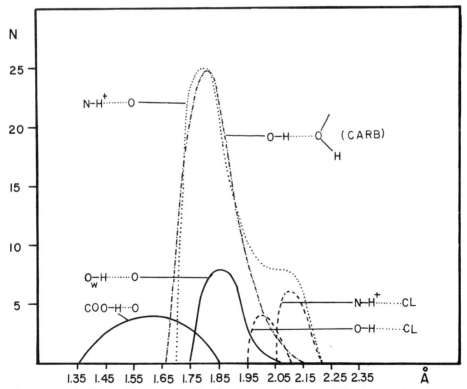

Figure 2.1.   Distribution of hydrogen-bond lengths observed in 32 neutron diffraction crystal structure analyses of amino acids (from Jeffrey and Maluszynska, 1982, reprinted with permission).

donor A atoms are electronegative relative to hydrogen and the acceptor B atoms have lone-pair unshared electrons. These are the most common hydrogen bonds both in chemistry and nature. They might be regarded as *normal* hydrogen bonds, with the other two categories being the minority exceptions. They are important and essential components of the structure and function of biological molecules.

Weak hydrogen bonds are formed when the hydrogen atom is covalently bonded to a slightly more electroneutral atom relative to hydrogen, as in C—H, Si—H, or when the acceptor group has no lone-pairs but has π electrons, such as C≡C or an aromatic ring. Although F is a very electronegative atom, F—C or F—S groups are only weak acceptors for reasons that will be discussed later. These interactions have similar energies and geometries to those of van der Waals complexes, and are distinguished from them by evidence of a directional involvement of the A—H bond.

Hydrogen bonds are group-pair properties. Unlike covalent bonds, hydrogen bonds are not atom-pair, but group-pair properties. For example, P—OH, H—O—H, and C—OH are distinguishably different in both their hydrogen-bond donor and acceptor properties. For this reason, it is not possible to separate hydrogen-bond distances into hydrogen-bond atomic radii, as Pauling did so successfully for covalent and ionic radii in *The Nature of the Chemical Bond*. Similarly, it is not possible to separate hydrogen-bond energies into atom-pair bond energy components. The nature of a hydrogen bond depends on the nature of the donor and acceptor groups. The most common donor and acceptor groups are given in Table 2.2 with the class of compounds in which they are observed.

P—OH, H—O$_W$—H, C—O—H, and to a lesser extent, ﹥N—H, can function as both donors and acceptors, which has important consequences as discussed in Chapter 6.

All types of hydrogen bonds can be *intramolecular* when donor and acceptor groups are on the same molecule or *intermolecular* when they are on different molecules. When A and B are the same, they are known as *homonuclear hydrogen bonds*, when different, *heteronuclear hydrogen bonds*. In the gas phase, homonuclear bonds lead to the formation of hydrogen bonded dimers. When A and B are different, the terms hydrogen bonded adducts or complexes are used. In liquids, hydrogen bonds can be between like molecules or between unlike molecules forming multicomponent systems. In solids, they can form hydrogen-bonded clusters, or one-, two-, or three-dimensional arrays of hydrogen-bonded molecules.

Three scalar quantities are necessary to define the geometry of a hydrogen bond. These are the A—H covalent bond length, the H----B hydrogen bond length and the A----B hydrogen bond distance. These quantities define the X—H----A hydrogen bond angle. Only in the strong hydrogen bonds is this angle ~180°. For moderate and weak hydrogen bonds in crystals, the angles are easily bent from linearity. As shown in Figure 2.2, *bent bonds are entropy-favored*; the probability that the angle θ is 180° is proportional to sinφ, so that the peak in the hydrogen bond angle distribution curve is ~155°. This is known as the *conic factor* or *correction*. Only when this correction is applied does the distribution of hydrogen bond angles cluster around 180°. In consequence, *it is only for strong bonds* that the A—(H)----B distances are a reliable measure of the hydrogen bond lengths. In structure analyses where hydrogen atom positions cannot be determined, as in proteins and nucleic acids, this feature is sometimes overlooked.

Table 2.2.
Functional Groups That Form Hydrogen Bonds.

Strong hydrogen bonds
  Donors and acceptors
    $[F\cdots H\cdots F]^-$ Symmetrical hydrogen bifluoride ion
    $[H\!-\!F\!-\!H]_n^-$ Anions in fluoride HF adducts
    $[O\!-\!H\cdots O]^-$ Organic hydrogen anions, hydrogen phosphates and sulfates,
                                hydrogen carboxylate ions
    $[\overset{+}{O}\!-\!H\cdots O]$ Hydroxonium ions, pseudo hydrates
    $[\overset{+}{N}\!-\!H\cdots N]$ Proton sponges
    $[N\!-\!H\cdots \overset{-}{N}]$

Moderate hydrogen bonds
  Donors and acceptors
    $O\!-\!H,\ P\!-\!O\!-\!H,\ H\!-\!O_W\!-\!H$ Water, hydrates, alcohols, carboxylic acids, phenols,
                                     carbohydrates, oligo- and polysaccharides, nucleosides,
                                     nucleotides, nucleic acids

    $\overset{C}{\underset{C}{>}}N\!-\!H,\ \overset{N}{\underset{C}{>}}N\!-\!H$ Secondary amines, amides, carbamates, hydrazides, purines,
                                     pyrimidines, barbiturates, nucleosides, nucleotides,
                                     peptides, proteins (main chain and side chain)

  Donors only
    $\overset{+}{N}(H_3)H$ Ammonium salts
    $-\overset{+}{N}(H_2)H$ Zwitterion amino acids
    $>\!\overset{+}{N}(H)H$

    $S\!-\!H$ Cysteine

    $\overset{C}{\underset{C}{>}}\overset{+}{N}\!-\!H$ Proteins (side chain, nucleic acids (low pH)

    $C\!-\!N(H)H$ Primary amines, pyrimidines, purines, barbiturates

  Acceptors only
    $\overset{C}{\underset{C}{>}}O$ Ethers, carbohydrates, oligo- and polysaccharides (ring and
                                     glycosidic oxygens)
    $>\!C\!=\!O$ Carboxylates, zwitterion amino acids
    $>\!C\!=\!O$ Carboxylic acids, ketones, esters, N-oxides, pyrimidines,
                                     purines, nucleosides, nucleotides, nucleic acids, peptides,
                                     proteins (main chain)
    $\overline{X}\!=\!O$ Oxyanions, nitrates, chlorates, sulfates, phosphates
    $\overset{}{\underset{/}{>}}N$ Tertiary amines
    $\overset{\backslash}{\underset{/}{}}N$ Purines, pyrimidines, barbiturates, nucleosides, nucleotides,
                                     nucleic acids
    $N\!=\!O$ Aromatic nitro compounds
    $>\!S$ Methionine

Weak hydrogen bonds
  Donors
    $C\!-\!H,\ Si\!-\!H$ (?)

  Acceptors
    $C\!\equiv\!C,\ \hexagon,\ F\!-\!C$ (?)

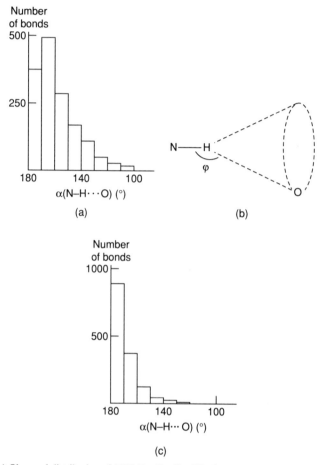

Figure 2.2.   (a) Observed distribution of 1509 N—H···O=C hydrogen bond angles in crystal structures; (b) the conic-correction factor; and (c) the distribution after correction (from Taylor, Kennard, and Versichel, 1984a).

An ab-initio molecular orbital calculation of the room temperature probability distribution of the O—H---O bond angle in a hydrogen-bonded methanol dimer by Newton, Jeffrey, and Takagi (1979) gave a maximum at 163°, in good agreement with the values from neutron diffraction studies of carbohydrates.

2.3 | INSIGHT FROM THEORY

What does theory say about the electronic nature of the hydrogen bond? The concept of the hydrogen bond evolves naturally from consideration of Pauling's atomic electronegativities. A consequence of the greater electronegativity of A relative to H in an A—H bond is that the hydrogen proton is stripped of some of its electron density, i.e., it is descreened. This results in a dipole at the terminus of the A—H bond which interacts with the monopole or dipole of the lone pairs on the acceptor atom and to a lesser degree more distant bond dipoles. Consequently hydrogen bond donor strengths are qualitatively proportional to these differences in electronegatives; F—H > O—H > N—H > C—H, and the hydrogen bond has a directional property, being strongest when A—H····B = 180°. In the days of valence-bond theory, Coulson (1952) and others were perhaps the first to state that this purely electrostatic interpretation was not sufficient. The other components of the hydrogen bond energy were identified as delocalization, repulsion, and dispersion. Coulson (1959) partitioned the hydrogen bond system X—H····A into contributions from the five valence bond structures shown below.

$$\Psi_{HB} = a\Psi_a + b\Psi_b + c\Psi_c + d\Psi_d + e\Psi_e$$

Ψ_a A—H····B   covalent A—H bond
Ψ_b A$^-$—H$^+$····B   ionic A—H bond
Ψ_c A$^-$—H····B$^+$ charge transfer, A····B bond
Ψ_d A$^+$—H$^-$····B   ionic A—H bond
Ψ_e A—H$^-$····B$^+$ charge transfer, H····B bond

His conclusion was that for O—H····O bonds at O····O = 2.8 Å, the electrostatic components, $\Psi_b + \Psi_d$, contributed about 65% of the hydrogen bond energy. As the O····O distance became closer, the quantum mechanical charge-transfer contributions became more important, while for longer weak bonds, the interaction became more electrostatic. Later ab-initio molecular orbital calculations support this general chemical concept. As a criterion, it is reasonable to assume that if the vibration frequency and the bond length of the covalent A—H bond are not significantly altered on hydrogen bond formation, the major component of the bonding is electrostatic. A very perceptive simple model of hydrogen bonding based on bond dipole moments and ionization potentials was presented by Allen (1975).[1]

Modern computers make it possible for ab-initio molecular orbital methods to explore the potential energy for hydrogen-bonded dimers with a variety of likely configurations and conformations. The lowest energy minima are sought with the structures and energies associated with them. Decomposing these energies into components, which is a valence bond concept, is somewhat artificial in terms of molecular orbital theory. Nevertheless, it is very useful from the chemical conceptual point of view. A way for doing this, known as the Morokuma (1971) decomposition method, decomposed the total bonding energy into electrostatic (es), polarization (pl), exchange repulsion (ex), charge-transfer (ct), and coupling (mix). The concept was applied to a number of hydrogen bonded adducts by Morokuma (1971), Kitaura and Morokuma (1976), and Umeyana and Morokuma (1977), as shown in Table 2.3. The electrostatic component includes monopole-monopole terms, r^{-1}, monopole-dipole, r^{-1}, dipole-dipole, r^{-3} and higher combinations of classical interactions between undisturbed

---

[1]This paper also contains perceptive comments and useful references.

Table 2.3.
The Kitaura and Morokuma (1976) Decomposition of Hydrogen-Bond Energies in kcal mol^{-1} for Some Hydrogen-Bonded Dimers.[a] **Taken from Morokuma (1977). Values in Parentheses are from Kollman (1977).**

Proton acceptor	Proton donor	$\Delta E = -E_b$	ES	EX	PL	CT	MIX
H$_3$N	HF	−16.3	−25.6	16.0	−2.0	−4.1	−0.7
		(−16.3)	(−25.0)	(15.2)	(−1.9)	(−4.6)	
H$_2$O	HF	−13.4	−18.9	10.5	−1.6	−3.1	−0.4
		(−13.2)	(−16.7)	(7.7)	(−1.3)	(−2.9)	
HF	HF	−7.6	−8.2	4.5	−0.4	−3.2	−0.3
		(−7.8)	(−8.8)	(4.6)	(−0.5)	(−3.1)	
H$_3$N	HOH	−9.0	−14.0	9.0	−1.1	−2.4	−0.4
H$_2$O	HOH	−7.8	−10.5	6.2	−0.6	−2.4	−0.5
H$_3$N	HNH$_2$	−4.2	−5.7	3.6	−0.6	−1.3	−0.2
H$_2$O	HNH$_2$	−4.1	−4.6	2.5	−0.3	−1.5	−0.2
H$_2$O	HNH$_3^+$	−37.3	−34.1	5.9	−4.1	−5.0	[b]
F$^-$	HF	−62.7	−86.1	67.5	−5.9	−27.9	−10.3[b]

[a]For neutral donors and acceptors, the electrostatic attractive and exchange repulsive terms dominate the interaction. For the ionic hydrogen bonds, the attractive polarization and charge-transfer terms become much more significant. Note that for F$^-$, CT is greater than ES—EX.
[b]Values from Vanquickenborne (1991).

monomer charge distributions.[2] The electron distributions of the molecules are disturbed by the close approaches due to the hydrogen bonding. This gives rise to polarization and the quantum mechanical interactions, exchange repulsion, charge transfer, and dispersion. The polarization is the effect of the distortion of the electron distributions of A—H by B and B by A—H. This is a stabilizing interaction. The exchange repulsion is the short-range repulsion of the electron distributions of the donor and acceptor groups. It accounts for the overlap of charges in occupied orbitals of both donor and acceptor. With the application of the Pauli principle, it is repulsive and is the major destabilizing term.[3] Charge-transfer is the result of the transfer of electrons between occupied orbitals on the donor to vacant orbitals on the acceptor and *vice versa*.[4] The coupling term allows for the fact that these four interactions are not strictly independent of each other. It is small, except for large bonding energies. The electrostatic, polarization, and charge transfer are attractive at equilibrium distances while the exchange repulsion is the balancing term.

As shown in Table 2.3, for moderate hydrogen bonds the electrostatic attraction (es) and the exchange repulsion (ex) terms account for more than 80% of the interaction energy. Since these terms partially cancel, the other terms, although small, can be

[2]Many, but not all, the empirical force-fields described in Chapter 11.8 model hydrogen bonding electrostatic interactions using only the monopole-monopole terms, for simplicity in parameterization. Since these interactions are nondirectional, should only those with appropriate directionality be regarded as hydrogen bonds?

[3]Since the greater electronegativity of A in an A—H bond withdraws electrons from the proton, the hydrogen bond could be regarded as a consequence of the reduction in the exchange repulsion term.

[4]The concept of a charge-transfer complex was developed by Mulliken (1952) to account for a characteristic transition band which appears in the ultraviolet and visible spectra of some molecular complexes. Ratajczak and Orville-Thomas (1980) consider that there is no fundamental difference between charge transfer and hydrogen bonding interactions; both are donor-acceptor interactions.

WATER DIMER LINEAR CONFIGURATION (0...0)

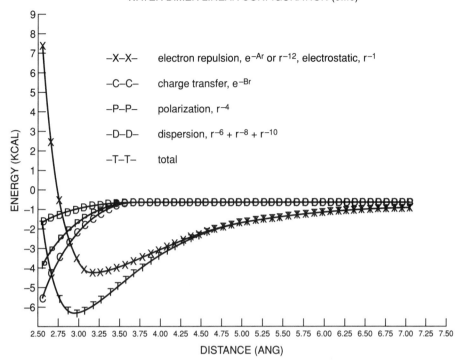

Figure 2.3. Morokuma decomposition of an ab-initio molecular orbital calculation for the water-dimer (from Singh and Kollman, 1985).

significant. The charge transfer term becomes important for the very strong [F—H—F]$^-$ hydrogen bond, accounting for 44% of the attractive energy, which is greater than the difference between the electrostatic attractive and repulsive terms.

This distribution is illustrated by the Morokuma component analysis of an ab-initio self-consistent field calculation of the water dimer potential by Singh and Kollman (1985). As shown in Figure 2.3, at the equilibrium O----O distance of 2.98 Å, the charge transfer, polarization, and dispersion components account for about 48% of the attractive energy. At 2.5 Å, they are 85% and at 3.2 Å only 33%, dropping to zero at about 4.0 Å. More recently a Morokuma decomposed calculation for the water dimer using a perturbation theory by Rybak, Zeriorski, and Szalewicz (1991) gave similar results. On the other hand, an ab-initio calculation by Szezesniak et al. (1993) suggests that for the weak H_3C—H----OH_2 hydrogen bond, almost equal contributions come from the various attractive components.

A simple classical electrostatic approach to hydrogen-bonding gives a reasonable answer. The importance of the electrostatic component to neutral hydrogen bonds is illustrated by the work of Spackman (1986). He extended an electrostatic model, which Buckingham and Fowler (1985) used for van der Waals complexes, to some hydrogen-bonded dimers. In this method the electrostatic energy is evaluated in terms of multiple moments of the monomer electron densities, while the charge-transfer and polarization are ignored.

Table 2.4.
Decomposition of Hydrogen-Bonding Energies For Some Dimers Calculated by the Method of Spackman (1986).

Dimer	E_{rep}	E_{per}	E_{es}	E_{disp}	E_{total}	E_{exp}
$(HF)_2$	1.9	0.5	-6.2	-0.6	-4.4	-4.6
$(HC\ell)_2$	0.7	0.2	-1.9	-0.5	-1.5	-2.3
$(HCN)_2$	3.2	0.7	-8.7	-1.3	-6.1	-4.4
FH----OC	1.5	1.3	-5.0	-0.6	-2.8	-2.8
FH----NCH	3.5	1.8	-12.2	-1.1	-8.0	-6.2
FH----C_2H_2	1.6	0.9	-5.3	-0.8	-3.6	—

The total energy is then comprised of four terms:

$$E_{tot} = E_{es} = E_{per} + E_{rep} + E_{disp}.$$

E_{es} is the classical electrostatic attraction between moments of atom-like fragments. E_{per} arises from the penetration of the fragment moments of one monomer inside the spherical charge distribution of the other monomer. It is classically electrostatic and a repulsion term. E_{rep} is a short-range repulsion term arising from the overlap of monomer charge distributions. It is given by the sum of exponential atom-atom potentials. E_{disp} is the attractive long-range dispersion energy term proportional to r^{-6}. Charge-transfer and polarization terms are ignored. Some results are given in Table 2.4. Application to the water dimer by Spackman (1987) gave for the configuration, (**I**), values of R = 2.67 Å, $\theta_1 = 52°$, $\theta_2 = 1°$, with $E_{tot} = 6.1$ kcal mol^{-1}. While the energy and orientation angles agree reasonably well with experiment and with ab-initio molecular orbital calculations, the equilibrium O----O distance is significantly shorter by 0.3 Å.

(**I**)

2.4 | CHARGE DENSITY STUDIES

The charge density studies using X-ray and neutron diffraction measurements, discussed in Chapter 11, have been applied to a number of crystal structures containing normal hydrogen bonds. In these studies, a deformation electron density is calculated by subtracting from the experimentally measured electron density that calculated by placing *isolated* spherical atoms at the nuclear positions. The results of a deformation density study for an O—H----O hydrogen bond is shown in Figure 2.4. It shows a conspicuous absence of electron density in the hydrogen bond. The most prominent feaa   b

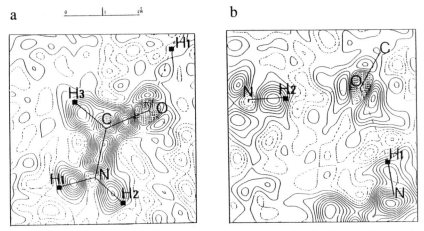

Figure 2.4.   The electronic structure of a hydrogen bond. The deformation density map of acetamide at 90K; contours at 0.05 eÅ^3. (a) in plane of molecule; (b) in plane of H_1, H_2, and O (from Stevens, 1978, reprinted with permission).

ture is a concentration of electron density about two-thirds along the A—H bond, arising from the covalent bonding, with a corresponding electron deficiency at the position of the hydrogen nucleus, and between the proton position and the acceptor lone-pair density. This gives rise to a dipole near the terminus of the A—H bond. There is a corresponding dipole at the acceptor atom, due to the concentration of lone-pair electron density away from the nucleus. These results suggest that the electrostatic component is primarily a dipole-dipole interaction. It is responsible for the common observation that the electronic A—H bond lengths in X-ray analyses are observed shorter than the corresponding internuclear distances from neutron diffraction analyses. This apparent shortening was predicted by Tomiie (1958) long before it was observed, from a valence bond calculation using the C—H bond as the example. This is consistent with the directional property of hydrogen bonds, since the dipole to dipole interactions are maximum when the dipoles are colinear and zero when orthogonal.

Charge density studies of strong hydrogen bonds are less common. A study of the short symmetrical hydrogen bond in sodium hydrogen diacetate by Stevens, Lehman, and Coppens (1977) showed a quite different deformation density, with no concentration in the O----H----O bonds and a barely significant deficiency at the center.

2.5   GEOMETRY IN CRYSTALS

X-ray and neutron crystal structure analysis provides the geometry of molecules and assemblies of molecules in exquisite detail. This information is computer accessible through the Cambridge Crystallographic Data Base (see Appendix I), which presently contains entries for more than 100,000 organic and organometallic crystal structures.

Of these, only about 900 are by neutron diffraction. A significant proportion of these molecules contain hydrogen-bonding functional groups from which it is possible to observe the properties of the hydrogen bond in the solid state. This database has formed the basis for a number of surveys of hydrogen bonding in crystal structures, which are described in subsequent chapters.

For inorganic crystal structures, i.e., those molecules not containing carbon, there is an Inorganic Structural Data Base. For gas-phase structural data, there is the MO-GADOC (molecular gas phase documentation) database. Information about access to these data bases is given in Appendix I.

Some Other Structural Definitions are Necessary

In the solid state,the packing of the molecules is determined by their shape and a variety of intermolecular forces, of which the hydrogen bonding is only one, not necessarily the dominant one. For this reason, more structural variety is observed than in gas phase hydrogen-bonded dimers and adducts.

While strong hydrogen bonds are close to being linear with a single acceptor, moderate hydrogen bonds can involve two acceptors, as in (**II**) or (**III**). These bonds are referred to as *three-centered*, since the hydrogen is bonded to three atoms: one by a covalent bond and two by hydrogen bonds. Since these are attractive forces, the hydrogen should lie close to the plane of A, B_1 and B_2, in which case the angles, $\alpha_1 + \alpha_2 + \alpha_3 \approx 360°$. That the H lies within 0.2 Å of the A, B_1, B_2 plane is an alternate criterion that has been used. Three-center bonds are also described as being *bifurcated*.[5] This description was first used when it was observed in the crystal structure of α-glycine by Albrecht and Corey (1939), and confirmed by a neutron diffraction analysis by Jönsson and Kvick (1972). In *The Hydrogen Bond* by Pimental and McClellan (1960), the term *bifurcated* was used to describe the configuration (**IV**) for water hydrogen bonding. This configuration is also possible for the amino group, (**V**), but in fact neither configuration is commonly observed in crystal structures, although they occur in some gas-phase hydrogen-bonded adducts.

$$
\begin{array}{ccccc}
(\mathbf{II}) & (\mathbf{III}) & (\mathbf{IV}) & (\mathbf{V}) & (\mathbf{VI})
\end{array}
$$

In this text, *three-centered* is used for (**II**), and *chelated* is used for (**III**), (**IV**), (**V**) and (**VI**). Jamvóz and Dobrowolski (1993) have recently used the term *double hydrogen bonds* for the configuration (**VI**) between $CH_2C\ell_2$ and CH_2Br_2 and strong hydrogen bond acceptor solvents, inferred from infrared data. Infrared spectroscopic evidence for three-centered intra- and intermolecular bonds of the type (**VII**) was presented by Gaultier and Hauw (1969) in some naththaquinone compounds.

---

[5]In some experimental and theoretical studies of gas-phase adducts, the terms *mono-dentate*, *bi-dentate*, and *multi-dentate* are used.

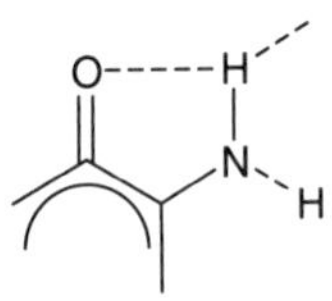

(VII)

An ab-initio molecular orbital calculation by Newton, Jeffrey, and Takagi (1979) predicted that a two-center O—H---O bond with H---O = 1.95 Å and O—H---O = 180°, and a three-center bond with H---O = 1.95 and 2.50 Å and both O—H---O angles = 150 and 110° would have comparable energies. The common occurrence of these bonds supports the view that two- and three-center bonds can have comparable energies. In certain crystal structures, such as the zwitterion amino acids, these three-center bonds account for about 75% of the hydrogen bonds with configurations shown in Figure 2.5. The rationale for this is that $\bar{C}{<}^O_O$ groups prefer to accept four hydrogen bonds, two per carboxyl oxygen, but the $\overset{+}{N}H_3$ group has only three protons. Some sharing must take place to satisfy the strong potential of C=O groups as hydrogen-bond acceptors. Three-center bonds are usually unsymmetrical with a major and minor component; with $r_1 > r_2$, as shown in Figure 2.6.

In carbohydrates, nucleosides, and nucleotides, the proportion of three-center bonds is about 25%. This arises because these molecules contain ether oxygens, i.e., C—O—C, which are hydrogen bond acceptors, but have no donor functionality. The proportion of ether oxygens to hydroxyl oxygens is also about 25%. An example where three-center hydrogen bonding satisfies both the donor and acceptor functionality is shown in Figure 2.7. A recent analysis of N-H---O=C bonds in 11 high-resolution X-ray protein crystal structure analyses by Pressner, Enger, and Saenger (1991) found that about 24% of the bonds were three-centered. These three-center bonds occurred mostly in the α-helix and β-sheet structure.

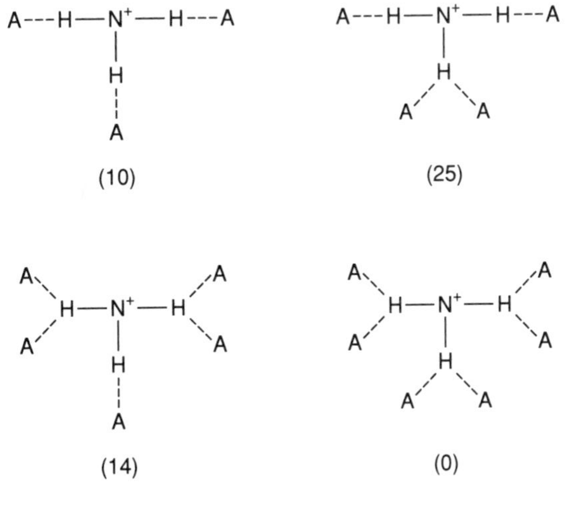

Figure 2.5. Two- and three-centered hydrogen bond configurations around the $\overset{+}{N}H_3$ in 49 zwitterion amino acid crystal structures (from Jeffrey and Mitra, 1984). Numbers of examples are given in parentheses.

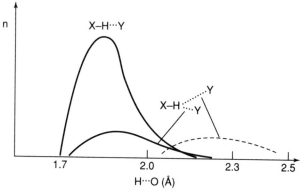

Figure 2.6. Distribution of hydrogen bond lengths in two- and three-center bonds in carbohydrate crystal structures.

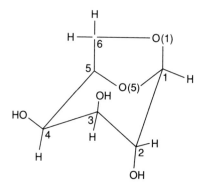

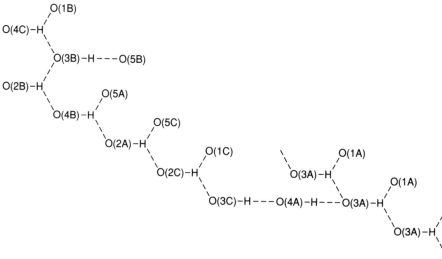

Figure 2.7. Hydrogen bonding scheme in the crystal structure of 1,6-anhydro-β-D-galactopyranose, by Ceccarelli, Ruble, and Jeffrey (1980). This illustrates the way in which three-center bonds incorporate the two rings oxygens O-1A,B,C and O-5A,B,C into the cooperative system of hydrogen bonds. A, B, and C refer to three symmetry independent molecules in the crystal structure.

24

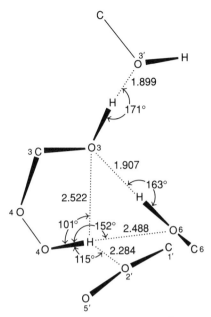

Figure 2.8. A four-center hydrogen bond in the crystal structure of sucrose by neutron diffraction (Brown and Levy, 1973). This was described as nonbonding in the original paper. (The H(O4) to H(O6) distance is comparable at 2.55 Å).

Three-center hydrogen bonding can therefore be viewed as a consequence of proton deficiency, i.e., greater acceptor functionality than number of hydrogen bonds available, hence the analogy of the three-center electron deficient bonds in the boron hydrides. (Some examples from neutron diffraction crystal structure analyses are given in Table 4.10.)

Four-center bonds, in which there are three acceptor groups, are observed in crystal structures, but are relatively rare, at <5%. The definition is that all the X–H----A angles must be greater than 90°. The H----B distances are generally longer than for three-center bonds and are sometimes regarded as nonbonding interactions as in the case of that in sucrose shown in Figure 2.8.

The configuration (**VIII**) is also referred to as *chelated hydrogen bonds. Tandem hydrogen bonds* (**IX**) are also reported.[6] Because of the close approach for the two hydrogens, these are disordered.

(**VIII**)   (**IX**)

---

[6]Frisch, Pople and Del Bene (1985) discuss the energies of configurations **IV** and **IX** for $(H_2O)_2$, $(NH_3)_2$, $(HF)_2$, $(H_2S)_2$, and $(HC\ell)_2$.

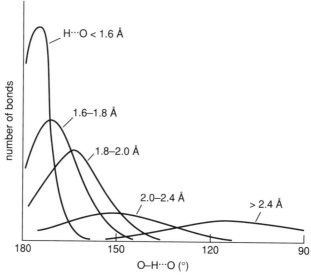

Figure 2.9. Distribution of O—H····O angles for different hydrogen bond lengths observed in carbohydrate crystal structures (*private communication*, T. Steiner).

A combination of three-center and chelated hydrogen bonding, (**X**) and (**XI**), is observed, albeit rarely, in the crystal structures of hydrates, discussed in Chapter 9.

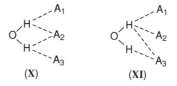

$$(X) \qquad\qquad (XI)$$

There is a correlation between bond lengths and angles. There is a qualitative relationship between hydrogen bond lengths and angles. The larger the angle, the shorter the bond, as shown in Figure 2.9. As a result the H----O distances can range from 1.6 to 3.0 Å, with a relatively small change in O----O distances. This is due to the constraints arising from the exchange repulsion of the electrons of the donor and acceptor atoms. This was pointed out by Savage and Finney (1986) and further elaborated by Savage (1986b) when discussing hydrogen bonding in hydrates. These repulsive interactions give rise to the *excluded region*, illustrated in Figure 2.10.

2.6 | THE VIBRATIONAL PROPERTIES

Molecules undergo thermal motion even in crystals at very low temperatures due to the zero-point energy. The motion of the hydrogen atoms is greater than that of the heavier atoms to which they are covalently bonded, as illustrated in Figure 2.11. The

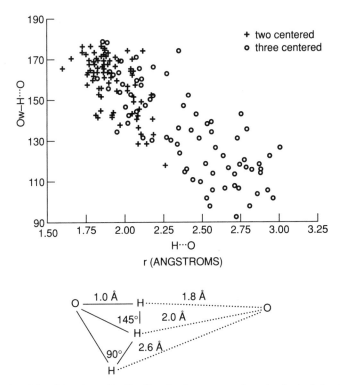

Figure 2.10. H····O bond lengths vs O—H····O angles in some organic molecular hydrates, illustrating the excluded regions.

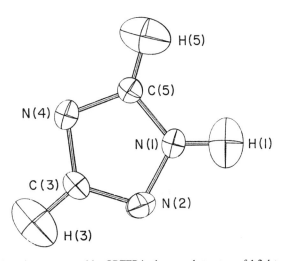

Figure 2.11. Thermal motion represented by ORTEP in the crystal structure of 1,2,4-triazole at 15 K by neutron diffraction. The ellipsoids are at 99% probability (from Jeffrey, Ruble, and Yates, 1983, reprinted with permission).

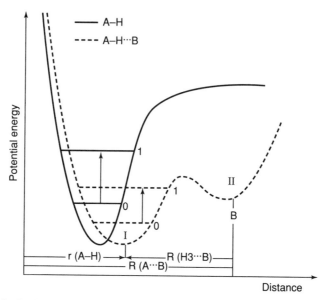

Figure 2.12. Qualitative potential energy curve for a free (——) and hydrogen-bond (----) A—H group (from Novak, 1974).

hydrogen, or deuterium, atom is said to be *riding* on the motion of the heavier atom to which it is covalently bonded. The formation of a hydrogen bond restricts the degree of vibrational motion, since the hydrogen atom is restrained by two bonds rather than one. A measure of the vibration of the A—H bonds is therefore a sensitive criterion of the formation and strength of hydrogen bonds. This is the basis for the extensive use of infrared spectroscopy for studying and classifying hydrogen bonds, discussed in Chapter 11.

The vibrational frequencies of both the A—H covalent bonds and the H----B hydrogen bonds, and of certain acceptor groups such as O=C, can be identified in favorable cases. The largest frequency change due to hydrogen bonding is observed in the A—H covalent bond stretching frequencies, and these changes are the most informative. Changes in the H----B hydrogen bond and acceptor frequencies are less dramatic and more difficult to interpret.

Infrared spectroscopy provides insight into the potential energy surfaces of hydrogen bonds. When an A—H bond becomes involved in hydrogen bonding, the potential energy curve becomes broader, a second minimum develops and the A—H bond stretching vibrational levels become closer, as illustrated in Figure 2.12. With moderate hydrogen bonds, an unsymmetrical double minimum develops, but with strong hydrogen bonds, the two minima become more equal. The barrier between them is reduced, forming *low-barrier hydrogen bonds*, as shown in Figure 2.13. In such bonds, the potential energy surface is quite flat, and the equilibrium position of the hydrogen atom becomes environmentally sensitive. In some cases, the *proton transfer* accompanies the formation of ionic hydrogen bonds, i.e., A—H----B → A⁺----H—B. In a few extreme cases, discussed in Chapter 3, there is a single potential energy minimum and the hydrogen bond is truly centric.

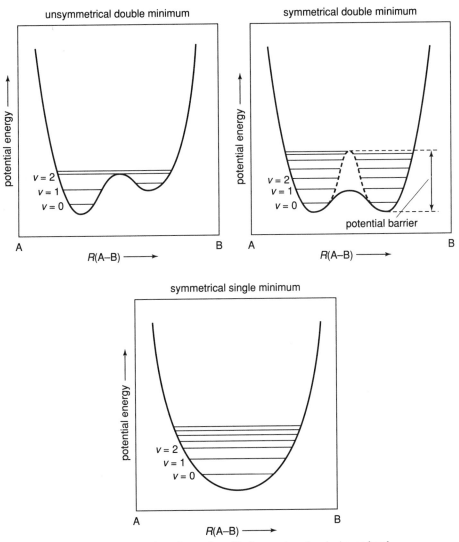

Figure 2.13.   Potential energy surfaces for motion of hydrogen atoms in a hydrogen bond.

The more recent development of Fourier transform infrared spectroscopy (FTIR) has increased the interpretative power of the method for studying hydrogen bonds. This is especially so when polarized infrared is used on single crystals, the crystal structure of which is known. An example is discussed in Chapter 11.

Proton NMR spectroscopy is a sensitive probe of hydrogen bonding, since the H^1 magnetic resonance is very sensitive to the electron environment around the proton. The chemical shift on hydrogen bond formation is one of the largest observed, sometimes exceeding 20 ppm. It provides the same type of general investigative tool for studying hydrogen bonding in liquids as does the X-ray and neutron crystallography in solids and infrared spectroscopy in gases, liquids and solids. While proton chemical shifts are useful for identifying hydrogen bonding, more detailed investigation is difficult because of the complex nature of solute-solvent interactions.

Solid-state ^{13}C NMR is more informative since it can be correlated with crystal structure analysis. However, solid-state proton NMR spectroscopy is not yet so well advanced, and ^{13}C solid-state NMR is more than one atom removed from the hydrogen bond in the important O—H····O and N—H····O bonds.

Hydrogen bonds are temperature sensitive. Increase of temperature increases the vibrational motion of the A—H bonds, which in turn increases the average A—H····B angles and weakens the hydrogen bonding, so that at a given temperature there can be an equilibrium A—H + B$\rightleftharpoons$A—H····B. This effect is most pronounced in gases and liquids. It could occur in solids, but this feature has not been systematically studied. From measurements of the equilibrium constants in liquids and solutions, the change in enthalpy ΔH_o and entropy ΔS_o on hydrogen formation can be determined, as described in Chapter 11.

Precise hydrogen bond energies are difficult to determine. Hydrogen bond energies can be measured by thermodynamics from the large number of properties of associated molecules which are temperature dependent. Although individual measurements quote precision to $\pm$ 0.5 kcal/mol^{-1}, the range for a particular type of bond may vary over several kcal/mol^{-1}, depending on the method used.

For gas-phase ion-molecule complexes, the more modern methods of ion-cyclotron resonance and mass spectroscopy provide the best experimental data. For neutral gas-phase adducts, the best source is now the advanced theoretical computations discussed in Chapter 11.

Qualitative estimates of hydrogen-bond energies have been obtained from the statistics of hydrogen-bond lengths. Since moderate hydrogen bonds have weak stretching force constraints, the bond lengths observed for any particular donor-acceptor combination may range over as much as 0.4 Å. If the mean values of their distributions from the crystal structures of particular classes of hydrogen bonds are compared, a qualitative estimate of relative strengths can be obtained, as discussed in Chapter 4. Bürgi and Dunitz (1988) have cautioned about using such statistical analyses to obtain quantitative energy relationships.

2.7 | ELECTROSTATIC POTENTIALS

The molecular electrostatic potential (MESP) is the potential energy of the electrostatic interaction of a molecule with a unit charge at points outside the van der Waals envelope of the molecule. It is an important concept, particularly in molecular biology, since it can be used to calculate the energetics of a variety of biological processes, as described for example by Warshel and Russell (1984). It is related to the charge density of the molecule $\bar{\rho}(r)$ by

$$\phi(r') = \int \bar{\rho}(r) |r' - r|^{-1} \, dr.$$

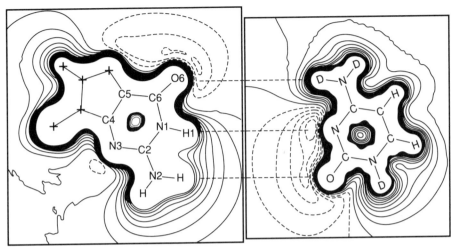

Figure 2.14. The complementarity of electrostatic potentials in a hydrogen-bonded guanine-cytosine base pair. Contours at 0.1 e/Å^3; full lines +ve, dotted lines −ve (0.2 e/Å^{-1}-33.2 kcal mole^{-1}).

It can be calculated theoretically or measured experimentally from the electron and nuclear structure factors from X-ray and neutron diffraction experiments, as described in Chapter 11.

The hydrogen bond donor and acceptor properties of molecules are clearly revealed by electropositive and electronegative regions in the electrostatic potential. The hydrogen bond can therefore be regarded as the consequence of a complementarity between electrostatic potentials, as illustrated in Figure 2.14. At hydrogen bonding distances, the electropositive and electronegative regions superimpose and cancel in the region between the molecules. This can be displayed by calculating electrostatic difference potentials as described by Ryan (1994).

2.8 | HYDROGEN-BOND LENGTHS VS VAN DER WAALS RADII SUMS

The suggestion made by Hamilton and Ibers (1968) that, a heavy atom distance less than the sums of their van der Waals radii is perhaps a sufficient, but not necessary, condition for hydrogen bonding, with variations on that theme, was adopted over-enthusiastically by crystallographers for many years, even after the hydrogen atoms could be observed. While this test can be applied to strong and moderately strong hydrogen bonds, it is totally inappropriate for the weaker bonds. Van der Waals interactions, attenuating as r^{-6}, are short-range forces relative to the electrostatic forces which are the major components of moderate or weak bonds, and attenuate as r^{-3} or r^{-1}.

In macromolecular crystal structures particularly, this raises the question, at what O----O, N----O or N----N distances is it appropriate to stop seeking evidence for hydrogen bonding?

Using the neutron diffraction data from six high-resolution crystal structures of cyclodextrin hydrates, Steiner and Saenger (1994) examined the reliability of using

O----O distances as a criterion for O—H----O hydrogen bonding. They examined the potential for hydrogen bonding out to H----O < 3.7 Å, O—H----O $> 90°$. The results are shown in Table 2.5.

While these results give the probability that the assignment of a hydrogen bond to a particular O----O is correct, it still does not solve the problem of identifying which are correct. With a cut-off at O----O $= 3.0$ Å, about 25% of the hydrogen bonds would not be identified. With a cut-off of 3.7 Å, about 35% of the O----O distances would be incorrectly associated with hydrogen bonds. With N—H----O hydrogen bonds, the problem is simpler since the hydrogen atom positions can be calculated with acceptable reliability.

2.9 | WHAT MAKES THE HYDROGEN BOND UNIQUE?

Umeyama and Morokuma (1977) tried to answer this question by comparing their decomposed energy components for FH----FCℓ versus HF----CℓF. Surprisingly, the energy components were rather similar. Morokuma (1977) had more success comparing $(HF)_2$ with $(LiF_2)_2$ and $(LiH_2)_2$.

The deformation electron density studies shown in Figure 2.4 reveal an expected accumulation of electron density between the atoms in the A—H covalent bond with a corresponding deficiency at the proton position. Part of this electron density comes from the hydrogen atom, leaving the proton partially deshielded. This gives rise to a dipole at the terminus of the A—H bond estimated by Craven (1987) to be about 1 D. For no other atoms is there an equivalent deshielding of the proton on formation of a covalent bond. The nearest might be A—Li where the deshielding could be about 20%. An interesting discussion of electrical interactions in relation to hydrogen bonding from a different point of view is provided by Dykstra (1988).

Table 2.5.
Percentage of O—H----O Hydrogen Bonds *vs* O----O
Distances Based on Cyclodextrin High-Resolution Neutron
Diffraction Studies by Steiner and Saenger (1994).[a]

O----O	N	Percentage of H-bonds
$2.6 \leq d < 3.0$	177	100
$3.0 \leq d < 3.1$	18	94(5)
$3.1 \leq d < 3.2$	25	84(7)
$3.2 \leq d < 3.3$	17	65(12)
$3.3 \leq d < 3.4$	28	57(9)
$3.4 \leq d < 3.5$	23	17(8)
$3.5 \leq d < 3.6$	27	11(6)
$3.6 \leq d < 3.7$	24	0

[a]N is number of O----O distances observed. (n) is the statistical uncertainty due to small number of data. A hydrogen bond is identified by an H----O distance $<$ 3.0 Å with the O—H----O angle $> 90°$. Of the 474 intermolecular O----O distances to < 4.0 Å, 193 are two-centered or major components of three-center bonds, 56 are minor components of three-center bonds, 225 are not hydrogen bonds.

3.1 | INTRODUCTION

Strong hydrogen bonds are the most interesting and intriguing category of hydrogen bonds. Research relating to strong hydrogen bonds started in the early days of infrared spectroscopy and X-ray and neutron diffraction crystal structure analysis. It has been stimulated more recently by studies of the $\overset{+}{N}$—H----N hydrogen bonds observed in the so-called *proton sponges*, and the discussion of the possible role of strong hydrogen bonds in enzyme catalysis. The strong hydrogen bond energies are in excess of 12 kcal/mol^{-1} and sometimes equal those of covalent bonds. They include configurations where the hydrogen atoms are at or close to the midpoint of the formal donor and acceptor atoms. They have low-energy barrier potential wells in which the position of the proton is very variable and appears to be sensitive to the molecular environment beyond its immediate vicinity. Strong bonds are characterized by large infrared ν_s stretching frequency shifts and by relatively large downfield proton NMR chemical shifts. They are formed when the proton is shared by two strong bases as in the [F—H—F]$^-$ ion or between ions and molecules when there is a deficiency of electron density on the donor group or an excess of electron density on the acceptor group as in O—H----$\overset{-}{O}$, $\overset{+}{O}$—H----O, $\overset{+}{N}$—H----N, or N—H----$\overset{-}{N}$. They are sometimes referred to as *ionic hydrogen bonds*, *positive-* or *negative-ion hydrogen bonds*, or *low-barrier hydrogen bonds*. Short and presumably strong intramolecular neutral hydrogen bonds are also formed when the configuration of the molecule brings the donor and acceptor atoms into much closer contact than the sum of their van der Waals radii, and when there is the *resonance assisted hydrogen bond* configuration discussed later in Chapter 6.

A review of theoretical calculations of the bond energies of binary complexes with ionic hydrogen bonds is given by Deakyne (1987).

Table 3.1.
Experimental Observed Properties of [F—H—F]⁻ and [F—D—F]⁻ Ions.

Cation	Bond lengths F····H····F (Å) (D)		IR and Raman (R) frequencies		
			ν_1 (cm^{-1})	ν_2 (cm^{-1})	ν_3 (cm^{-1})
Na$^+$	1.132	1.132	630(R)	1210 1220 1240	~1500
Na$^+$	1.132(D)	1.132		893	
K$^+$	1.138	1.138	595(R) 604(R)	1222 1225 1233	1450
K$^+$	(D)		596(R) 603(R)	885(R) 894(R) 911(R)	1045
pCH₃C₆H₅NH₂⁺	1.025	1.235	450(R)	1080 1230	1720

3.2 THE HYDROGEN BIFLUORIDE ION: A PROTOTYPE STRONG BOND

Hydrogen bifluoride was one of the first systems to be studied by X-ray diffraction by Bozorth (1923) and revealed a short F----F distance of 2.25 Å; a value which is valid today. It was the object of early studies by infrared spectroscopy by Ketelaar (1941), by NMR by Waugh, Humphrey, and Yost (1953), and by neutron diffraction by Peterson and Levy (1952). Some of the experimental properties of the [F—H—F]⁻ and [F—D—F]⁻ ion are given in Table 3.1. There has always been debate as to whether the ion is truly centrosymmetric with the hydrogen atom at the midpoint of the bond. Even in neutron diffraction analysis, a single site for the proton or two thermally overlapping sites is not easily distinguished. A good discussion of the difficulty of deciding whether a strong hydrogen bond is truly centric from crystal structural data is given in *Hydrogen Bonding in Solids* by Hamilton and Ibers (1968).

The bond energies reported for [F—H—F]⁻ range from 36 to 60 kcal/mole⁻¹ depending upon the experimental method used or theoretically calculated. To quote Emsley (1980), "There are many reasonable answers . . . but none sufficiently reliable to be quoted in preference to the rest." A value of 39.0 ± 1 kcal/mol⁻¹ from ion cyclotron resonance spectroscopy by McMahon and Larson (1982) is presently considered the most reliable value.[*] Other hydrogen bond energies involving F⁻ given by ion cyclotron resonance are given in Table 3.2. A more complete table including hydrogen-bonded complexes with Cℓ⁻, CN⁻, Br⁻, I⁻, OH⁻, and CH₃O⁻ is given by Larson and McMahon (1982, 1983), from which a relationship between binding energies and electronegativity is derived. The question whether the F—H—F⁻ bond is the strongest hydrogen bond is discussed in an excellent review by Hibbert and Emsley (1990). A theoretical ab-initio calculation by Frisch et al. (1986) for a symmetrical ion gave H—F distances of 1.134–1.164 Å with energies between 43 and 58 kcal/mol⁻¹, depending upon the level of approximation of the calculation.

---

[*]50 years ago Pauling (1927) obtained a value of 49.5 kcal/mol⁻ for the (F—H—F)⁻ bond by a simple calculation.

Table 3.2.
Strong Hydrogen-Bond Energies For Binary Ionic Adducts
Determined By Ion Cyclotron Resonance Spectroscopy (icr).
Estimated Error ± 1 kcal/mol^{-1} (from Hibbert and Emsley,
1990).

		Hydrogen bond energy
Species	Method	kcal/mol^{-1}
F$^-$----H—A and FH----B$^-$		
F$^-$----HF	icr	39
F$^-$----HO$_2$CH	icr	19
F$^-$----HO$_2$CMe	icr	21
F$^-$----HOMe	icr	30
F$^-$----HOEt	icr	32
F$^-$----HOPh	icr	20
F$^-$----HOH	icr	23
F$^-$----H$_2$NPh	icr	27
FH----Cℓ^-	icr/ir	22
FH----Br$^-$	icr/ir	17
FH----I$^-$	icr/ir	15
FH----CN$^-$	icr/ir	21

The ^{1}H NMR chemical shifts reported for the [F—H—F]$^-$ ion range from 7.6 to 16.4 ppm relative to (CH$_3$)$_4$Si, and the ^{19}F shifts are from 119 to 476 pm relative to CF$_6$. This short and possibly symmetrical hydrogen bond is particular to fluoride since the [Cℓ—H—Cℓ]$^-$ bond is asymmetrical with Cℓ—H and H----Cℓ bond lengths of 1.368 and 1.850 Å.

3.3 | OTHER H----F̄ BONDS

HF forms crystalline adducts with a number of fluoride salts which contain (F.nHF)$^-$ anions having strong H----F hydrogen bonds. Early X-ray analyses of KH$_2$F$_3$ by Forrester et al. (1963) and KH$_4$F$_5$ by Coyle, Schroeder, and Ibers (1970) revealed that the anions had short F----(H)----F distances. Mootz and his collaborators (1981–1991) have extended these studies by means of phase diagrams and low-temperature X-ray crystal structure analyses which located the hydrogen atoms. The crystal structures studied and the H—F----H hydrogen bond geometries are given in Table 3.3 and Figures 3.1–3.4. The H----F hydrogen bond lengths range from 1.13 to 1.70 Å with no obvious correlation between the position of the proton in the H—F----H bonds and the config- uration of the anion. This is shown by comparing (H$_2$F$_3$)$^-$ ions in the K salts and in the (CH$_3$)$_4$N$^+$ salt, or the (H$_3$F$_4$)$^-$ ions in (CH$_3$)$_4$N$^+$ and NO$^+$ salts. This suggests that the proton lies in a flat potential surface in which its position is determined not by the donor and acceptor atoms, but by the more remote crystal field. Pyridine forms par- ticularly interesting complexes with HF, C$_6$H$_5$N.nHF with n = 1–4. The C$_6$H$_5$N.HF is

Table 3.3.
Hydrogen Bond Distances in $(H_nF_{n+1})^-$ Anions.

Salt	Anion	F—H (Å)	H---F (Å)	F---F (Å)	F—H---F (°)
$H_2O \cdot 2HF^a$	HF_2^-	0.85	1.52	2.361	171
$H_2O \cdot 4HF^a$	$H_3F_4^-$	0.77	1.61	2.377	174
		0.90	1.47	2.367	173
		0.80	1.67	2.460	173
$NH_3 \cdot 4HF^b$	$H_3F_4^-$	0.78	1.63	2.406	178
$NH_3 \cdot 5HF^b$	$H_4F_5^-$	0.78	1.70	2.444	177
$C_5H_5N \cdot 2HF^c$	HF_2^-	0.98	1.34	2.326	179
$C_5H_5N \cdot 3HF^c$	$H_2F_3^-$	0.80	1.55	2.345	173
		0.83	1.52	2.343	180
$C_5H_5N \cdot 4HF^c$	$H_3F_4^-$	0.87	1.51	2.361	164
		0.87	1.52	2.389	174
$KF \cdot 2.5HF^d$	$H_2F_3^-$	0.73	1.64	2.352	165
		0.96	1.37	2.323	178
	$H_3F_4^-$	0.88	1.58	2.441	163
		0.71	1.70	2.402	168
		1.08	1.21	2.281	172
$KF \cdot 3HF^d$	$H_3F_4^-$	0.77	1.63	2.401	180
$NOF \cdot 3HF^e$	$H_3F_4^-$	0.80	1.59	2.389	172
$NOF \cdot 4HF^e$	$H_4F_5^-$	0.81	1.63	2.437	178
$(CH_3)_4NF \cdot 2HF^f$	$H_2F_3^-$	0.93	1.39	2.316	175
		0.94	1.38	2.302	168
$(CH_3)_4NF \cdot 3HF^f$	$H_3F_4^-$	0.90	1.45	2.351	170
		0.82	1.62	2.357	162
$(CH_3)_4NF \cdot 5HF^f$	$H_5F_6^-$	0.77	1.72	2.484	178
		1.13	1.13	2.266	180

esd's: F—H and H---F 0.02 to 0.06 Å, F---F 0.001 to 0.004 Å, F—H---F 1 to 5°.

[a]Mootz and Poll (1982).
[b]Mootz and Poll (1984a).
[c]Boenigk and Mootz (1988).
[d]Mootz and Boenigk (1986).
[e]Mootz and Poll (1984b).
[f]Mootz and Boenigk (1987).

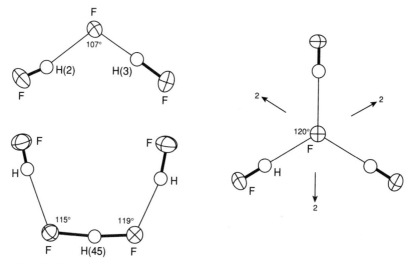

Figure 3.1. Configurations of $(H_2F_3)^-$ and $(H_3F_4)^-$ anions in the crystal structure of $KF_2.5HF$ (from Mootz and Boenigk, 1986).

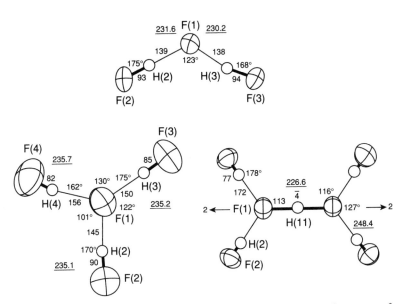

Figure 3.2. Configurations of $(H_2F_3)^-$, $(H_3F_4)^-$, and $(H_4F_5)^-$ anions in the crystal structures of $(CH_3)_4N)^+$ salts (from Mootz and Boenigk, 1987).

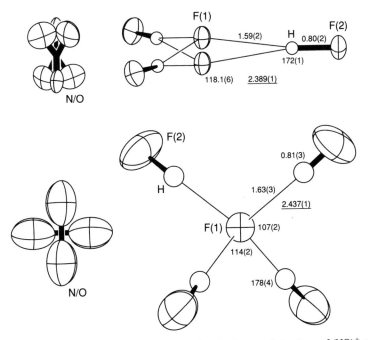

Figure 3.3. Configurations of $(H_3F_4)^-$ and $(H_4F_5)^-$ anions in the crystal structures of $(NO)^+$ salts. The NO^+ ions are disordered, as in the $(H_3F_4)^-$ ion (from Mootz and Poll, 1984b).

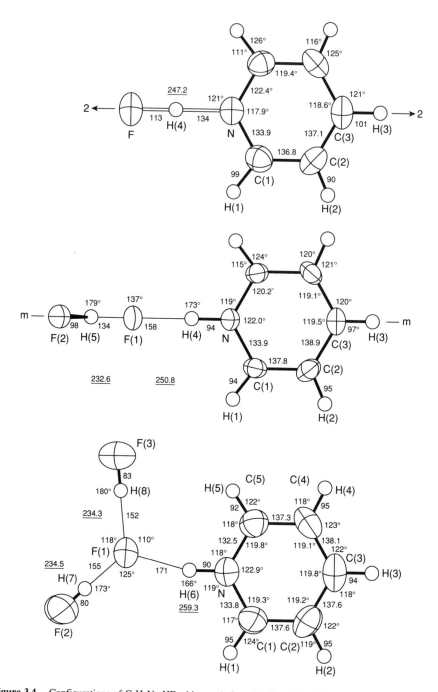

Figure 3.4. Configurations of $C_5H_5N \cdot nHF$ with $n = 1$, 2, and 3 (from Boenigk and Mootz, 1988).

a molecular complex with an F—H----N$\langle$ hydrogen bond, shown in Figure 3.4. The higher complexes are proton transfer complexes with $C_6H_5NH^+$ and $F(HF)_n^-$ ions. The H----F$^-$ bonds are strong, while the $\overset{+}{N}$—H----F bonds are normal with N—H = 0.94, 0.90 Å, and H----F = 1.58 and 1.71 Å. Even with the relatively high uncertainty in the hydrogen positions of $\sim \pm$ 0.05 Å from the X-ray data, it is clear that the position of the hydrogen atoms in the hydrogen bond is sensitive to the crystal field environment beyond the configuration of the $(H_nF_{n+1})^-$ anions. This work, if repeated with neutron diffraction data to provide more precise proton nuclear positions with corrections for thermal motion effects, could form a sound basis for a theoretical study of what determines the proton position in strong hydrogen bonds.

3.4 | O—H----Ō BONDS

These occur as intramolecular hydrogen bonds in organic hydrogen anions, some carboxylic acids, and the salts of hydrogen oxyanions. The H----O hydrogen bond lengths are in the range 1.2–1.6 Å, with significant lengthening of the covalent O—H bond length. These short hydrogen bond lengths are accompanied by the large shifts in the infrared stretching frequencies which are characteristic of strong hydrogen bonds, as discussed by Novak (1974) and in Chapter 11. As with the (H—F----H)$^-$ bonds, the bond angles are close to 180°. There are similar variations in the position of the protons. The correlation between the O—H and H----O distances is only qualitative in the region < 1.6 Å, as shown in Figure 3.5.

Examples of bond lengths from neutron diffraction studies of organic hydrogen anions are shown in Table 3.4. All these results correspond to unsymmetrical O—H----Ō hydrogen bonds with the exception of those of the maleate ions in imidazolium and potassium hydrogen maleate, (**I**), by James and Matsushima (1976); and Currie and Speakman (1970); potassium hydrogen chloro maleate, (**II**), by Ellison and Levy (1965); and one of the two-symmetry independent anions in the crystal structure of lithium hydrogen phthalate methanolate (**III**), by Küppers, Kvick, and Olovsson (1981).

(I)   (II)   (III)

In the crystal structures containing the anions (**I**), (**II**), and (**III**), which were determined by neutron diffraction analyses, the difference in the O—H and H----Ō bond lengths is less than 3 σ.

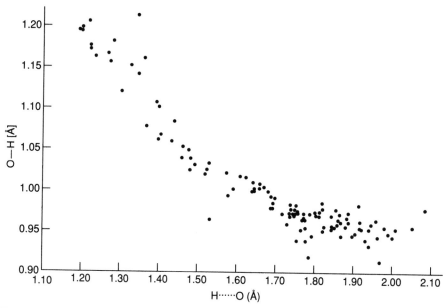

Figure 3.5.   Plot of O—H vs H⋯O based on neutron diffraction data (from Jeffrey and Saenger, 1991).

As candidates for symmetrical hydrogen bonds, these structures prompted an X—N deformation density analysis of the imidazolium hydrogen maleate by Hsu and Schlemper (1980), which proved to be inconclusive. There was also an early ab-initio molecular orbital calculation at the 4-31G basis set level for (**I**) by George, Boch and Trachtman (1983) which favored an unsymmetrical hydrogen bond. However, the question now appears to be resolved in favor of a single minimum symmetrical bond by a recent polarized infrared study of single crystals of potassium hydrogen and deuterium maleate at 310 K and 15 K by Bezyszyn et al. (1992). The evidence is based on the observation that the out-of-plane ν_r O—H⋯O deformation vibration, at 1190 cm^{-1}, obeys the C_s site symmetry selection rules and appears at almost the same frequency in spectra polarized parallel to the a and b crystallographic axes. This interesting paper illustrated both the advantage of combining infrared and neutron diffraction data and the complexity of attempting to completely identify the various vibration modes associated with hydrogen bonding in a relatively simple crystal structure.

The Hesitating Proton is a Characteristic of Strong Hydrogen Bonds

The observation that symmetrical, or almost symmetrical, hydrogen bonds are observed in the anions in imidazolium and potassium hydrogen maleate, in potassium chloromaleate and one of the two hydrogen phthalate ions in the lithium salt but not in the other, or in the other crystal structures shown in Table 3.4, raises the same question as with the (F—H—F)$^-$ ion complexes; what determines the position of the hydrogen atom in these strong hydrogen bonds? Disturbing the constitutional symmetry

Table 3.4.
Intramolecular O—H····Ō Hydrogen Bonds in Some Organic Hydrogen Anions; Neutron Diffraction Data.

Crystal structure	H····O (Å)	O—H (Å)	O····O (Å)	O—H····O (°)
Imidazolium H maleate (N)	1.196(4)	1.195(4)	2.393(3)	176.8(4)
K H maleate	1.204(4)	1.199(4)	2.437(4)	
K H chloromaleate	1.206(5)	1.199(5)	2.403(3)	175.4(4)
Li H phthalate CH$_3$OH[a]	1.203(5)	1.195(5)	2.393(4)	171.3(5)
	1.226(5)	1.172(5)	2.388(4)	169.3(6)
Mg (H maleate)$_2$ 6H$_2$O	1.228(7)	1.186(7)	2.414(5)	176(1)
Ca (H maleate)$_2$ 5H$_2$O (N)	1.305(3)	1.121(3)	2.424(2)	175.9(3)
Na H maleate 3H$_2$O (N)	1.367(2)	1.079(2)	2.445(2)	176.1(2)
Li H phthalate H$_2$O[a]	1.216(2)	1.181(2)	2.390(2)	171.7(1)
	1.296(2)	1.114(2)	2.400(2)	168.9(1)
Cu (H phthalate)$_2$ 2H$_2$O	1.290(4)	1.108(4)	2.391(2)	172.1(4)
K (H nitrosodisulphonate)[b]	1.23(1)	1.22(1)	2.45(1)	177.7(9)
	1.235(8)	1.211(7)	2.446(7)	178.6(7)
	1.249(8)	1.166(8)	2.415(8)	178.8(7)
K H formate (120 K)	1.270(1)	1.167(1)	2.437(1)	179.3(1)
(295 K)	1.281(3)	1.163(3)	2.444(1)	179.4(4)
K H oxydiacetate	1.328(3)	1.152(3)	2.476(2)	174(2)
K H crotonate	1.348(2)	1.141(2)	2.488(2)	178.0(3)
K H (dichloroacetate)$_2$	1.392(4)	1.107(4)	2.498(3)	177.8(3)
K H oxalate	1.463(5)	1.054(5)	2.518(3)	174.6(4)
Na H oxalate H$_2$O	1.531(2)	0.964(2)	2.566(2)	176.8(2)

[a]Two symmetry independent ions.
[b]Three symmetry independent ions. NO—H—ON bonds.

of the ion, as in the chloromaleate, does not seem to be as significant as disturbing the symmetry of the crystal field environment of the anion.

In the lithium hydrogen phthalate methanolate crystal structure there are two symmetry independent molecules in the unit cell of the crystal structure. One has a centered O—H····Ō bond within the experimental uncertainty with O—H distances 1.195(5) and 1.205(5) Å, while the other is definitely asymmetrical with O—H = 1.172(5) Å and H····O = 1.225(5) Å. There is a similar difference between the hydrogen bonds in the two symmetry independent anions in the hydrate, as shown in Table 3.4. The difference is ascribed by Küppers, Kvick, and Olovsson (1981) to differences in the ionic environment of the two hydrogen phthalate ions, but the difference is a very subtle one.

Short intramolecular O—H····O bonds are observed in the X-ray structure analyses of ammonium hydrogen caronate (**IV**) and potassium hydrogen caronate (**V**) by Küppers and Jessen (1993). The geometric conditions for the formation of such bonds is discussed and the relationship to the first and second dissociation constants.

(**IV**)

Table 3.5.
Strong O—H···Ō Hydrogen Bonds Observed in Hydrogen Phosphates and Sulphates From Neutron Diffraction Data.

Compound	H···O (Å)	O—H (Å)	O···O (Å)	O—H···O (°)
$H_2PO_4^- \cdot COH \cdot (NH_2)_2^+$	1.223(6)	1.207(6)	2.421(3)	169.9(4)
$Na_3H(SO_4)_2$	1.276(3)	1.156(3)	2.432(4)	179.1(3)
KD_2PO_4	1.397(5)	1.063(5)	2.455(6)	172.6(5)
$CaHPO_4$	1.283(3)	1.182(3)	2.461(2)	173.5(3)
NaH_2PO_4	1.459(8)	1.040(9)	2.485(5)	167.7(7)
	1.524(8)	1.026(8)	2.550(5)	177.6(7)
	1.590(7)	1.003(7)	2.591(5)	175.6(6)
	1.644(8)	1.001(9)	2.644(5)	177.0(7)
RbH_2PO_4	1.430(5)	1.061(16)	2.489(11)	175.4(25)

Strong anionic O—H···Ō hydrogen bonds are observed in the crystal structures of salts of hydrogen oxyanions. Data from the crystal structures of acid phosphates and sulfates which have been studied by neutron diffraction are given in Table 3.5. A more comprehensive review by Ferraris and Ivaldi (1984) reports H····Ō and O····Ō distances for more than 500 hydrogen bonds in borates (HBO_4^{4-}), $H_2BO_4^{3-}$, HBO_4^{2-}, HBO_3^{2-}, H_3BO_3); silicates ($HSiO_4^{3-}$, $H_2SiO_4^{2-}$); arsenates ($HAsO_4^{2-}$, $H_2AsO_4^-$, H_3AsO_4); phosphates ($H_2PO_4^{2-}$, $H_2PO_4^-$, H_3PO_4, $HP_2O_7^{3-}$, $H_2P_2O_7^{2-}$, $H_3P_2O_7^-$); carbonates (HCO_3^-); selenates ($HSeO_3^-$, H_2SeO_3, $HSeO_4^-$, H_2SeO_4); sulphates (HSO_4^-, H_2SO_4), HNO_3, and HIO_3. The range of H····Ō and O····Ō distances observed is shown in Table 3.6. All these hydrogen bonds are unsymmetrical and cover a range from strong to moderate with 1.5 Å > H····O > 1.5 Å.

Some Strong O—H···Ō Bonds are Disordered by Crystallographic Symmetry

When there is a crystallographic center or mirror plane of symmetry between the oxygen atoms, the hydrogen atoms half-occupy sites across the symmetry element, i.e., O—($\frac{1}{2}$H)····($\frac{1}{2}$H)—O, with very short O····O distances, as shown in Table 3.7. From diffraction experiments, it cannot be determined whether the disorder is spatial, i.e., O—H···O in one unit cell and O····H—O in another, or dynamic O—H···O ⇆ O···H—O. When the separation of the two half hydrogen electron or Fermi densities is small, it is very difficult to ascertain whether these bonds have single minimum potentials or double potentials, as discussed by Hamilton and Ibers (1968).

These symmetrical hydrogen bonds were referred to as type A by Speakman (1972), as distinct from the unsymmetrical type B. Those in potassium hydrogen ditrifluoroacetate and sodium and potassium hydrogen diacetate are candidates for single potential minima on the basis of their low-temperature infrared and Raman spectra from Novak (1974) and Hadzi, Orel, and Novak (1973). The evidence is

1. low infrared ν_s—OH stretching frequencies; from infrared, 720 cm^{-1} and Raman 320 cm^{-1},

Table 3.6.
Range of Hydrogen Bond Geometries Observed in Crystal Structures of Oxyanions.

Anions	No. of bonds	Range H----O	Range O----O
$HC\ell O_3$	8	1.331	2.560
HNO_3	3	1.339	2.535
HBO_3^{2-}, H_2BO_3	46	1.364–1.372	2.717–2.736
HBO_4^{4-}, $H_2BO_4^{3-}$, $H_3BO_4^{2-}$	43	1.465–1.485	2.790–2.847
$HP_2O_7^{3-}$, $H_2P_2O_7^{2-}$, $H_3P_2O_7^{-}$	24	1.538–1.564	2.507–2.551
HSO_4^{-}, H_2SO_4	34	1.530–1.551	2.569–2.595
HPO_4^{2-}, $H_2PO_4^{-}$, H_3PO_4	209	1.544–1.581	2.574–2.597
$HSiO_4^{3-}$, $H_2SiO_4^{2-}$	13	1.673–1.685	2.628–2.743
$HAsO_4^{2-}$, $H_2AsO_4^{-}$, H_3AsO_4	76	1.697–1.731	2.610–2.651
$HSeO_4^{-}$, H_2SeO_4	12	1.710–1.736	2.459–2.565
$HSeO_3^{-}$, H_2SeO_3	53	1.736–1.759	2.572–2.639

2.   high ν_s—OH/ν_s—OD ratio; and

3.   for $KH(CF_3COO)_2$ the O----O distance does not increase on deuteration.

Neutron and X-ray studies of tetramethyl ammonium hydrogen succinate monohydrate with charge density and electrostatic potential analyses by Flensburg, Larsen, and Stewart (1995), together with an H^2 NMR study by Kalsbeck, Schaumberg, and Larsen (1993), provide strong evidence for a single minimum potential.

Table 3.7.
Symmetrically Related O—($^1/_2$H)----($^1/_2$H)—$\bar{\text{O}}$ Bonds in Some Hydrogen Carboxylate and Phosphate Anions.

Crystal structures[a]	0.5[O—(1/2H)----($^1/_2$H)—O] (Å)	Symmetry
NH_4 H dichloroacetate (X)	1.216	$\bar{1}$
K H di-trifluoroacetate (N)	1.218	$\bar{1}$
K H dibenzoate (X)	1.220	$\bar{1}$
K H biphenylacetate (X)	1.221	1
K H succinate (N)	1.222	$\bar{2}$
K H di-aspirinate (N)	1.224	$\bar{1}$
Na H diacetate	1.237	2
Rb H oxydiacetate (N)	1.224	$\bar{2}$
hydrazinium hydrogen oxalate (N)	1.224	$\bar{1}$
K H di-p-hydroxybenzoate hydrate (X)	1.119	$\bar{1}$
K H malonate (N)	1.229	$\bar{1}$
$Ca(H_2PO_4)_2$ (X)	1.217, 1.211	1
(NH_4) H maleate (X)	1.218	m
Cu H maleate (X)	1.205	m
K H maleate (X)	1.218	m
K H difluoromaleate (X)	1.208	$\bar{m}$
(CH_3NH_3) H succinate H_2O (X, N)	1.221	$\bar{1}$

[a](X) = X-ray study, (N) = neutron study.

Table 3.8.
Hydrates With Oxonium Cations

H_3O^+	Monohydrates of HF, HCℓ, HBr, H_2SO_4, HCℓO_4, HNO_3, H_2SeO_4, CF$_3$HSO$_3$, C$_7$H$_8$HSO$_3$, Sb$_2$HSO$_4$SO$_4$)$_3$, Dihydrates of H$_4$P$_2$O$_6$, H$_2$SO$_4{}^a$
$H_5O_2^+$	Dihydrates of HF, HCℓ, HBr, HNO$_3$, HCℓO_4, Mn(HSO$_4$)$_2$ and HCℓO_4·2.5H$_2$O
	ZnCℓ_2·0.5HCℓ·H$_2$O, HCℓ·3H$_2$O, MnH$_2$SO$_4$·4H$_2$O, HAuCℓ_4·4H$_2$O, C$_4$H$_3$O$_5$N$_3$·3H$_2$O,
	nitro compounds:
	2-nitro-1,3-indandione·2H$_2$O C$_9$H$_9$O$_6$N,
	3-nitronanilic acid·2H$_2$O,
	5-nitrobarbituric acid·3H$_2$O C$_4$H$_3$O$_5$N·3H$_2$O
	cyanonitranilic acid·2H$_2$O
	nitronic acid·2H$_2$O
$H_7^+O_3$ ($H_5O_2^+$ + H_2O)	Trihydrates of HCℓ, HBr, HCℓO_4, HBr·4H$_2$O, C$_6$H$_7$Br$_2$HSO$_3$, nitranilic acid and HBr·4H$_2$O^b
$H_9^+O_4$ (H_3^+O + $3H_2O$)	Tetrahydrates of HBr^b, nitranilic acid, F$_3$CSO$_3$H, F$_3$CSO$_3$·5H$_2$O, and HCℓ·6H$_2$O
$H_{11}^+O_5$ (H_3^+O + $4H_2O$)	

^{a}Contains two H_3^+O ions H-bonding to $SO_4^=$.
^{b}Contains both $H_7O_3^+$ and $H_9O_4^+$ ions.

3.5 $\overset{+}{O}$—H----O BONDS

The strong $\overset{+}{O}$—H----O hydrogen bonds observed in the hydrates of strong acids are asso-ciated with the concept of the *hydrated proton*. That the H$^+$ ion in water or aqueous so-lutions occurs as a hydrated H$_3$O$^+$ species (the oxonium ion) dates back to the early part of the century as a mechanism for acid catalysis and to explain the high mobility of hy-drogen ions. The diaqua oxonium in H$_5$O$_2^+$ was suggested by Huggins (1936a). Infrared spectroscopic studies of the hydrates of strong acids by Bethell and Sheppard (1953) sug-gested that the H$_3^+$O is a symmetrical pyramid. A very comprehensive review of the sub-stantial literature relating to the hydrated proton is provided by Ratcliffe and Irish (1985).

Species of hydrated protons are observed in the crystal structures of the hydrates of strong acids. A large number of these, shown in Table 3.8, have been identified by phase diagrams, infrared spectra, or crystal data. These hydrates generally have melt-ing points between 0 and $-50°C$, but there are some notable exceptions, e.g., H$_2$SO$_4$·H$_2$O at 81.5°C and HCℓO_4·H$_2$O at 50°C.

3.6 THE HYDRATED PROTON

In a review of the hydrated proton in solids, Lundgren and Olovsson (1976) postulated a series of *hydrated proton complexes* or *oxonium water clusters* in mono- to hexahy-drates of strong acids from H$_3$O$^+$ to H$_{13}$O$_6{}^+$. This work was based mainly on the O----O distances, since many of the analyses were before hydrogen atom positions could be determined by X-ray analysis. They showed that the spread of the O----O distances in these crystal structures systematically increases with the size of the complex or cluster from 2.41 to 2.50 Å for monohydrates, 2.42 to 2.52 Å for dihydrates, and 2.44 to 2.89 Å for trihydrates and higher. In the monohydrates or dihydrates the oxonium cations are hydrogen-bonded to the anions. In higher hydrates, oxonium water clusters are formed which can either be isolated from other waters, i.e., surrounded by anions, or they can form infinite chains or layers with additional water molecules.

In the first neutron diffraction study of $HC\ell O_4 \cdot H_2O$, the oxonium ions were orientationally disordered. In the monohydrates of F_3CSO_3H and $H_3CC_6H_4SO_3H$, the H_3^+O cation is a rather flat pyramid with H—O—H angles of 109–115°, but the O—H and H----O bond lengths are different in the two structures, as shown in Figure 3.6. In the trifluoromethane sulfonate, one hydrogen bond is longer than the other two, and there is a corresponding difference in the covalent O—H bonds, i.e., there are two strong bonds and one moderate hydrogen bond. In the p-toluene sulfonic acid hydrate, both hydrogen bond and covalent bond lengths are similar, indicating three strong $\overset{+}{O}$—H----O bonds. With dihydrates and higher, the H_3O^+ cation is replaced by $H_5O_2^+$. This di-aqua ion has a linking O—$\overset{+}{H}$----O hydrogen bond which is short but unsymmetrical, except when there is a center of symmetry as shown by the neutron diffraction data given in Table 3.9.

With higher hydrates, the $H_5O_2^+$ ion is hydrogen bonded to other water molecules to form $(H_7O_3)^+$ and $(H_9O_4)^+$ clusters and chains, as shown in Figure 3.7, and Figure 3.8. In the crystal structure of yttrium deuterium oxalate trideuterate, analyzed by neutron diffraction by Brunton and Johnson (1975), $D_5O_2^+$ ions are hydrogen bonded within a deuterium oxalate cage, shown in Figure 3.9.

Hydronium ions also occur in the crystal structures of $HF.nH_2O$ with n = 1, 2, and 4 studied by Mootz, Ohms, and Poll (1981) and Mootz and Poll (1982), in $H_2SO_4.6.5H_2O$, and $8H_2O$ studied by Mootz and Merscherz-Quack (1987), and in $FSO_3H.H_2O$ and $SbF_6CF_3SO_3H$ by Mootz and Bartman (1991a,b). These crystal struc-

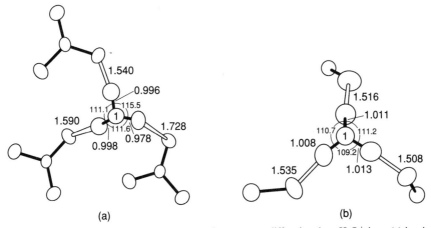

Figure 3.6. Geometry of hydrated proton clusters from neutron diffraction data; H_3O^+ ions. (a) in trifluorsulphonic acid monohydrate, $F_3CSO_3H \cdot H_2O$; and (b) in p-toluenesulphonic acid monohydrate, $H_3CC_6H_4SO_3H \cdot H_2O$.

Table 3.9.
O—H⋯O Hydrogen Bond Lengths (Å) in Some Crystal Structures With $H_5O_2^+$ Ions From Neutron Diffraction Analyses.

Structure	O⋯H	O⋯H⋯O		H⋯O
Picryl sulphonic acid $4H_2O^a$	1.609			1.776
		1.128	1.301	
	1.634			1.720
Yttrium oxalate trideuterate[b]	1.652			1.652
		1.221	1.221	
	1.711			1.711
5-Sulfosalicylic acid $3H_2O^c$	1.501			1.732
		1.095	1.340	
	1.689			1.725

[a]Lundgren and Tellgren (1974).
[b]Brunton and Johnson (1975).
[c]Williams, Peterson, and Levy (1972).

tures contain layers and ribbons of hydrogen bonds with H_3O^+ and F^- ions. As with the other strong hydrogen bonds, the position of the proton is variable, giving a variety of hydrogen bond lengths. The $\overset{+}{O}$—H⋯F hydrogen bond lengths range from 1.29 to 1.61 Å, the $\overset{+}{O}$—H⋯O from 1.48 to 1.81 Å and the F—H⋯$\overset{-}{F}$ from 1.40 to 1.61 Å.

The $HClO_4.nH_2O$ hydrates have been studied by X-ray diffraction, giving the hydrogen bond geometries shown in Table 3.10. As the hydration increases, the hydrogen bond tends to lengthen. The potential energy surfaces and vibrational spectra of

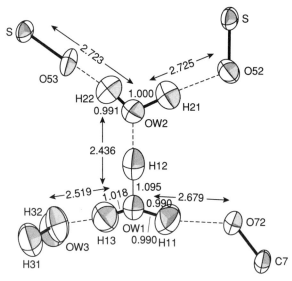

Figure 3.7. The $H_7O_3^-$ ion in 5-sulfosalicyclic acid trihydrate (from Williams, Peterson, and Levy, 1972).

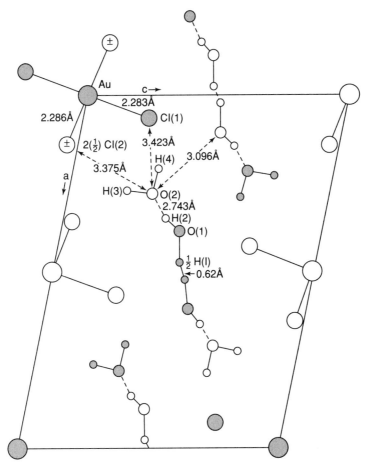

Figure 3.8. $(H_9O_4^+)_n$ chains from a neutron diffraction analysis of $HAuCl_4 \cdot 4H_2O$ by Williams and Peterson (1969). Note the three-centered $O_WH\begin{smallmatrix} O \\ Cl \end{smallmatrix}$ bonds.

Table 3.10.
Hydrogen Bond Lengths Less Than 2.0 Å and Angles in $HClO_4 \cdot nH_2O$ Hydrates.[a]

		O—H⋯O—Cl⁻ (Å)	O—H⋯O (Å)	O—H⋯O_W (Å)	O⋯H⋯O_W (°)
n = 1	range (3)	1.62–1.69	179–169		
	mean	1.65	174		
n = 2	range (3)	1.84–1.86	174–173	(1) 1.21	180
	mean	1.85	174		
n = 2.5	range (4)	1.82–1.93	177–154	(5) 1.56–1.77	171–149
	mean	1.86	164	1.66	163
n = 3.0	range (3)	1.90–1.92	165–147	(3) 1.54–1.77	175–175
	mean	1.91	158	162	175
n = 3.5	range (4)	1.87–1.99	162–148	(10) 1.54–1.94	176–153
	mean	1.93	155	1.77	165

[a]The H⋯O distances were obtained by normalizing the O—H bond lengths to 1.10 Å for O⋯O < 2.7 Å and 0.95 Å for O⋯O > 2.7 Å (from Brown, 1976).

47

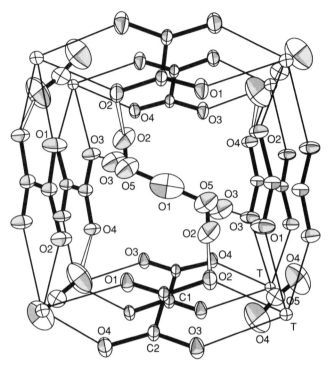

Figure 3.9. $D_5O_2^+$ ions in an enclosure of hydrogen oxalate ions in yttrium deuterium oxalate trideuterate (from Brunton and Johnson, 1975).

$H_5O_2^+$ and larger hydrated complexes were recently calculated by ab-initio methods by Ojamäe, Shavitt, and Singer (1995).

$H_3O^+ n(H_2O)$ charged clusters are believed to be present in the cold environment of the earth's upper atmosphere and in outer space. They can be produced in the laboratory, identified, and their infrared spectra studied by a method known as mass analyses molecular ion consequence spectroscopy (MAMICS) as described by Crofton, Price, and Lee (1994) and Castleman (1994). Binding energies for $H_3O^+ nH_2O$ with n = 1–7 from gas-phase ion-molecular equilibria measurements give $-\Delta H$ values ranging from 36 kcal/mol^{-1} for $H_3O^+ \cdot H_2O$ with a 10 kcal/mol^{-1} increment for each progressive hydration to n = 7. Similar data have been obtained for the hydrogen-bonded $NH_4^+ n(H_2O)$ and $NH_4^+ n(NH_3)$ clusters.

3.7 | OTHER POSITIVE ION HYDROGEN BONDS

A series of binary positive ion complexes was studied by Desmeules and Allen (1980) at the HF/4-31G level and by Del Bene (1989) at the MP4/6-31+G level. Some representative values for hydrogen bond distances and energies are given in Table 3.11.

Table 3.11.
Geometries and Energies of Some Linear Positive-ion Hydrogen Bonds. Top line,
Desmeules and Allen (1980), bottom line Del Bene (1989).

Complex	H⋯B (Å)	A⋯B (Å)	A—H⋯B (°)	E	ΔH
$(H_3N$—H⋯$NH_3)^+$	1.616	2.716	180	−32	−25
	1.750	2.816	171	−26	
$(H_3N$—H⋯$OH_2)^+$	1.617	2.671	180	−28	−17
	1.738	2.772	180	−20	−19
$(H_3N$—H⋯$FH)^+$	1.708	2.735	180	−18	
	1.778	2.809	171	−12	
$(H_3N$—H⋯$SH_2)^+$	2.334	3.367	180	−14	
	2.425	3.451	175	−12	
$(H_3N$—H⋯$C\ell H)^+$	2.377	3.397	180	−9	
	2.462	3.480	171	−7	

The higher level calculations significantly lengthened the hydrogen bonds and reduced the calculated energies.

3.8 | O—H⋯O BONDS

Strong neutral intramolecular O—H⋯O hydrogen bonds are observed in the crystal structures of β-diketo enols and some dibasic acids. In compounds containing the 1-hydroxy-3-keto configuration, (**V**), the planar configuration forces the formation of a short intramolecular O—H⋯O hydrogen bond. The crystal structures of a series of these compounds have been studied, two by neutron diffraction and 11 by X-ray diffraction. As shown in Table 3.12, the hydrogen bond lengths range from 1.32 to 1.70 Å, with the corresponding lengthening of the covalent O—H bonds for the shorter hydrogen bonds, consistent with low-barrier potential wells between the oxygen centers. The O⋯O distances determined by the molecular configuration are relatively constant. Similar intramolecular hydrogen bonds are observed in the crystal structures of pyridine 2,3-dicarboxylic acid (quinolinic acid), (**VI**), and 6-hydroxy-1-fulven carboxaldehyde, (**VII**), shown in Table 3.13. Resonance assisted hydrogen bonding, discussed in Chapter 6, contributes to the greater strength of these bonds, as compared to the moderate O—H⋯O bonds discussed in the next chapter.

(**V**)   (**VI**)   (**VII**)

Table 3.12.
Strong Intramolecular O—H····O Hydrogen Bonds in 1,3-Diaryl-Hydroxy-Keto Propanes (1,3-Propane Dione Enols).

Compound[a]	H····O (Å)	O—H (Å)	O····O (Å)
1-mesityl-3-(o-nitrophenyl)-	1.70	0.91	2.554
1-(p-methoxyphenyl)-3-			
(m-nitrophenyl)-	1.53	0.98	2.432
1-(p-nitrophenyl)-3-mesityl-	1.49	1.07	2.512
1-mesityl-3-(p-methoxyphenyl)-	1.42	1.16	2.499
1-(p-methoxyphenyl)-3-			
(m-nitrophenyl)-	1.39	1.10	2.434
1-phenyl-3-(p-nitrophenyl)-	1.38	1.12	2.465
1-phenyl-3-(p-methoxyphenyl)-	1.37	1.15	2.471
1-phenyl-3-phenyl-[b]	1.36	1.16	2.463
1-mesityl-3-(m-nitrophenyl)-	1.35	1.20	2.492
1-phenyl-3-mesityl-	1.33	1.20	2.461
1-phenyl-	1.32	1.24	2.489
3-(4-nitrophenyl-)[c]	1.38	1.14	2.445
3-(4-phenoxyphenyl-)[c]	1.42	1.11	2.443
3-(4-isopropenylphenol-)[c]	—	—	2.419
3-(p-chlorophenol)-phenol-	1.44	1.09	2.471

[a]1-Phenyl- and 1-phenyl-3-phenyl- from Jones (1976a, 1976b).
[b]By neutron diffraction.
[c]From Emsley et al. (1989). This paper also discusses differences in the chemical shifts of the enol proton and the in-frared vibration modes and provides evidence that these bonds are both short and strong. All other compounds from Bertollasi et al.(1991). σ(O—H) and (H····O) $\approx$ 0.01 Å; σ(O····O) $\approx$ 0.004 Å.

Potentially symmetric intramolecular O—H····O hydrogen bonds have been observed in a number of organometallic complexes. A symmetric O—H····O bond is reported by a neutron diffraction analysis of an organometallic complex [3,3′-dimethyl—3,3′-(2-ni-tropropanol diylidenediamono)-bis(2-butanone oximato-N,N′,N″,N‴]nickel II with the O—H····O bond lengths equal at 1.20(1) Å. In contrast, the organic complexes of nickel and platinum, [Ni and Pt $(C_5H_{11}N_2O)_2H]^+ \cdot C\ell^- \cdot 3.5H_2O$, contain short unsymmetrical $\overset{+}{O}$—H····O hydrogen bonds. The hydrogen bond is shorter in the nickel complex with H····O = 1.242(5) Å and O—H = 1.187(5) Å than in the platinum salt, where the bond

Table 3.13.
Strong Intramolecular O—H····O and O—D····O Hydrogen Bonds Observed in Quinolinic Acid (Pyridine 2,3-Dicarboxylic Acid) From Neutron Diffraction Data.

	Temp	H····O (Å)	O—H (Å)	O····O (Å)	O—H····O (°)
(H)[a]	(120 K)	1.227(3)	1.176(3)	2.400(2)	174.5(2)
	(298 K)	1.238(5)	1.163(5)	2.398(3)	174.4(2)
(D)[a]	(120 K)	1.253(2)	1.150(2)	2.401(2)	175.2(1)
	(298 K)	1.257(2)	1.138(2)	2.393(2)	175.2(1)
(D)[b]	(100 K)	1.227(3)	1.176(3)	2.400(2)	174.5(4)
	(293 K)	1.238(5)	1.163(5)	2.398(3)	174.4(4)

[a]Kvick et al. (1974).
[b]Measurements were made at 35, 80, 100, and 298 K and sophisticated thermal motion corrections were applied by Takusagawa and Koetzle (1979). They found that at every temperature the deuterium atom was farther from the midpoint of O····O than was the hydrogen atom.

lengths are H----O = 1.389(7) Å and O—H = 1.087(8) Å. In tetraaqua bis(hydrogen maleate)Ni II, $[Ni(C_4H_3O_4)_2 \cdot (H_2O)_4]$, the hydrogen bond is also unsymmetrical with O----O = 2.438 Å and H----O = 1.49 Å. In $[Pt\{(SO_3)_2H\}C\ell_2]^{-3}$ there is a short O—($^1/_2$H)----($^1/_2$H)—O distance of 2.382(6) Å across a mirror plane of symmetry.

3.9 $\overset{+}{N}$—H----N BONDS

The hydrogen bonding in binary complexes $(H_3N—H—NH_3)^+$ was studied by high-level ab-initio molecular orbital calculations by Del Bene (1989) giving H----N = 1.756 Å, with a binding energy of -26 kcal/mol^{-1}, in good agreement with the experimental ΔH value of -24 kcal/mol^{-1} of Payzant, Cunningham, and Kebarle (1973).

Strong hydrogen bonds $\overset{+}{N}$—H----N are observed in the crystal structures of the *proton sponges*. These compounds are fused-ring aromatic inside-protonated diamines with exceptionally high basicity which is attributed to the close intramolecular proximity of the basic centers. The molecules readily gain a proton in an acidic aqueous solution to become cations with the formation of $\overset{+}{N}$—H----N hydrogen bonds. Their high basicities were discovered by Alder et al. (1968), but did not attract much synthetic, spectroscopic, or crystallographic attention until nearly 20 years later by Staub, Saupe, and Krüger (1983), Saupe, Krüger, and Staub (1986), Staub and Saupe (1988), and Pawelka and Zeegers-Huyskens (1989). Typical examples are the protonated cations of 4,5-bis(dimethylamine) fluorine, DMAFH (**VIII**), 4,5-bis(dimethylamine) phenantrines, DMAPH (**IX**), and 1,8-bis(dimethylamine) naphthalene, DMANH (**X**).

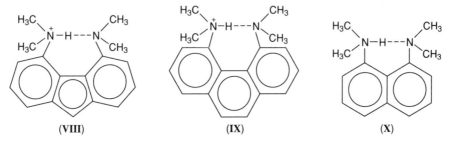

(**VIII**)   (**IX**)   (**X**)

In each case, the methyl groups provide sufficient steric interaction to force the nitrogen lone-pairs inwards to face each other. In the crystal structure of the unprotonated DMAN by Einspar et al. (1973), the naphthalene ring is distorted such that the two nitrogen atoms are 0.4 Å on either side of a mean plane giving a nonbonding N----N distance of 2.79 Å.

The protonation results in shorter N----N distances and relieves the necessity for an aromatic fused-ring distortion. A large number of salts of DMANH$^+$ have been studied by X-ray crystal structure analysis and by infrared and NMR spectroscopy in solution. The crystal structural data are shown in Table 3.14. These hydrogen bonds range from possibly single-minimum hydrogen bonds, as in DMAHN tetrazolate hydrate, to normal hydrogen bonds where there is no evidence of significant lengthening

Table 3.14.
Hydrogen-bond Lengths in the [DMANH]$^+$ Cations.

Anion	N····N	H····$\overset{+}{\text{N}}$	N—H	N—H$\overset{+}{\text{N}}$
BF$_4^-$	2.562	(1.31	1.31)^a	159
Br$^-$	2.554(5)	(1.31	1.31)^a	153(3)
[tetrazole]$^-$ H$_2$O	2.573(2)	(1.312)	1.312(5)	157(2)
1,8-bis(trifluoroacetamide)				
naphthalene (TFAN)	2.588(4)	1.42(3)	1.22(3)	256(3)
OTeF$_5^-$ (triclinic)	2.574	1.46	1.17	159
(orthorhombic)		(1.37	1.37)^a	140(3)
[tris(hexafluoracetato)]$^{-3}$ Cu^{2+}	2.65(2)	1.48(17)	1.27(21)	148(12)
NO$_3^-$	2.642(3)	1.39(3)	1.28(3)	164(2)
pentachlorophenolate	2.609(5)	1.38(5)	1.24(5)	700(4)
CℓO$_4^-$	2.675(5)	1.51	1.118(4)	167(4)
[2,4-dinitroimidazolate]$^{-1}$	2.606(3)	1.48(3)	1.18(3)	160(3)
[pentachlorophenolate]$^-$	2.555(3)	1.49(2)	1.11(2)	162(2)
[pentachlorophenol] at 100 K				
[1-oxo-2-phenyl 1,2-dicarbo-	2.577(3)	1.50(3)	1.22(3)	140(3)
dodecaborate]$^-$				
[chloranilate]$^-$ 2H$_2$O at 295 K	2.589(3)	1.51(3)	1.14(3)	155(2)
at 150 K	2.588(2)	1.59(3)	1.07(3)	152(2)
[pentafluorophenolate]$^-$	2.565(3)	1.56(6)^b		154(6)
[pentafluorophenol] at 100 K		1.84(7)		141(6)
[hydrogen squarate]$^-$	2.583(2)	1.59(2)	1.08(2)	157(2)
picrate	2.716(3)	1.68(3)	1.07(3)	162(3)
[3,4-furan dicarboxylate]$^-$	2.621(3)	1.62(2)	1.06(2)	155(2)
[tris(hexafluoracetato)]$^{-3}$ Mg^{2+}	2.60(1)	1.63(11)	1.25(11)	134(8)
1,8-bis(4-toluenesulphonamido)-	2.610(5)	1.63(5)	1.05(5)	152(5)
2,4,5,7-tetranitronaphthalene				
[hydrogen squarate]$^{-2}$ 4H$_2$O	2.594(3)	1.66(6)^b	0.97(6)	162(5)
at 100 K	2.574(3)	1.69(6)^b	0.94(6)	156(5)
[dihydrogen hemimellitate]$^-$ 0.5H$_2$O	2.604(2)	1.72(2)^c	0.94(3)	155(2)
at 100 K		1.72^c	0.94(6)	164(5)
[D-hydrogentartrate]$^-$ 3H$_2$O	2.610	1.75(5)^d	0.91(5)	157(5)
at 100 K		1.8^d	0.8(1)	160(10)
8-dimethylaminomethyl-	2.629(2)	1.61(3)^e	1.04(3)	165(3)
1-dimethylammoniomethyl-				
naphthalene nitrate				
at 100 K		1.83(5)^e	0.83(6)	161(5)

^{a}Mirror symmetry.
^{b}H disordered over two sites in 0.54:0.46 ratio.
^{c}H disordered over two sites in 0.7:0.3 ratio.
^{d}H disordered over two sites in 0.71:0.29 ratio.
^{e}H disordered over two sites in 0.66:0.34 ratio.
The references to individual crystal structure analyses are given in reviews by Llamas-Saiz Foces-Foces, and Elguero (1994) and by Jeffrey (1995).

of the covalent N—H bond length. As with the O····O distances in the 1,3-hydroxy ke-topropanes, the N····N distances are relatively constant, while there is a varying degree of lengthening of the N—H covalent bond and the hydrogen bond lengths range from 1.4 to 1.8 Å. Some are centrosymmetrical while others are unevenly disordered over two sites. As with the F—H····F bonds discussed previously, the hydrogen atom appears to be located on a flat potential energy surface, the nature of which is influenced by the crystal field environment.

An intermolecular $\overset{+}{N}$—H----N hydrogen bond is observed in the crystal structure of phthalazine semifluoroborate between the cation $C_8H_6N_2H^+$ and the molecule $C_8H_6N_2$, (**XI**), but it is only moderately strong with lengths N—H = 1.07(1) Å and H----N = 1.63(2) Å, illustrating the importance of the intramolecular steric compression in the DMANH$^+$ salts.

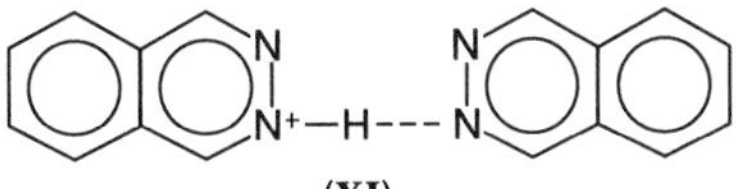

(**XI**)

Pulsed beam ion source mass spectrometry by Payzant, Cunningham, and Kebarle (1973) on gas phase NH_4^+ $(H_2O)_n$ clusters gave ΔH values for the $(H_3N$—H----$OH_2)^+$ bond of -24 kcal mol^{-1}.

3.10 | N—H----$\overline{\text{N}}$ BONDS

The protonation of DMAN to form a cation has also been observed with compounds with acidic N—H groups to form intramolecular N—H----N $anions$. Examples are salts of DMANH$^+$ with 1,8-bis(trifluoroacetamido) naphthalene (TFAN), (**XII**) and 1,8-bis(4-toluenesulphonamido)-2,4,5,7-tetranitronaphthalene (TSATNN), (**XIII**).

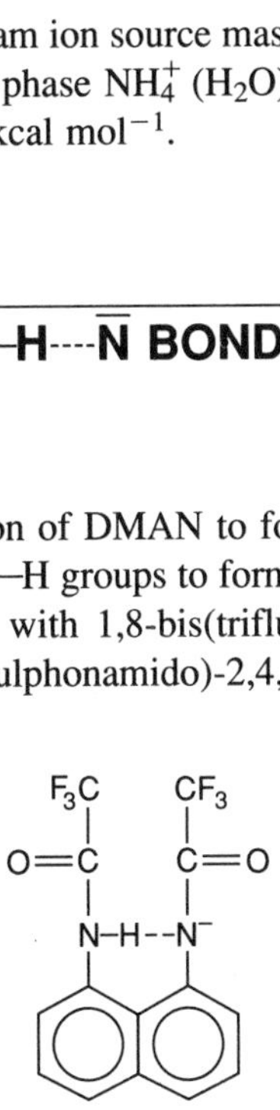

(**XII**)	(**XIII**)
TFAN(H)$^-$	**TSATNN(H)**$^-$

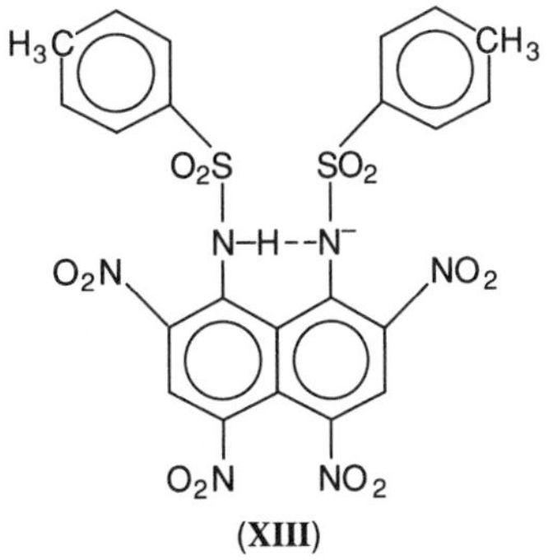

Ab-initio calculations by Del Bene (1989) on the $(H_3N$—H—$NH_3)^-$ complex indicated a weaker bond than in the positive ionic bond with H----N = 1.90 Å and a binding energy of -14 kcal/mol^{-1} versus 1.76 Å and -26 kcal/mol^{-1} for the positive ion bond.

In the DMAH$^+$—TSATNH$^-$ salt, the N—H---- $\overline{\text{N}}$ bond length in the cation is barely within the strong category, and the lengthening of the $\overset{+}{N}$—H bond is much less.

In both crystal structures, the N—H----N bonds are almost in the normal category with H----N = 1.85(5), N—H = 0.82(5), N----$\overline{\text{N}}$ = 2.600(5) Å, N—H----$\overline{\text{N}}$ = 153(5)° in TRANS, and 1.63(5), 1.05(5), 2.610(5) Å, N—H----$\overline{\text{N}}$ = 152(5)° in TSATNH$^-$.

In contrast, in DMAH$^+$—TFAN$^-$, the $\overset{+}{N}$—H····N bond length in the cation is one of the shortest observed and the N—H bond length is significantly extended.

All the evidence so far available points to the hydrogen bifluoride, hydrogen maleate, hydrogen phthalate, and hydrogen succinate anions as the only candidates for symmetrical single minimum hydrogen bonds. In all other examples of strong hydrogen bonds, the hydrogen, or deuterium, atoms appear to be located off-center. However, how much off-center varies with the crystal structure. The potential surface is clearly very flat and the position of the hydrogen atom appears to be determined by its long-range electrostatic environment.

This raises the question of existence of centered hydrogen bonding in anions or cations in solutions, where at least statistically, the environment will be symmetrical. There is evidence of symmetrical bonds in solution. An early example was in the crystal structures of nickel and copper bis(2,3-butanedione-dioxime), where it was suggested from IR spectral data that the hydrogen bonding was much stronger in solution. Other evidence comes from the infrared spectra and large downfield ^{1}H NMR shifts reported by Hibbert and Emsley (1990). A method for determining the relative strength of normal and low barrier hydrogen bonds O—H····O and O—H····N in solution is given by Schwartz and Drueckhammer (1995) using the cis-trans isomerism of the monotetrabutyl ammonium salts of litraconic, metaconic, fumaric, and maleic acids.

3.11 | HETERONUCLEAR BONDS

The candidates for strong heteronuclear hydrogen bonds are O—H····$\bar{F}$, F—H····$\bar{O}$, $\overset{+}{O}$—H····N, $\overset{+}{N}$—H····O, N—H····$\bar{O}$, and N—H····$\bar{F}$.

Theoretical calculations of the geometries and binding energies of $(H_2N$—H····OH$)^-$ and $(H_2N$—H····F$)^-$ are given by Del Bene (1989), with H····O$^-$ and H····F$^-$ having bond lengths of 1.72 Å and 1.60 Å and binding energies of 16 kcal/mol^{-1}.

O—H····F$^-$ hydrogen bonds occur in the crystalline adducts of carboxylic acids and fluoride salts. Data are available for KF and CsF succinic acid and KF malonic acid. In the succinic acid adducts, which are isostructural, the F$^-$ is on a mirror plane of symmetry, as in (**XIV**).

(**XIV**)

The O····F$^-$ distances are short in both structures, 2.441 and 2.449 Å, but the hydrogen bond lengths are different. In the potassium salt adduct O—H = 1.000(1) Å, H····F$^-$ = 1.411(2) Å from neutron diffraction data. In the cesium salt, O—H = 1.10(6) Å and H····F$^-$ = 1.35(6) Å. The H····F$^-$····H angle is bent, 125(1)°, 133(6)° in both structures.

Adducts With Strong Acids Form Strong Hydrogen Bonds

A strong F—H---O=C bond of 1.40 Å occurs in the crystalline adducts of $CH_3COOH.HF$ determined by Bartmann and Mootz (1991). In the same paper, proton transfer occurs in the adduct of CH_3COOH with the super-acid CF_3SO_3H forming a strongly hydrogen-bonded adduct with acetinium cations and the $CF_3SO_3^-$ anions.

An ab-initio molecular orbital calculation of the HOH---F^- dimer by Del Bene (1988a) gave a hydrogen bond length of 1.379 Å, with an energy of 27.3 kcal/mol[-1].

O_W—H---F^- hydrogen bonds occur in the hydrates of metal fluorides. There are many examples of these with O---F^- distances ranging from 2.608 Å (in $ZnF_2 \cdot 4H_2O$) to 2.820 Å (in $KMnF_4 \cdot H_2O$). The hydrogen bond distances based on normal O—H values of 0.96 Å range from 1.617 Å (in $ZnF_2 \cdot 4H_2O$) to 1.938 Å (in $MnFeF_5 \cdot 2H_2O$). Clearly, these are normal hydrogen bonds, with the difference reflecting the difference in donor strength between a carboxylic acid hydroxyl and a water hydroxyl.

Moderate Hydrogen Bonds 4

4.1 | INTRODUCTION

Moderate, conventional, or normal hydrogen bonds are formed between neutral donor groups A—H and neutral atoms containing lone-pair electrons. The exceptions are $\overset{+}{N}H_4$ and $\underset{}{\overset{+}{N}}$—H and O$\overset{-}{=}$C which do not generally form strong bonds. The strongest bonds are P—OH····O=P and the weakest are S—H····S=C. The hydrogen bonds involving oxygen and nitrogen atoms are particularly important in biological small molecules where they determine molecular packing and solvation, and in biological macromolecules where they determine packing and solvation and can influence conformation. In solids, they may link molecules into finite clusters or one-, two-, or three-dimensional infinite arrays. The flat potential energy surface which is characteristic of the strong hydrogen bonds and responsible for the variable location of the proton is replaced by a well-defined minimum in which the proton is located close to that of a nonhydrogen-bonded A—H covalent bond, as shown in the O—H vs H····O plot in Figure 3.1.

Tables of thermodynamic data are to be found in Pimental and McClellan (1960) and Joesten and Schaad (1974). The latter contains a table of thermodynamic and spectroscopic data for over 2000 hydrogen-bonded acid-base adducts which include the reported values for $K_{25°C}$, $-\Delta G$ and ΔH and ν_s X—H shifts when available. Although the error limits quoted are small—generally less than 0.5 kcal/mol^{-1}—the variations in values obtained by the application of different methods to the same systems are large enough to obscure any differences due to different molecule constitutions or configurations. For example, gas-phase measurements on the carboxylic acid dimers from formic to heptanoic acid gave $-\Delta H$ values from 6.4 to 8.6 kcal/mol^{-1} per hydrogen bond with estimated errors from 0.1 to 1 kcal/mol^{-1}. Six different experimental methods were used, and no correlation with the constitution of the carboxylic acid could be observed. A table for some 200 various one-, two-, and three-component systems using a wide variety of thermodynamic methods gave $-\Delta H$ values ranging from 3 to 12 kcal/mol^{-1}. For ethanol gas as donor, for example, the values ranged from 3.4 to 6.2 kcal/mol^{-1} (Appendix A in Pimental and McClellan, 1960). This lack of precision was a factor inhibiting development of the Badger and Bauer (1937) rule linking energies with infrared stretching frequencies. This is unfortunate because such a relationship could establish a useful link between the energies and stretching frequencies as well as to hydrogen-bond lengths in crystalline solids for which there is much information, as described later in this chapter.

4.2 | IN GAS-PHASE ADDUCTS

High-resolution rotational spectroscopy in the microwave wavelength region provides a wealth of information about hydrogen bonding in the gas-phase binary complexes. As described in Chapter 11, this information includes distances between centers of mass, dipole moments, stretching and bending force constants, and dissociation energies. An excellent review is provided by Legon and Millen (1986), with references to the complexes studied up to that date.

Legon and Millen (1987a) also observed an interesting relationship between the ratios of the hydrogen bond stretching force constants. These ratios are approximately constant between pairs of donor molecules irrespective of the acceptors. Using the force constants given in Table 4.1, for example,

$$K_\sigma(\text{FH----B})/K_\sigma(\text{C}\ell\text{H----B}) \approx 2$$

$$K_\sigma(\text{FH----B})/K_\sigma(\text{NCH----B}) \approx 2.5.$$

This suggested that the force constants could be expressed by the product

$$K_\sigma = c \cdot N \cdot E$$

where E is the *electrophilicity* of the donor molecule and N is the *nucleophilicity* of the acceptor, with c a constant. Arbitrary values of E = 10 for HF and N = 10 for H_2O gave a value of c = 0.25 N m^{-1}. This provided the basis for the table of electrophilicities and nucleophilicities shown in Table 4.2 which could be used as a measure of the rel-

Table 4.1.
Observed (and Calculated) Intermolecular Stretching Force Constants (Kg.Nm^{-1}) For Gas Phase Complexes AH----B.

B	AH				
	HF	HCℓ	HCN	HBr	HC≡CH
H_2O	24.9	12.5	11.1	(10.5)	6.5
CH_3CN	20.1	10.7	9.8	(8.9)	4.7
HCN	18.2	9.1	8.1	7.3	(4.4)
H_2S	12.0	6.8	4.7	5.9	(2.9)
PH_3	10.9	5.9	4.3	5.0	(2.6)
CO	8.5	3.9	3.3	3.0	(2.0)
N_2	5.5	2.5	2.3	—	—

Table 4.2.
Nucleophilicities, N, and Electrophilicities, E, for Gas-phase Dimer Components (from Legon and Millen, 1992).

Molecule	N	E
H_2O	10.00	5.0
HF	4.8	10.0
HCℓ	3.1	5.0
HCN	7.3	4.25
HC≡CH	5.1	2.4

ative strengths of hydrogen bonds. If $E_A \cdot N_B > E_B \cdot N_A$, the A—H----B bond is stronger than B—H----A. Thus for HF and H_2O, the stronger hydrogen bond is FH----OH_2. Since all the molecules in Table 4.2 can be donors or acceptors, this method predicts a sequence of bond strengths shown in Table 4.3.[1]

4.3 | GEOMETRIES FROM CRYSTAL STRUCTURE DATA

A major source of information about hydrogen bonds in the solid state comes from crystal structure analysis. With the introduction of direct methods for solving the phase problem and the major advances in instrumentation and computing, the number of crystal structure analyses has dramatically increased in recent years.[2] With the emphasis

---

[1]This is reminiscent of an empirical relationship proposed by Drago and Weyland (1965) for predicting gas-phase enthalpies. Each Lewis acid and each base are assigned E, electrostatic, and C, covalent, parameters and $-\Delta H = E_A E_B + C_A C_B$.

[2]See Appendix I for Databases.

Table 4.3.
Relative Strengths of Hydrogen Bonding in Gas Phase Adducts Predicted
From Values of $E_{A—H} \cdot N_B$.

Complex	$E_{A—H} \cdot N_B$	Complex	$E_{A—H} \cdot N_B$
FH----OH_2	100	CℓH----NCH	36.1
FH----NCH	73	NCH----NCH	31
FH----HC≡CH	51	FH----CℓH	31
HOH----OH_2	50	HOH----HC≡CH	25.5
CℓH----OH_2	50	CℓH----HC≡CH	25.5
FH----FH	48	NCH----HC≡CH	21.7
NCH----OH_2	42.5	CℓH----CℓH	15.5
		HC≡CH----HC≡CH	12.2

on the small molecule components of the biological macromolecules, many of these crystal structures involve molecules with hydrogen-bonding functional groups.

The development of the computer accessible Cambridge Crystallographic Data Base and the Inorganic Data Base, described by Allen, Kennard, and Watson (1994), has prompted many surveys directed specifically to the exploration of hydrogen-bond geometries. These are listed in Table 4.4. They fall into two categories, those aimed at specific classes of compounds and those directed towards specific types of hydrogen bonds.

O—H----O Bonds are the Most Studied Hydrogen Bonds

These are the hydrogen bonds for which most structural and spectroscopic information is available. As shown in Table 4.5, they occur in several different classes of compounds. Those in the carboxylic and amino acids and their hydrates extend over a wide range of hydrogen bond lengths. They demonstrate the effect of the different combinations of O—H and O_WH donor and acceptor groups. As shown in Table 4.6, the hydrogen bond lengths fall into four distinct categories due to the marked difference between the donor and acceptor properties of the water molecules. Water is a strong acceptor, i.e., base, and a relatively weak donor, i.e., acid, as also observed in the sequence of gas-phase hydrogen bond strengths shown in Table 4.3.

The hydrogen bonds in carboxylic acid hydrates have, in fact, some of the properties of strong bonds and are often described as such.[3] There is an observable lengthening of the covalent O—H bond and the O—H----O_W angles are close to being lin-

---

[3]Some authors prefer to use the categories very strong, with H----O < 1.5 Å and strong with H----O $= 1.5$–1.7 Å.

Table 4.4.
Statistical Surveys of Hydrogen Bond Geometries.

Carbohydrates
1. A survey of O—H----O hydrogen bond geometries observed in neutron diffraction crystal structure analyses of carbohydrates (Jeffrey and Takagi, 1978).
2. 100 O—H----O bonds from 24 neutron diffraction analyses of monosaccharides and maltose and sucrose (Ceccarelli, Jeffrey, and Taylor, 1981).
3. Hydrogen-bonding patterns in 43 monosaccharide and 15 disaccharide crystal structures (Jeffrey and Mitra, 1983).
4. A survey of O—H----O hydrogen bond and nonbonding geometries out to H----O distances less than 5.0 Å from 17 neutron diffraction analyses including two high-resolution studies of hydrate β-cyclodextrins (Steiner and Saenger, 1992a)
5. A survey of the effect of different patterns of O—H----O hydrogen-bonding on the covalent O—H bond length (Steiner and Saenger, 1992b).
6. A survey of the evidence for C—H----O hydrogen-bonding from carbohydrate neutron diffraction analyses (Steiner and Saenger, 1992c).
7. A statistical survey between O----O distances and occurrence of hydrogen-bonding in six high-resolution neutron crystal structure analyses of cyclodextrin complexes (Steiner and Saenger, 1994).

Amino acids
8. Analysis of data from neutron diffraction studies of 23 amino acids (Ramanadham and Chidambaram, 1978).
9. 168 $\overset{+}{N}(H_2)H$----O=$\overset{-}{C}$, O=C, O=$\overset{-}{S}$ bonds from 32 neutron diffraction analyses (Jeffrey and Maluszynska, 1982).
10. Hydrogen-bonding patterns in 56 crystal structures to illustrate the predominance of three-center hydrogen bonding (Jeffrey and Mitra, 1984).

Table 4.4. (continued)

Nucleosides and nucleotides

11. Hydrogen bonding in 86 X-ray crystal structure analyses classified according to functional groups: donors, —OH, O_WH, $\diagup$N—H P—OH, —N(H)H, $\overset{+}{N}$—H, $=\overset{+}{N}$(H)H; acceptors, O=C, $O\diagdown^{C}_{C}$, O_W, N$\diagup$, $C\ell^-$, O=P (Jeffrey, Maluszynska, and Mitra, 1985).

Barbiturates, purines and pyrimidines

12. A survey of 832 hydrogen bonds in 45 barbiturate crystal structures and 214 purine and pyrimidine crystal structures. Donor groups are $\diagup\overset{+}{N}$—H, $\diagup$N—H, —N(H)H, O_WH. Acceptor groups are O_W, O=C, OH, O$\diagdown$, N$\diagdown$, (Jeffrey and Maluszynska, 1986).

13. Hydrogen-bonding in crystal structures of nucleic acid components: purines, pyrimidines, nucleosides, and nucleotides. A comprehensive survey of the structural data available up to 1988 (Jeffrey, 1989).

Water in hydrates

14. The first review of water structure in organic hydrate crystals based on 62 crystal structure analyses. Most based on X—O_W and O_W---A distances (Clark, 1963).

15. A survey of O—H---O hydrogen bonds in salt hydrates to explore the effect of cationic charge on water coordination and geometry, from neutron diffraction data (Ferraris and Franchini-Angela, 1972; Chiari and Ferraris, 1982).

16. Geometrical data of O_WH---O_W two- and three-center hydrogen bonds from neutron diffraction analyses of most inorganic salt hydrates (Falk and Knop, 1973). This paper also contains the fundamental frequencies of uncoupled HDO bonds in 38 inorganic salt hydrates.

17. A survey of the hydrogen-bonding coordination of 207 water molecules in the hydrates of carbohydrates, amino acids and peptides, purines and pyrimidines, and nucleosides and nucleotides (Jeffrey and Maluszynska, 1990a).

18. A survey of C—H---O_W hydrogen bonds for 100 water molecules in 46 neutron diffraction analyses, showing how these bonds increase the coordination around the water molecules (Steiner and Saenger, 1993a).

General

19. O—H---O hydrogen bonds. A statistical survey based on 196 O—H---O hydrogen bonds from 45 crystal structures of polyalcohols, saccharides and related compounds, with a quantum mechanical analysis (Kroon et al. 1975).

20. O—H---O hydrogen bonds. An analysis of 356 hydrogen bonds from amino acids, peptides, and oligosaccharides (Mitra and Ramakrishnan, 1977).

21. An analysis of 1352 N—H---O=C hydrogen bonds in 889 crystal structures. Donors include $R_3\overset{+}{N}H$, $R_2\overset{+}{N}(H)H$, $\overset{+}{R}N(H_2)H$, $\overset{+}{N}(H_3)H$, $\diagup\overset{+}{N}$—H, $\diagup$N—H. Acceptors include O=C=O, O=C—NH$_2$, O=C$\diagdown^{OH}$, O=C$\diagup$ (Taylor, Kennard, and Versichel, 1984a).

22. Comparison of X-ray and neutron diffraction results for NH---O=C hydrogen bonds (Taylor, Kennard, and Versichel, 1983).

23. Survey of three- and four-centered NH---O=C hydrogen bonds in 889 crystal structures (Taylor, Kennard, and Versichel, 1984b).

24. Evidence for C—H---O, C—H---N and C—H---$C\ell$ hydrogen bonds from 113 crystal structures (Taylor and Kennard, 1982).

25. O—H---N(sp^2) hydrogen bonds. Survey with a theoretical analysis which discussed bond lengths and geometries (Llamas-Saiz et al., 1992).

26. N—H---N(sp^2) hydrogen bonds. Survey of N---N, H---N, and N—H---N distances for R—N(H)H and $\overset{R_1}{\underset{R_2}{\diagdown}}$N—H as donors varying the values of N---N$_{max}$ from 3.20 Å to 2.95 Å (Llamas-Saiz and Foces-Foces, 1990).

27. N—H---O hydrogen bonds in carboxylates, sulfonates and monohydrogen phosphonates as revealed by geometrical similarities (Piraud, Baudoux, and Durant, 1995).

60

Table 4.5.
O—H···O Bond Lengths (Å) From Different Classes of Small Molecules.[a]

	Carboxylic acids	Amino acids	Carbohydrates	Inorganic salt hydrates	Purines and pyrimidines	Organic hydrates	Nucleosides and nucleotides
Number of data	26	26	255	296	66	46	322
min	1.40	1.44	1.74	1.74	1.60	1.60	1.55
max	2.01	2.06	1.96	2.26	2.46	2.25	2.18
mean	1.71	1.74	1.82[b]	1.82	1.83	1.90	1.92

[a]These data are based on neutron diffraction analyses, except for the nucleosides and nucleotides.
[b]1.84 based on 450 combined X-ray and neutron values.

ear. The relationship of O—H to H···O is somewhat more regular than with the strong bonds discussed in Chapter 3, as shown in Figure 3.5. The ν_s-OH stretching frequencies lie in the 500–1500 cm^{-1} range. The crystal structure of ammonium tetraoxalate dihydrate, for example, contains short O—H···O_W, intermediate O_W···O=C, and long O_WH···O=C hydrogen bonds, all in the same crystal structure.

In the carbohydrate crystal structures determined by neutron diffraction, which are mostly monosaccharides, the O—H···O bond lengths lie in a narrower range. The only donors are C—O—H and the only acceptors hydroxyl oxygens and ether oxygens, i.e., ring or glycosidic, with water as donors and acceptors in the hydrated crystals. A survey by Ceccarelli, Jeffrey, and Taylor (1981) showed a small difference between the O—H···O—H bonds, with a mean H···O value of 1.81 Å, and the O—H···O⟨ bonds with a mean H···O value of 1.88 Å. Hydrates are relatively rare in monosaccharides, but become increasingly common in disaccharides and higher oligosaccharides. The O—H···O_W and O_W—H···O_W hydrogen bonds tend to be shorter and longer, respectively, than the O—H···O and O_WH···O bonds, illustrating again the weak donor and stronger acceptor property of water.

Normal Hydrogen Bonds are Soft Interactions

Normal hydrogen bonds have vibrational force constants that are an order of magnitude less than those for covalent bonds. Single C—C bonds, with bond energies around 80 kcal/mol^{-1} and stretching force constants of around 10^6 dynes cm^{-1}, can have lengths ranging from 1.50 to 1.57 Å due to conformational constraints, e.g., the necessity to close rings. Hydrogen bonds such as O—H···O or N—H···O with energies of ~5 kcal/mol^{-1} and X—H stretching force constants of less than 10^5 dynes cm^{-1} can have hydrogen bond lengths in crystal structures which range over more than 0.5 Å and angles 178–140°, due to conformation or packing constraints.

For statistical reasons, the most probable hydrogen-bond angle is not 180°, although this may correspond to the conformation of strongest energy. As discussed in Chapter 2, the probability of having an X—H···A angle of θ is proportional to $\sin\theta$. Statistical analyses of O—H···O bond angles show that while the most probable angle is about 155°, it becomes 180° when the *conic correction* is made.

Table 4.6.
Some O—H····O$_W$ and O$_W$H····O=C Hydrogen-Bond Lengths (Å) Observed in Carboxylic Acid Crystal Structures by Neutron Diffraction.

	H····O	O—H	O—H····O	Crystal structure

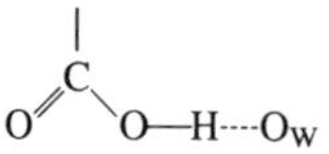

O=C—O—H····O$_W$

	1.403	1.069	178	Ammonium tetroxalate 2H$_2$O
	1.439	1.085	179	L-cysteic acid H$_2$O
	1.480	1.026	179	α-Oxalic acid H$_2$O
	1.483	1.040	173	Ammonium tetraoxalate 2H$_2$O
	1.492	1.031	177	Deuterated oxalic acid 2D$_2$O
	1.520	1.020	174	Deuterated oxalic acid 2D$_2$O

O=C—OH····O=C—OH

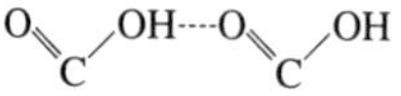

	1.642	1.011	165	Acetic acid
	1.640	1.002	176	Glycollic acid
	1.646	1.003	175	Glycollic acid
	1.658	1.009	177	Trifluoroacetic acid
	1.687	0.995	174	Succinic acid
	1.694	0.986	175	Acetic acid/phosphoric
	1.699	0.992	175	Acid complex

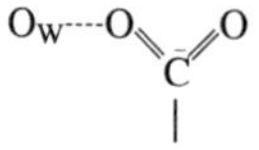

O$_W$····O=C

	1.764	0.954	171	Glycyl-glycine HCℓ H$_2$O
	1.766	0.966	170	Ammonium tetroxalate 2H$_2$O
	1.787	0.973	167	Ammonium tetroxalate 2H$_2$O
	1.770	0.974	172	L-arginine H$_2$O
	1.817	0.950	174	L-histidine HCℓ H$_2$O
	1.857	0.960	169	L-lysine HCℓ H$_2$O
	1.868	0.946	173	L-serine H$_2$O
	1.879	0.957	164	L-asparagine H$_2$O
	1.888	0.962	171	L-asparagine H$_2$O
	1.965	0.915	169	L-arginine H$_2$O

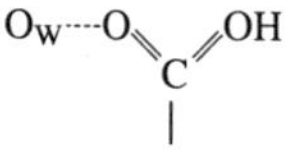

O$_W$····O=C—OH

	1.917	0.964	167	α-Oxalic acid H$_2$O
	1.979	0.956	157	α-Oxalic acid H$_2$O
	1.906	0.949	165	Ammonium tetroxalate 2H$_2$O
	2.000	0.945	158	Ammonium tetroxalate 2H$_2$O
	1.939	0.954	168	Deuterated α-oxalic acid 2H$_2$O
	2.008	0.954	156	Deuterated α-oxalic acid 2H$_2$O

C—OH····O=C, O=C—HO—C gas phase dimers

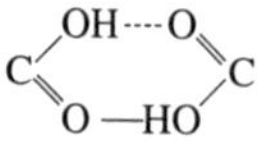

	1.66	1.04	180[a]	Formic acid
	1.64	1.04[a]	180[a]	Acetic acid
	1.65	1.04[a]	180[a]	Propionic acid

[a] Assumed.

Using neutron diffraction data from the crystal structures of some monosaccharides and from hydrated β-cyclodextrins, Steiner and Saenger (1992a) recently produced the interesting scatter plots for O—H----O angles *vs* hydrogen bond lengths and O----O distances, shown in Figure 4.1. They identified six regions.

1. the region of directional two-center bonds and the major components of three-center bonds.

2. the region of the minor components of three-center bonds with a few four-center bonds.

3. a region of nonbonding next neighbors.

4. a region of nonbonding second neighbors.

5. a poorly populated region of stretched two-center bonds.

6. the excluded region, due to O----O repulsions.

The solid line at O----O $= 2.65$ Å, $75° \langle\langle$O—H----O$\rangle\rangle$ $155°$ defines the *exclusion boundary* where no hydrogen bonds are observed, proposed by Savage and Finney (1986) and illustrated in Figure 2.9. Between (2) and (4) is a region of nondirectional monopole-monopole interactions (hydrogen bonds?).

There are few neutron diffraction data available from the crystal structures of small peptides. (The neutron diffraction data from the amino acids is not appropriate since the molecules in these crystal structures are zwitterions and the hydrogen bonds are $\overset{+}{N}$—H----O$=\overset{-}{C}$.) But a similar study of N—H----O$=$C bonds is feasible from good quality X-ray studies of peptides since the hydrogen positions of $>$N—H groups can be calculated from those of the heavy atoms with an acceptable degree of precision.

The Hydrogen Bond Data from the Crystal Structures of the Small-molecule Nucleic Acid Components Provide a Broader Perspective on Hydrogen Bonding

The crystal structures of the purines, pyrimidines, nucleosides, and nucleotides provide data on six different donor groups and six different acceptor groups. Although relatively few of these data are from neutron diffraction studies, they do provide a comparison of hydrogen-bond lengths for a wide variety of bonds within a limited class of compounds. The data obtained by using these analyses, where the positions of the hydrogen atoms were reasonably reliable, are given in Table 4.7. These results indicate the relative donor strengths P—OH $> \overset{+}{>N}$—H $> >$N—H $>$ C—OH $>$ O$_W$H $>$ N(H)H, and acceptor strengths O$=\overset{-}{P} >$ O$_W >$ O$=$C $>$ N$\langle$N(H)H $>$ O$\langle^C_C$.

In the nucleoside and nucleotide class, the short hydrogen bonds are P—OH----O$=\overset{-}{P}$, 1.55–1.69 Å; P—OH----OH, 1.65–1.89 Å; O$_W$—H----O$=\overset{-}{P}$, 1.66–1.88 Å; and P—O—H----O$_W$, 1.59–1.68 Å. All other O—H----O bonds lie in the range of 1.74–2.18 Å.

A more recent and more extensive analysis of $>$N—H----N$\langle$, C—O—H----N$\langle$, and O$_W$H----N$\langle$ bonds, irrespective of crystal structure type, by Llamas-Siaz and Foces-

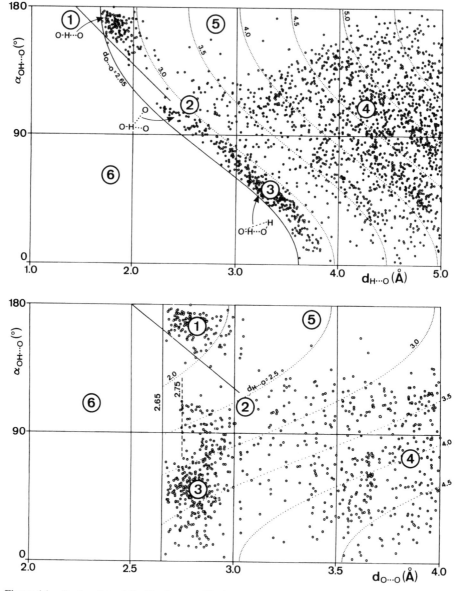

Figure 4.1. Scatter plots of O—H····O versus H····O bond lengths (top) and O····O bond distances (bottom) from neutron diffraction data of carbohydrate crystal structures (from Steiner and Saenger, 1992a, reprinted with permission).

Foces (1990) and Llamas-Siaz et al. (1992) gave comparable values, as shown in Table 4.8.

Another survey included 1352 $\overset{+}{>}$N—H····O=C and $>$N—H····O=C hydrogen bonds in 889 crystal structures irrespective of type of structure. These results are shown in Table 4.9. This large database permitted subtle distinctions between the donor and acceptor groups. For donors the relative donor strengths are $R_3\overset{+}{N}H > R_2\overset{+}{N}(H)H >$

Table 4.7.
Ranges and Mean Values (Å) of Hydrogen Bonds in Crystal Structures of Purines and Pyrimidines, Nucleosides, and Nucleotides (from Jeffrey, 1989).[a]

Donors	Acceptors									
	O=P	O$_w$	O=C	O<(C,H)	N<	NH$_2$	O<(C,C)	⟨m⟩[b]	Cℓ⁻	N<[c]
P—OH	5	5		2	1				1.65	
	1.55	1.59		1.65	1.76					
	1.69	1.68		1.89						
	1.58	1.64		1.77						
$\overset{+}{N}$—H		10						1.69	21	
		1.65							1.98	
		1.74							2.22	
		1.69							2.06	
N—H	9	22	119	24	47		5	1.80	8	284
	1.58	1.59	1.69	1.72	1.73		1.77		2.10	1.75
	1.89	2.06	2.32	2.16	2.23		2.25		2.17	2.32
	1.76	1.86	1.87	1.88	1.90		2.00		2.13	1.96
C—OH	6	29	76	78	54	1	5	1.85	23	120
	1.70	1.57	1.63	1.69	1.71		1.95		2.02	1.76
	1.95	2.04	2.65	2.57	2.62		2.97		2.43	2.19
	1.78	1.79	1.83	1.86	1.88	1.83	2.19		2.18	1.96
O$_w$H	11	35	29	41	23		3	1.90	26	
	1.66	1.60	1.72	1.76	1.78		1.92		2.13	
	2.11	2.25	2.23	2.18	2.19		2.10		2.40	
	1.86	1.90	1.86	1.90	1.92		2.03		2.24	
N(H)H	13	30	67	40	69	3	9	2.01	19	307
	1.67	1.71	1.68	1.86	1.85	2.01	1.94		2.14	1.83
	2.07	2.13	2.73	2.44	2.76	2.58	2.60		2.64	2.30
	1.86	1.95	2.00	2.07	2.04	2.20	2.26		2.29	2.06
⟨m⟩	1.80	1.85	1.89	1.92	1.95	—	2.12		2.19	

[a]For each entry, row 1 is the number of data; row 2 the minimum value; row 3 the maximum value; and row 4 the mean value.
[b]⟨m⟩ is the weighted mean.
[c]From survey 25 and 26 in Table 4.4.

$R_3\overset{+}{N}(H_2)H > \overset{+}{N}(H_3)H > {>}N—H$. For acceptors it is $O{=}\overset{|}{C}{=} > O{=}\overset{|}{C}—NH_2 > O{=}C\overset{C}{\underset{C}{<}} > O{=}\overset{|}{C}—OH$.

While these surveys gave a good qualitative view of the relative strengths of the various donor and acceptor groups, it was pointed out by Bürgi and Dunitz (1988) that they do not provide a theoretically sound basis for attempting to calculate bond energies.

Using O----O Distances as a Criterion of Hydrogen Bonding

As discussed in Chapter 2, repulsive interactions tend to maintain the distance of the O----O separation at ~2.8 ± 0.1 Å in hydrates and carbohydrates, while the H----O

Table 4.8.

Comparison of N—H····N and N—H····O Hydrogen Bond Lengths From Two Independent Surveys.[a]

	From Llamas-Siaz and Foces-Foces (1990)	Purines and pyrimidines 2-center and components of 3-center Jeffrey (1989)	Nucleosides and nucleotides 2-center and components of 3-center Jeffrey (1989)
		R—NH$_2$····N	
H····N (Å) mean	2.06	2.02[b]	2.13
min	1.83	1.85	1.89
max	2.30	2.28	2.76
		N—H····N	
H····N (Å) mean	1.96	1.93[c]	1.88
min	1.75	1.73	1.78
max	2.32	2.23	1.99
		C—O—H····N	
H····N (Å) mean	1.87	1.83	1.89
min	1.59	1.71	1.77
max	2.18	2.02	2.62
		O$_W$H····N	
H····N (Å) mean	1.96	1.99[d]	1.94
min	1.76	1.78	1.85
max	2.19	2.84	2.16

[a]All N—H bonds normalized.
[b]Excluding three three-center bonds with N—H····N < 140°.
[c]Excluding one three-center bond with N—H····N < 140°.
[d]Excluding one three-center bond with O$_W$—H····N < 140°; including one major component of three-center bond = 2.84 Å.

bond lengths have a much larger range from 1.8 to 2.6 Å. In consequence, O····O distances are a relatively insensitive measure of hydrogen bond length and angles, as shown in Figure 2.10.

Unlike the $>$N$^+$—H and $>$N—H bonds, the hydrogen atom positions in P—O—H and C—O—H bonds cannot be deduced from the covalent bonding geometry due to the rotational freedom around the P—O or C—O bonds. This makes it difficult to identify O—H····O hydrogen bonds in the low-resolution crystal structure analyses common with biological macromolecules, such as proteins, nucleic acids, and polysaccharides. The reliability of identifying O—H····O hydrogen bonds from O····O distances has been explored by Steiner and Saenger (1994), as discussed in Chapter 2.

About a Quarter of Normal Hydrogen Bonds are Three-centered

The survey of O—H····O hydrogen bonds out to O····O distances of 3.0 Å from carbohydrate neutron structure analyses by Ceccarelli, Jeffrey, and Taylor (1981)

Table 4.9.
Survey of Intermolecular N—H⋯O=C Hydrogen Bond Lengths (Å) (from Taylor, Kennard, and Versichel, 1984b).

Donor	Acceptor				Weighted row mean
	O=C=O	O=C—NH$_2$	O=C(C)(C)	O=C—OH	
R$_3$NH$^+$	11 / 1.72	2 / 1.84	1 / 1.94	0	1.755
R$_2$NH$_2^+$	47 / 1.796	3 / 1.79	3 / 1.97	6 / 1.89	1.805
RNH$_3^+$	226 / 1.841	15 / 1.89	8 / 1.87	68 / 1.936	1.865
>N$^+$—H	36 / 1.869	12 / 1.86	2 / 1.84	11 / 1.98	1.887
$\overset{+}{N}$H$_4$	56 / 1.886	4 / 1.99	2 / 1.99	13 / 1.92	1.900
>N—H	74 / 1.928	597 / 1.934	38 / 1.970	117 / 2.002	1.945
Weighted column mean	1.855	1.931	2.024	1.972	

revealed that about 25% of the hydrogen bonds were three-centered.[4] These three-centered bonds frequently included a ring or glycosidic ether oxygen as one of the acceptors. In this way, it was possible to retain an energetically favored cooperative chain of ⋯O–H⋯O–H⋯O–H⋯ bonds, while incorporating the ether oxygens, as illustrated by the example in Figure 2.6.

In the carbohydrates and amino acid crystal structures, the close proximity of donor and acceptor groups favors three-center bonding. In addition, there is an excess of acceptor functionality over the number of donor hydrogen atoms available. In a disaccharide, for example, there are eight OH groups which can be donors and acceptors, and three ether oxygens which can only be acceptors. In these crystal structures, the three-center bonds as defined in Chapter 2 ranged from symmetrical, i.e., $r_1 \approx r_2$, $\alpha_1 \approx \alpha_2$, to quite asymmetrical, with $r_1 > r_2$, $\alpha_1 > \alpha_2$, as shown in the neutron diffraction data in Table 4.10, but the asymmetrical configuration is generally more common. The major components of three-center bonds have a bond length distribution not very different from that of the two-center bonds, as shown in Figure 2.6.

A similar proportion of about 25% is observed in the nucleic acid small molecule components, but as shown in Table 4.11, the percentage does depend on the donor-acceptor combination involved.

A more extensive review of N—H donors and C=O acceptors based on a data set of 889 structures by Taylor, Kennard, and Versichel (1984a) gave a proportion of 20%. They found the tendency to form three-center bonds decreased in the order

[4]The data in Table 2.8 indicate that a cut-off of 3.0 Å excluded 25% of the hydrogen bonds.

Table 4.10.

The Geometry of Three-Center Bonds Observed in the Neutron Diffraction Analyses of Carbohydrates and Amino Acids.

$$A{\rm —H}\begin{matrix} \theta \\ \alpha \\ \theta' \end{matrix}\begin{matrix} r \\ \diagup{\rm —B} \\ \diagdown \\ r' \diagdown{\rm B} \end{matrix} \qquad \Sigma = \theta + \theta' + \alpha$$

	r (Å)	r' (Å)	θ (°)	θ' (°)	Σ (°)
Carbohydrates					
Methyl α-D-altropyranoside	2.085	2.140	130	140	358
β-L-Arabinopyranose	2.201	2.839	150	147	360
	2.088	2.579	170	129	356
β-D-Fructopyranose	1.977	2.593	169	114	358
Methyl α-D-galactopyranoside	1.983	2.636	160	124	359
	2.127	2.210	137	146	360
β-L-Lyxopyranose	1.957	2.300	147	107	360
	2.114	2.634	146	123	359
Methyl α-D-glucopyranoside	1.998	2.391	152	106	360
α-L-Rhamnose H_2O	1.981	2.608	160	107	359
Methyl β-D-ribopyranoside	1.958	2.568	139	96	360
	1.989	2.495	148	135	360
α-L-Sorbopyranose	1.953	2.584	163	120	356
β-Maltose H_2O	1.927	2.613	150	97	360
3-Amino-1,6-anhydro-β-D-glucopyranose	2.031	2.713	139	90	356
D-Glucitol	2.219	2.328	165	107	358
Sucrose	1.908	2.506	168	94	360
Amino acids					
α-Glycine	2.119	2.364	154	115	359
Glycyl glycine HCℓ H_2O	2.070	2.361	155	97	360
Triglycine SO_4	2.044	2.537	146	111	357
	2.044	2.345	140	111	356
L-Glutamic acid HCℓ	1.977	2.504	148	117	358
$NH_4NH_3CH_2COOHSO_4$	1.910	2.630	158	101	358
	1.898	2.553	156	114	359
α-L-Glutamine	1.853	2.366	163	97	360
	1.752	2.836	164	109	359
β-L-Glutamine	1.844	2.578	168	96	359
L-Histidine	1.840	2.327	160	98	358
	1.786	2.520	158	107	360
L-Asparagine H_2O	1.833	2.291	157	99	356
L-Histidine HCℓ H_2O	1.741	2.656	168	94	359

$R_3\overset{+}{N}{\rm —H} < R_2H\overset{+}{N}{\rm —H} < RH_2\overset{+}{N}{\rm —H} < H_3\overset{+}{N}{\rm —H} < {\rm \diagup}\overset{+}{N}{\rm —H}$, which is the reverse of the order of mean bond lengths. This is understandable because weak bonds tend to have smaller X—H----A bond angles, leaving more space for a second acceptor group.

A much higher proportion of three-center bonds was observed in the zwitterion crystal structures of the amino acids surveyed by Jeffrey and Mitra (1984). In a sample of 49 crystal structures, 39 had three-center bonds with the configurations shown in Figure 2.6. This high proportion was attributed to *proton deficiency*. Carboxylate groups prefer to accept four hydrogen bonds, two per $\bar{C}{=}O$ group, but the $\overset{+}{N}H_3$ group has only three protons.

Table 4.11.
Percentages of Three-Center Bonds Observed in the Purine and Pyrimidine, Nucleoside and Nucleotide Crystal Structures According to Donor and Acceptor Groups.

Donors	O=P	O_W	O=C	O⟨C/H	N⟨	NH_2	O⟨C/C	Cl^-
						Acceptors		
P—OH			0					
$\overset{+}{N}$—H		8^a						5^a
N—H	14^a	28	18	23	22	—	—	14^a
C—OH	0^a	18	29	33	28	—	25^a	48
O_WH	10^a	14	37	19	21	—	—	4
N(H)H	31	29	21	20	36	—	25^a	9

^{a}Less than 20 observations.

Three-center chelated hydrogen bonds are also observed in the zwitterion crystal structures of the amino acids. These types of bonds are generally very asymmetrical, as is shown in Table 4.12.

Some examples of three-centered/bifurcated hydrogen bonds observed in purine, pyrimidine, nucleoside and nucleotide crystal structures are shown in Table 4.13 and Figure 4.2. More examples formed by water molecules are described in Chapter 9.

Four-center bonds are relatively rare. The general survey of 1509 N—H---O=C bonds found less than 4%. Generally, four-center bonds have a major component with a hydrogen-bond length of about 2.2 Å and two other components with distances about 2.5 Å, all with angles greater than 90° as illustrated in Figure 2.8.

Table 4.12.
Some Three-Center Chelated Hydrogen Bonds Observed in the Amino Acid Crystal Structures (Neutron Diffraction Data).

	r_1 (Å)	r_2 (Å)	θ_1 (°)	θ_2 (°)	α (°)	Σ (°)	Acceptors
Triglycine SO_4	2.001	2.225	161	126	69	356	O=S=O
$NH_4NH_3CH_2COOHSO_4$	1.936	2.713	166	129	60	355	O=S=O
	1.817	2.713	177	120	60	357	O=S=O
L-Tyrosine	1.789	2.625	170	133	57	360	O=C=O
L-Alanine	1.779	2.521	168	133	59	360	O=C=O
γ-Glycine	1.763	2.641	171	132	57	360	O=C=O
L-Lysine HCl 2H$_2$O	1.740	2.710	174	128	55	357	O=C=O
α-Glycine	1.728	2.648	170	131	57	358	O=C=O
Triglycine	1.723	2.510	177	117	66	360	O=S=O
L-Cysteine	1.710	2.711	174	130	55	359	O=C=O

Table 4.13.

Three-Centered/Bifurcated and Four-Centered Hydrogen-Bond Configurations in Purine and Pyrimidine (P & P) and Nucleoside and Nucleotide (N & N) Crystal Structures.

Donor	Acceptor A$_1$	A$_2$	A$_3$	Distances (Å) r$_1$	r$_2$	r$_3$	r$_4$	Angles (°) θ$_1$	θ$_2$	θ$_3$	θ$_4$
P & P											
—NH$_2$[a]	O–N–O (nitrate)		O=N	1.89	2.61	2.72	1.82	157	105	98	170
O$_w$H$_2$[b]	O=C	O–N–O (nitrate)		1.79	2.86	2.50	2.07	171	95	119	172
O$_w$H$_2$[c]	O=C	O=N	O=N	1.87	2.31	2.51	2.07	148	125	109	143
O$_w$H[d]	O$_w$	Br⁻	NH$_2$	2.00	2.33	2.65		104	159	95	
$\overset{+}{N}$(H$_3$)H[e]	O=C	O$_w$	O=C	2.17	2.34	2.44		131	120	104	
>N—H[f]	O=C	N	N	2.04	2.94	2.90		115	143	105	
N & N											
—NH$_2$[g]	OH	O=C	O<	2.00	2.41	2.75	2.16	149	115	94	141
—NH$_2$[h]	O=N̄	OH	O=N̄	2.01	2.74	2.58	1.94	171	94	103	169
—NH$_2$[i]	Cℓ⁻	OH	Cℓ⁻	2.35	2.79	2.49	2.68	164	88	105	145
—NH$_2$[j]	Cℓ⁻	OH	OH	2.54	2.38	2.78	1.86	140	114	89	160
O$_w$H$_2$[k]	N	O$_w$	O=P̄	1.93	2.75	2.53	1.85	152	104	120	151
N(H)H[l]	OH	OH	N	2.50	2.56	2.51*		157	105	9	
N(H)H[m]	NH$_2$	N	OH	2.53	2.79*	2.88		140	94	15	
OH[n]	O=C	O<	O	1.87	2.76	2.78*		158	111	91	

*Intramolecular bond.
[a] Adeninium dinitrate.
[b] Guanine picrate H$_2$O.
[c] C$_{12}$H$_{17}$N$_4$OS$^+$ · C$_{10}$H$_7$N$_4$O$_5^-$ · 2H$_2$O.
[d] Adeninium hydrobromide hemihydrate.
[e] Ammonium uracil-5-carboxylate monohydrate.
[f] 6-Methyl uracil-5-acetic acid.
[g] 3-Amino-6-(β-D-ribofuranosyl)-6H-1,2,6-thiadiazine 1.1 dioxide.
[h] Cytidinium nitrate.
[i] 1-(β-D-Arabinofurosyl)cytosine HCℓ.
[j] 1-D-Xylofuranosyl-cytosine HCℓ.
[k] 8,2-Anhydro-8-mercapto-9-β-D-arabinofuranosyl-adenine 5'-monophosphate 3H$_2$O.
[l] 5-Amino-1-β-D-ribofuranosyl-imidazole-4-carboxamide.
[m] 1-β-D-Arabinofuranosyl-cytosine.
[n] 2,2'-Anhydro-1-β-D-arabinofuranosyl-2-thio-uracil.

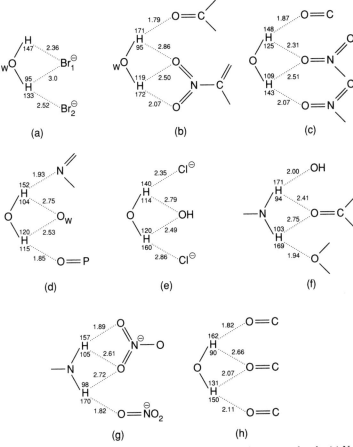

Figure 4.2. Some hydrogen-bonding configurations involving two three-center bonds. (a) N-[3-(aden-9-yl)propyl]-3-carbamoyl-pyridinium bromide hydrobromide dihydrate [ADENIC]; (b) guanine picrate monohydrate [GUNPIC10]; (c) thiamine picronolate dihydrate [THPROL]; (d) 8,2′-anhydro-8-mercapto-β-D-arabinofuranosyladenine-5′-monophosphate trihydrate [AMAFAP]; (e) α-D-xylofuranosyl cytosine·HCℓ [XFURCC10]; (f) 3-amino-6-(β-D-ribofuranosyl)-6H-1,2,6-thiadiazino-1,1-dioxide [RBFROX]; (g) adenium dinitrate [BIDRUBIO]; (h) perdeuterated violuric acid. [   ] Refcode in Cambridge Crystallographic Database (from Jeffrey and Saenger, 1991).

4.4 INTRAMOLECULAR BONDS

Intramolecular hydrogen bonding was recognized by Sidgwick and Callow (1924) as being responsible for significant differences in physical properties between ortho vs meta and para hydroxy and amino benzene derivatives. One of the early successes of infrared spectroscopy was the identification of the intramolecular hydrogen bonds in a variety of compounds such as o-nitrophenol, (**I**), and salicylaldehyde, (**II**), from the characteristics of stretching frequencies.

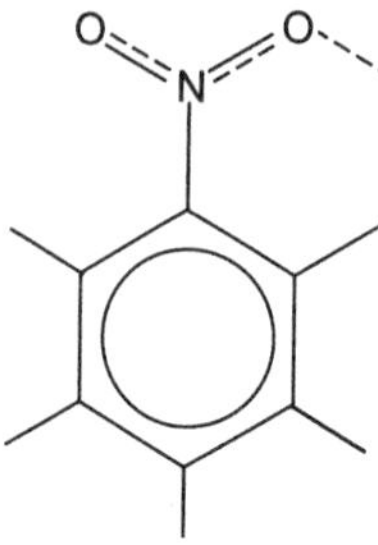

(I) 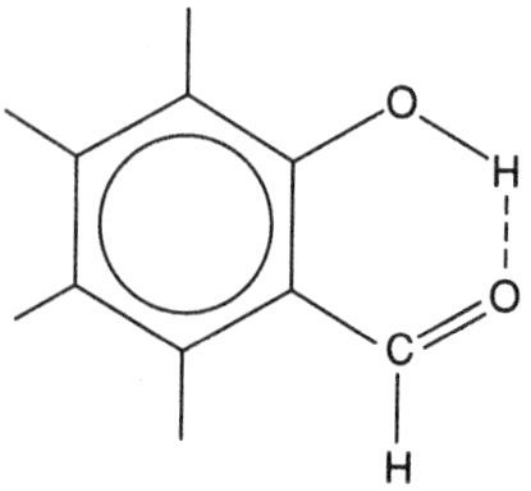   (II)

Intramolecular hydrogen bonds can be formed between donor and acceptor groups in the same molecule when the molecular configuration and conformation brings them within hydrogen bond geometry. When the A—(H)----B distance is very short, i.e., O----O < 2.5 Å and N----N < 2.6 Å, the strong hydrogen bonds discussed in Chapter 3 are formed. When the distances are between 2.5 and 3.2 Å, normal hydrogen bonds may be formed.

Intramolecular hydrogen bonds can occur in all three phases, but in solution with polar solvents and in crystals, they compete with intermolecular hydrogen bonds and generally lose. This is an inconvenience in theoretical calculations of gas-phase isolated molecules, which favor conformations with intramolecular hydrogen bonding, which are seldom observed in the world of solutions and crystals.

Carbohydrates are good candidates for studying intramolecular hydrogen bonding because of the high concentration of hydroxyl groups attached to a framework of adjacent carbon atoms. They can be on adjacent carbon atoms, i.e., *vicinal*, (**III**), or alternate carbon atoms, *syndiaxial*, (**IV**), or across rings, (**V**).

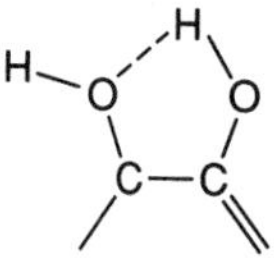

(III) 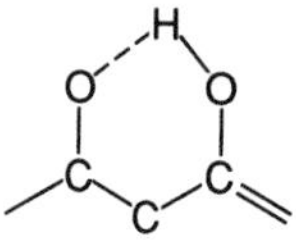   (IV) 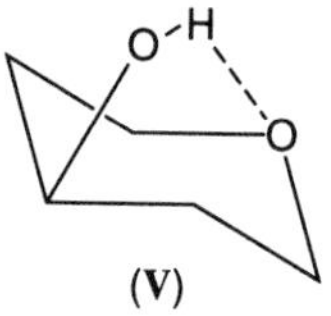   (V)

Vicinal intramolecular hydrogen bonding is rare except as the minor component of a three-center bond, because the angular geometry is unfavorable with O—H----O < 120°. Syndiaxial intramolecular hydrogen bonding has more favorable geometry, as shown by the examples in Table 4.14. In methyl β-D-ribopyranoside, methyl 1-thio-α-ribopyranoside, and methyl 1,5-dithio-α-ribopyranoside 0.25 H_2O, (**VI**), the unusual $^1C_4$-D chair conformation is stabilized by the syndiaxial intramolecular hydrogen bond. (In (**VI**), there are two symmetry-independent molecules with different bonding directions.)

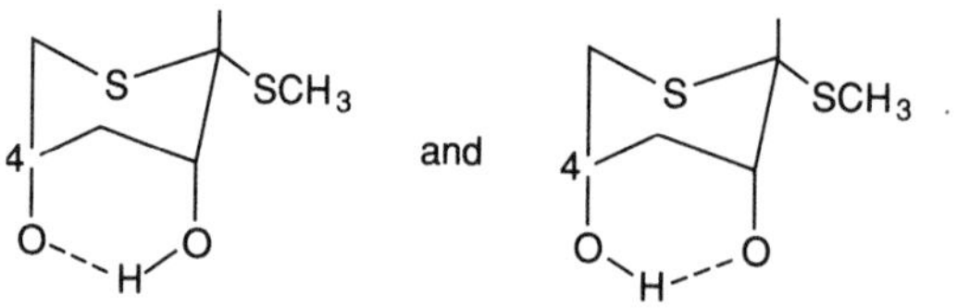

and

(VI)

Table 4.14.
Some Examples of Syndiaxial Intramolecular Hydrogen Bonding Observed in Carbohydrate Crystal Structures.

Molecule	Bond	H----O (Å)	O—H----O (°)
Methyl-β-D-ribopyranoside (neutron data)	O(2)H----O(4)	1.959	139
Methyl 1-thio-α-D-ribopyranoside	O(2)H----O(4)	2.09	142
Methyl 1,5-dithio-α-D-ribopyranoside	O(2)H----O(4)	2.32	135
	O(4)H----O(2)	2.20	129
Potassium D-gluconate Form A (neutron data)	O(4)H----O(2)	1.737	147

In the form A of potassium hydrogen gluconate monohydrate, which was studied by neutron diffraction, a syndiaxial O—H----O hydrogen bond stabilizes a straight carbon chain conformation, whereas in form B, without the intramolecular bond, the carbon chain conformation is bent, as shown in Figure 4.3.

Inter-residue intramolecular hydrogen bonding is commonly observed between the monomer components of disaccharides and higher oligosaccharides. Frequently these bonds are the major components of three-center bonds with minor components to nearby ring or linkage oxygen atoms. Two well-known examples of inter-residue intramolecular hydrogen bonding are shown in Figure 4.4. The N–H----O=C bonds which determine the conformation of the helices and sheets in protein crystals structures, discussed in Chapter 10, are also examples of inter-residue hydrogen bonds linking peptide units.

In hydrates, *indirect intramolecular hydrogen bonding* can occur involving a water molecule as in methyl β-maltose monohydrate in Figure 4.4(b). Some examples are given in Table 4.15. These also occur in proteins as discussed in Chapter 10.

These inter-residue intramolecular hydrogen bonds stabilize particular conformations in the crystals. *The degree to which they persist in solution is controversial.* Recent

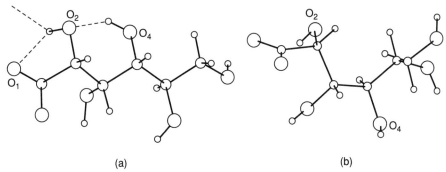

(a)   (b)

Figure 4.3. The straight chain and bent chain conformations of the gluconate ions as they appear in the A (a) and B (b) forms of potassium D-gluconate monohydrate (neutron data).

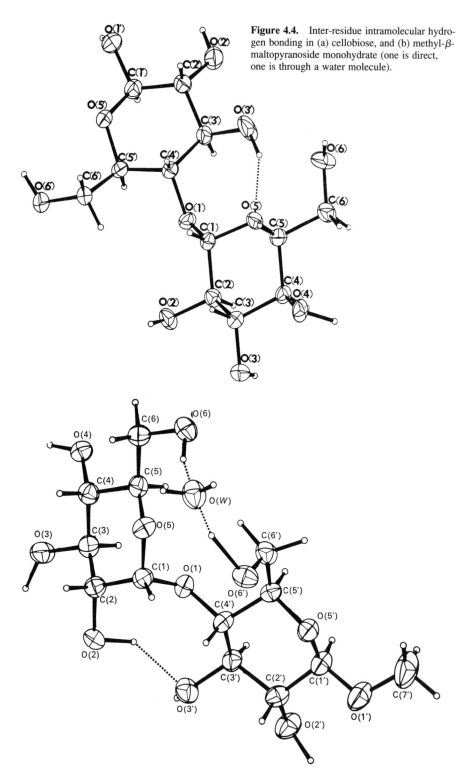

Figure 4.4. Inter-residue intramolecular hydrogen bonding in (a) cellobiose, and (b) methyl-β-maltopyranoside monohydrate (one is direct, one is through a water molecule).

Table 4.15.
Indirect Inter-residue Intramolecular Hydrogen Bonds in Some Oligosaccharides.

Compound	Bonds	$OH\cdots O_W$ (Å)	$O_WH\cdots O$ (Å)	$O\!-\!H\cdots O_W$ (°)	$O_W\!\!-\!H\cdots O$ (°)
Methyl β-maltoside	O(6')$\cdots$O$_W$H$\cdots$O(6)	1.99^a	1.90	148	147
α,α-Trehalose dihydrate	O(2)H$\cdots$O$_W$H$\cdots$O(6')	1.77	1.81	170	167
α,β-Melibiose trihydrate	O(6')H$\cdots$O$_W$H$\cdots$O(4)	1.88	1.85	172	170
Raffinose pentahydrate	O(2')H$\cdots$O(W4)H$\cdots$O(5)	1.80	1.96	158	157
Nystose trihydrate	O(3'')H$\cdots$O(W3)H$\cdots$O(3''')	1.73	2.18	154	155
α,β-Maltose	O(6)H$\cdots$O(4')H$\cdots$O(6')	1.81	1.79	176	157
1-Kestose	O(2)H$\cdots$O(4'')H$\cdots$O(3'')	2.13^a	1.88^a	148	166

^{a}Major component of three-center bond.

NMR studies of sucrose in water and acetone solutions by Adams and Lerner (1992) showed no distinction between the hydroxyl protons with regard to the temperature coefficients of the chemical shifts, scalar coupling constants, and exchange rates which might distinguish the hydroxyl groups involved in intra- and intermolecular hydrogen bonding.

In nucleotide and nucleoside crystal structures, intramolecular hydrogen bonding can occur between the ribofuranose moiety and the purine or pyrimidine base, stabilizing particular conformations. Some examples are shown in Table 4.16. As with the carbohydrates, these intramolecular hydrogen bonds are often the major or minor components of three-center bonds.

Intramolecular N—H$\cdots$O=C bonding is responsible for the helical and sheet structures of proteins which are discussed in Chapter 10. In the amino acids, the proximity

Table 4.16.
Some Important Intramolecular Hydrogen Bonds Between Bases and Riboside Observed in Nucleoside and Nucleotide Crystal Structures.

Compound	Donor	Acceptor	$H\cdots A'$ (Å)	$X\!-\!H\cdots A'$ (°)
6-Methyl 2'-deoxyuridine	O(5')H	O=C(4)	1.79	158
		O(4')	2.38	110
9-b-Arabinofuranosyl- 8-morpholineadenine dihydrate	O(5')H	N(3)	178	168
		O(4')	2.49	105
8(α-Hydroxyisopropyl- adenosine) hydrate	O(5')	N(3)	1.82	169
		O(4')	2.55	104
Inosine (orthorhombic)	O(5')$\cdots$	N(3)	1.95	165
	O(5'')$\cdots$	N(7')	2.61^a	90
5-Hydroxyuridine	O(5)H$\cdots$	O=C(4)	2.37^a	107
Xanthosine dihydrate	N(3)H	O(5')	1.86	167
		O(4')	2.56	113
Oxoformycin B	N(3)H	O(4')O(5')	1.85	157
		O(4')	2.37	117

^{a}Minor component of three-center bond where major component is intermolecular.

of the $\overset{+}{N}H_3$ and $O\text{=}C\text{=}O$ groups in **(VII)** will result in one component of the many three-center bonds being intramolecular.

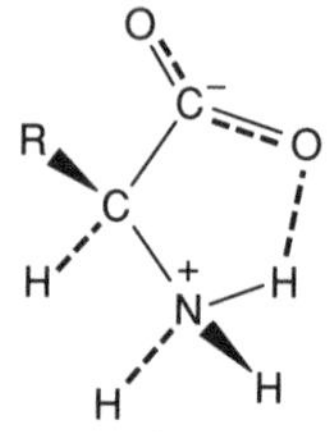

(VII)

Unfortunately there have been no systematic studies of the hydrogen bonding in the many X-ray crystal structure analyses of the linear and cyclic oligopeptides. Although these compounds do not give the high quality crystals for diffraction generally available from carbohydrates, there are sufficient good quality analyses of peptides that some valuable information could be derived. This is particularly true of the cyclic peptides where the molecular configurations place constraints on the formation of hydrogen bonds, and a significant number of nonbonding functional groups can be anticipated.

4.5 | BOND ACCEPTOR GEOMETRIES

Both experimental data and theoretical calculations show that in the gas-phase water dimer the hydrogen bond makes an angle of about 55° to the bisector of the H—O—H angle of the acceptor molecules, as in **(I)**.

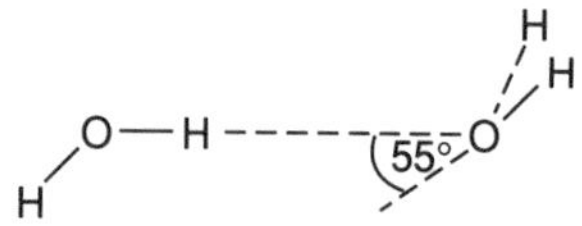

(I)

Legon and Millen (1987b) proposed as a definitive rule that in hydrogen-bonded gas phase dimers, the direction of the hydrogen bond is towards the axis of one of the lone-pair orbitals of the acceptor atom with a low energy barrier for a flip to the other lone pair. This suggests that the distribution of the oxygen lone-pairs rather than the direction of the dipole moment governs the acceptor geometry.

Two studies using crystal structural data were made of the directional properties of acceptor oxygen atoms, one by Taylor, Kennard, and Versichel (1983) and another by Murray-Rust and Glusker (1984). Both studies showed weak directionality in crystals consistent with the diffuse nature and weak directional properties of the lone-pair electron densities. For carbonyl groups, the H---O=C angles lie over the whole range from 90 to 180° with a maximum at ~130° as shown in Figure 4.5. The directionality observed in gas-phase dimers is not strong enough to persist over packing considerations in the solid state.

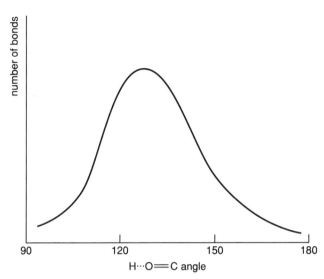

Figure 4.5. Typical distribution of angles at O in X—H···C=O in hydrogen bonds free of steric constraints. Observed for water-backbone hydrogen bonds in proteins (private communication, T. Steiner).

Table 4.17.
Two-Center Hydrogen Bond Lengths to Halide Ions Observed in Nucleic Acid Component Small Molecule Crystal Structures.

| | | Purines and Pyrimidines | | | Nucleosides and Nucleotides |
		$C\ell^-$	Br^-	I^-	$C\ell^-$
$\overset{+}{N}(H_2)H$	n	2			
$\overset{+}{N}(H_3)H$	r(min)	2.12			
	r(max)	2.15			
	$\langle r \rangle$	2.13			
$\overset{+}{N}-H$	n	17	2		4
	r(min)	1.98	2.26		2.11
	R(max)	2.09	2.28		2.22
	$\langle r \rangle$	2.04	2.27	2.67	2.14
$N-H$	n	8			
	r(min)	2.10			
	r(max)	2.17			
	$\langle r \rangle$	2.13			
—N(H)H	n	12	2	2	5
	r(min)	2.14	2.27	2.63	2.22
	r(max)	2.35	2.38	2.67	2.35
	$\langle r \rangle$	2.25	2.32	2.65	2.28
O—H	n	4			17
	r(min)	2.08			2.02
	r(max)	2.25			2.24
	$\langle r \rangle$	2.14			2.15
$O_W H$	n	26	4	2	1
	r(min)	2.13	2.31	2.54	2.28
	r(max)	2.40	2.42	2.61	
	$\langle r \rangle$	2.24	2.36	2.57	2.28

Halide ions can more easily occupy minimum energy sites on the hydrogen bond potential energy surfaces and the hydrogen bonds do not extend over the broad ranges observed in other intermolecular hydrogen bonds, as shown in Table 4.17. The mean increments in hydrogen bond lengths from $C\ell^-$, to Br^- to I^- is 0.10 Å and 0.32 Å. Halide ions as hydrogen bond acceptors give relatively large infrared spectral shifts, rather surprisingly in the order $C\ell^- > F^- > Br^- > I^-$ as pointed out by Allerhand and Schleyer (1963b).

4.6 TRANSITION METALS AS HYDROGEN BOND ACCEPTORS

Hydrogen forms complexes with transition metals. These can be terminal H—M bonds, C—H----M or $\overset{H}{\underset{C}{|}}$----M or bridging M----H----M. Many crystal structures of these hydrides have been reported and are reviewed by Teller and Bau (1981) and more recently by Heinekey and Oldham (1993).

It is only recently that crystal structures have been determined where the bonding corresponds to a hydrogen bond. These are the neutron diffraction analyses at 20 K of $[NPr^n_4]_2[PtC\ell_4]$ cis $[PtC\ell_2(NH_2Me)_2]$ by Brammer et al. (1991) and $Et_3NH^+Co(CO_4)^-$ by Brammer et al. (1992). A brief review of hydrogen bonds involving transition metals is by Brammer et al. (1995).

Infrared spectroscopy evidence is provided by Kazarian, Hamley, and Poliakoff (1993) for O—H----M hydrogen bonding in solution between fluoro alcohols and the neutral transition metal complexes $(\eta^5\text{-}C_5R_5)ML_2$, where R = H, CH_3, M = Co, Rh, Ir, and L = CO, C_2H_4, N_2 and PMe_3.

Two examples have been reported in which an Ir—H bond functions as the base. One is in $[IrH(OH)(PMe_3)]PF_6$ by Stevens et al. (1990) with a H----H distance of 2.4 Å. The other is the compound, (**I**), with H----H = 1.8 Å, by Lee et al. (1994).

(**I**)

R = Me, a; n—Bu, b; p-tolyl, c; Ph, d; p-FC$_6$H$_4$, e; 3,4-F$_2$C$_6$H$_3$, f.

Weak Hydrogen Bonds 5

5.1 | IN GAS-PHASE ADDUCTS

As with other hydrogen bonds, microwave spectroscopy and ab-initio calculations provide complementary methods for studying weak hydrogen bonding in the gas phase. The energies of weak hydrogen bonded binary complexes are comparable to those formed by van der Waals interactions. The distinguishing feature depends on the directionality of the donor A—H bond. Typical examples are the gas-phase adducts of HF, HCℓ, HBr, and HCN with N_2, CO, OCS, and CO_2 having binding energies of about 3 kcal/mol^{-1} by microwave spectroscopy, the references for which are given in a review by Legon and Millen (1986). In the theoretical calculations, a variety of possible configurations with comparable energies have to be explored of which only those with the correct configurational properties are hydrogen bonded. In the theoretical calculations of the 1:1 adducts of acetylene with HF, H_2O, and NH_3, by Frisch, Pople, and Del Bene (1983) for example, the configurations shown in Figure 5.1 were explored of which only 2, 4 and 7 were hydrogen bonded. A similar variety of hydrogen bonded and van der Waals complexes was explored in a study of the complexes of HF with the chloromethanes by Del Bene and Mettee (1993), shown in Figure 5.2.[1] The experimental measurements of Hunt and Andrews (1992) using Fourier Transform Infrared matrix isolation spectroscopy revealed a number of weak 1:1 and 2:1 complexes between HF and CCℓ_4, CH$_2$Cℓ_2 and CHCℓ_3. The van der Waals HF----CH$_2$Cℓ_2 is referred to as an *anti-hydrogen bonding* complex.

Acetylene was recognized as a weak hydrogen bond donor and weak acceptor through the triple bond, from early spectroscopic studies as reported by Green (1974).

---

[1]The chelated hydrogen bond configuration for CH$_2$Cℓ_2----F—H corresponds to the bifurcated bond of Pimental and McClellan (1960) and to the double hydrogen bonds reported from infrared data by Jamvóz and Dobrowolski (1993).

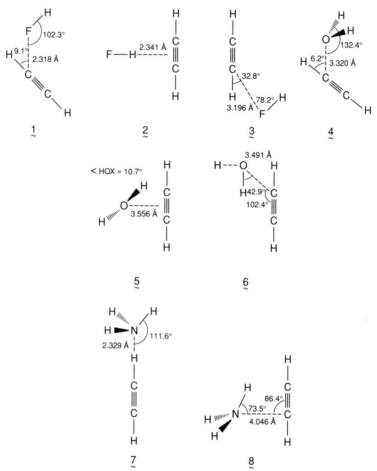

Figure 5.1. Calculated 1:1 adduct complexes of acetylene with HF, H_2O and NH3. Only for HF is the acetylene molecule the acceptor. The lowest energy configurations, and those predicted to be observed, are **2** with an estimated binding energy of ~3 kcal/mol; **4** also about 3 kcal/mol, and **7** at about 3.6 kcal/mol. Lower energy values were expected for larger basis sets and basis set superposition corrections (from Frisch, Pople, and Del Bene, 1983).

Crystalline acetylene has two polymorphs, one orthorhombic and one cubic. A neutron diffraction study at $-140°C$ by McMullan, Kvick, and Popelier (1992) of the orthorhombic form showed that the familiar herringbone packing shown in Figure 5.3 is stabilized by a $\equiv$C—H----π bond of 2.67 Å.

A microwave study of the HC$\equiv$CH----H_2O complex by Peterson and Klemperer (1984) gave the acetylene as the donor with $\equiv$C—H----OH_2 = 2.23 Å. Ab-initio calculations by Turi and Dannenberg (1993a) explored the weak complexes between HCN and HC$\equiv$CH with water and formamide giving the results shown in Table 5.1. A similar calculation by Turi and Dannenburg (1993b) of the acetic acid dimer gave an energy of about 0.5 to 1.0 kcal/mol^{-1} for the C—H----O$=$C bond, as compared with 4.7 kcal/mol^{-1} for the C—O—H----O$=$C hydrogen bond.

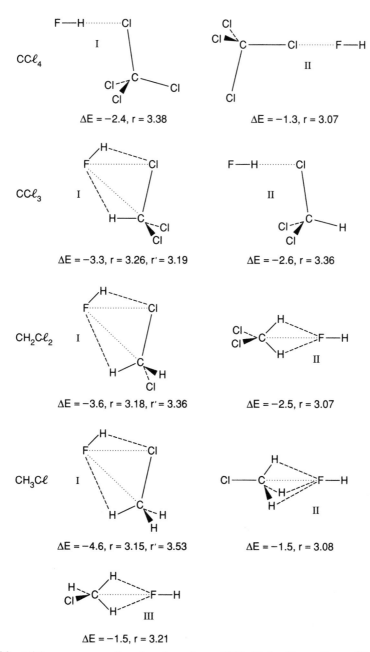

Figure 5.2. Minimum energy configuration of complexes of HF with the chloromethanes. ΔE in kcal/mol^{-1}, r is F—Cℓ in Å, r′ is F—C in Å (from Del Bene and Mettee, 1993).

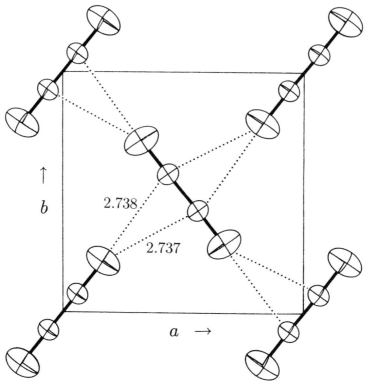

Figure 5.3. Arrangement of C_2D_2 molecules on the orthorhombic (001) face at 15 K, showing the shortest intermolecular separations in Å. Thermal ellipsoid surfaces are at 75% probability levels (from McMullan, Kvick, and Popelier, 1992, reprinted with permission).

The acceptor properties of multiple bonds and cyclopropane and benzene with $HC\ell$ as the donor were examined by microwave spectroscopy by Legon and Millen (1982) giving the T-shaped configurations shown in Figure 5.4.

Allerhand and Schleyer (1963a) expressed doubts that methane would form a hydrogen bonded adduct with water without substituent electron withdrawing groups. Recent ab-initio molecular orbital calculations suggest that, in fact, weakly bonded complexes might be formed. Szezesniak et al. (1993) examined seven configurations for CH_4----H_2O and found the binding energies in the sequence shown in Figure 5.5, ranging from 0.8 kcal/mol^{-1} at C----O = 3.5 Å for configuration F—H to 0.2 kcal/mol^{-1} at 3.5 Å for configuration F—C. Configuration F—H was favored over E—H by 0.02 kcal/mol^{-1}

On the other hand, Latajka and Scheiner (1987), Novao et al. (1991), and van Mourik and Diuyneveldt (1995) found the configuration E—H to be the more stable, but the energy differences are very small and within the limits of precision of the methods. The latter authors also examined H_2NCH_3----H_2O and $H3\overset{+}{N}CH_3$—H_2O. The substitution of H_2N made little difference to the interaction energy, but the ion-water adduct formed a strong hydrogen bond with an energy of 9.3 kcal/mol^{-1}.

Table 5.1.
Calculated Energies and Hydrogen Bond Geometries For N≡CH····OH₂,
N≡CH····O=CH₂, HC≡CH····OH₂ and HC≡CH····O=CH₂ Dimers By Ab-initio
MP2/D95++(dp) Calculations From Turi and Dannenberg (1993a).

Dimer	C—H (Å)	H····O (Å)	C····O (Å)	$\beta(°)$	$\gamma(°)$	Energy (kcal/mol⁻¹)
N≡CH····OH₂	1.076	2.051	3.126	180	180	−3.79
HC≡CH····OH₂	1.074	2.191	3.264	180	179	−2.19
N≡CH····O=CH₂		2.106	3.176	128	174	−2.74
HC≡CH····O=CH₂		2.251	3.229	101	151	−1.15

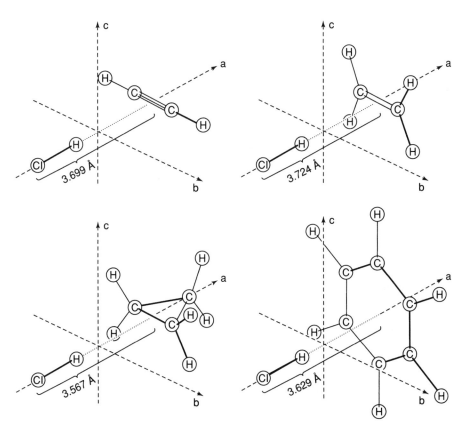

Figure 5.4. Weak hydrogen bonds with T configuration in gas-phase dimers formed from HCℓ and acetylene, ethylene, cyclopropane and benzene (from Legon, 1983).

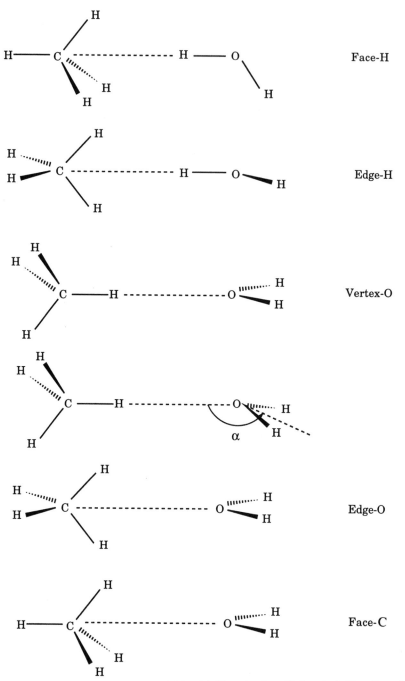

Figure 5.5. CH₄····H₂O configurations studied by ab-initio molecular orbital methods (from Szezesiak et al., 1993, reprinted with permission).

5.2 | C—H----B BONDS IN CRYSTALS

The most controversial and most discussed of the weak hydrogen bond donors in crystals is the C—H group. When attached to multiple bonds or to electron withdrawing groups, the spectroscopic and theoretical evidence provides clear evidence of hydrogen bonding. But the evidence is less convincing in the solid state, where these weak bonds are competing with stronger hydrogen bonds in addition to van der Waals interactions of comparable energies.

Under what circumstances are C—H----B hydrogen bonds formed in crystals and how significant are they in determining molecular configuration or packing? This is a particularly important question in molecular biology where the molecules involved have so many intra- and intermolecule C—H----O and C—H----N contacts.

Although spectroscopists recognized C—H----O hydrogen bonds, the crystallographers were skeptical. The C—H group qualifies as a hydrogen bond donor according to Pauling's table of atomic electronegativities. Pauling's electronegativity for carbon is greater than that of hydrogen and by that criterion C—H groups should be hydrogen bond donors. A charge density study by Craven (1987) derived a dipole moment of about 1 Debye at the C—H proton. The role of C—H bonds as potential hydrogen-bond donors was recognized from the beginning of interest in hydrogen bonding in a review by Hunter (1947) and later by Allerhand and Schleyer (1963a). Proton magnetic resonance evidence for hydrogen bonding by chloroform in solution was presented by Huggins and Pimental (1955). Strong spectroscopic evidence was presented by Green (1974) for the donor properties of R_3C—H groups when the proton is deshielded by the electron withdrawing properties of a substituent or a multiple covalent bond. Infrared studies by Krumm, Kurowa, and Rebare (1967) provided evidence of C—H----O≡C hydrogen bonding in polyglycine. Infrared matrix isolation of hydrogen bonding by Jeng and Ault (1991) in F_3CH, $(F_2CH)_2O$, and F_3COCF_2H also gave evidence of C—H hydrogen bonding. NMR studies of the chemical shift of acetylenic proton due to hydrogen bonding is reviewed by Foster and Fyfe (1969). The effectiveness in promoting C—H----O hydrogen bonding was found to be in the sequence H—C≡C—H, N≡C—H, $C\ell_3CH$, $C\ell_2CH_2$. Desiraju and Murty (1987) correlated ≡C—H----O distances from 3.2–3.4 Å observed in crystal structures with infrared C—H stretching frequencies from 3190 to 3280 cm^{-1}. As shown in Figure 5.6, there is a trend, but the correlation is poor.

There are a number of isolated references in the literature to C—H----O hydrogen bonds in crystal structures based on short C----O distances that were seriously considered in some early crystal structure analyses. In the crystal structure of dimethyloxalate by Dougall and Jeffrey (1953), C—H----O hydrogen bonds were regarded as responsible for the relatively high melting point, 54°C, and the difference in IR spectra between the solid and liquid. In the crystal structure of uracil by Parry (1954), one carboxyl oxygen accepts two N—H----O bonds, while the other has two C—H----O contacts. Ferguson and Tyrrell (1965) identified a C—H----O hydrogen bond in the crystal structure of O′-bromobenzoyl acetylene with a C----O distance of 3.26 Å. A neutron diffraction study of acetic acid by Jönsson (1971) showed intermolecular C—H----O≡C hydrogen bonding with H----O = 2.409 Å. Later C—H hydrogen bonds were largely

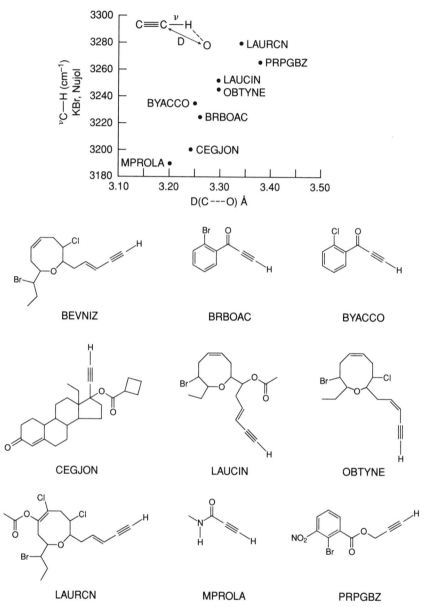

Figure 5.6. Correlation between ν_{C-H} stretching frequencies and intermolecular $\equiv$C—H···O=C and $\equiv$C—H···O$\big\langle$ distances. Letters refer to Cambridge Database Refcodes (from Desiraju and Murty, 1987). (A better correlation might be with H···O distances.)

ignored by crystallographers due to their reliance on comparison with van der Waals radius sums. An argument for C—H···O hydrogen bonding by Sutor (1962, 1963) was strongly refuted by Donohue (1968)[2] and the ensuing dogma amongst crystallographers that C—H···O bonds in crystals do not exist prevailed until a study by Taylor and

Kennard (1982). This survey examined 661 C—H contacts to O, N, Cℓ, and S atoms in 113 neutron diffraction crystal structures of organic compounds. They found 59 H----O distances between 2.05 and 2.4 Å with C—H----O angles greater than 90° as evidence of C—H hydrogen bonding. There were 11 C—H----N distances between 2.52 and 2.72 Å, 13 C—H----Cℓ⁻ distances between 2.57 and 2.94 Å, and 4 C—H----S distances between 2.5 and 2.9 Å, all considered less than the van der Waals distance criterium. Of the H----B distances 0.3 Å less than the sum of the H and B van der Waals radii, 90% were C—H----O. Although this relied on the dubious comparison with van der Waals radii, it was concluded that this was positive evidence for C—H----O hydrogen bonds, when the molecular packing provided appropriate geometry. This occurred in about 10% of the interactions studied.

In a study by Jeffrey and Maluszynska (1982) of hydrogen bond geometries from the neutron diffraction data on the amino acids, similar geometrical evidence for C—H----O=C̄ and C—H----Cℓ⁻ hydrogen bonds was presented. More than half of these interactions were from the more acidic α carbon C—H groups.

A number of examples of two- and three-center C—H----O bonds in the crystal structures of nucleosides with H----O distances between 2.1 and 2.5 Å and angles greater than 90° are reported by Jeffrey (1989), examples of which are shown in Table 5.2.

A recent study by Steiner and Saenger (1992c) gives an interesting perspective on the problem of identifying hydrogen bonds from C—H----O geometries in crystal structures. They used neutron diffraction data from 30 carbohydrate crystal structures including cyclodextrin hydrates. They identified 395 potential C—H donors, of which 61% were to OH, 27% to O$\langle$, and less than 2% to O=C.[3] The results are shown in the scatter plot of H----O vs C—H----O angles shown in Figure 5.7. The cluster of intramolecular bonds with angles less than 110° corresponds to syndiaxial interactions demanded by the configuration of the molecules. These are *forced contacts*, which may even be repulsive interactions, since with angles so close to 90°, the dipole-dipole and dipole-monopole components to the electrostatic energy will be very small. The other hydrogen bond data points scatter rather evenly. There is no evidence of directionality such as observed in Figure 4.1, where the shorter H----O hydrogen bond distances cluster in area ① where the O—H----O angles are > 150°. The even distribution either side of the van der Waals separation line suggests that the electrostatic components are nondirectional monopole-monopole with the contact distances determined by the crystal packing rather than the reverse. Some of the shorter H----O distances may in fact be repulsive forced interactions.

C—H----O$_W$ Hydrogen Bonds Increase the Coordination of Water Molecules

Water molecules in hydrates are commonly observed to accept one or two hydrogen bonds, or are coordinated to one or two cations. As will be described in Chapter

---

[2]Donohue corrected an earlier hydrogen-bonding base-pair concept of Watson and Crick (1953) and is described by Watson (1968) as knowing more about hydrogen bonds than anyone else in the world except Linus himself.

[3]These percentages are a function of the number of acceptor groups present, rather than the acceptor strength.

Table 5.2.
C—H····O Hydrogen Bonds Observed in Nucleoside and Nucleotide Crystal Structures (From Jeffrey, 1989).

Two-center C—H (θ, r to A)

Compound	A	r (Å)	θ (°)
5-nitro-1-(β-D-ribosyluronic acid uracil H_2O (neutron)	O<	2.08	156
	O=C	2.31	168
3-deaza-4-deoxyuridine	OH	2.22	166
uridine 3'-monophosphate H_2O	O<P (H)	2.23	149
	O—H	2.29[a]	171
	O=C	2.41	157
5-amino uridine	O<	2.53	151

Three-center C—H ($θ_1, r_1$ to A_1; $θ_2, r_2$ to A_2)

Compound	A_1	A_2	r_1 (Å)	r_2 (Å)	$θ_1$ (°)	$θ_2$ (°)
5-methoxy uridine	O<	O=C	2.17[a]	2.37	162	98
cytidinium nitrate	O—H	O<	2.19[a]	2.38[a]	153	96
inosine (monoclinic)	O—H	OH	2.25	2.28	131	139
2'-O-methyl cytidine	O—H	O<	2.25[a]	2.40[a]	166	104
	O<	OH	2.33[a]	2.53	154	90
5-nitro-1-(β-D-ribosyluronic uracil H_2O (neutron)	O=N	OH	2.33	2.36	132	134
	O=C	O=N	2.45	2.56	146	117
uridine 3'-monophosphate H_2O	O=C	O<	2.56	2.55[a]	156	116

[a]intramolecular.

9, there are a few examples of zero, four, and five acceptors. From a survey of 101 water molecules in 47 neutron crystal structure analyses, Steiner and Saenger (1993a) suggest that the *free acceptor potential* of the three-coordinated water molecules acts as an acceptor for C—H····O_W hydrogen bonds. In a sample of 101 waters, the inclusion of C—H····O_W bonds reduced the number of single acceptors from 44 to 18 and increased the number of double acceptors from 46 to 58. There was also a large increase in triple acceptors and a few examples of four and five acceptors. The hydrogen bond criteria used in this analysis were H····O_W < 3.0 Å for O—H····O_W, N—H····O_W; C—H····O_W < 2.8 Å, A—H····B > 90°; M^+····O_W < 3.0 Å. Steiner and Saenger (1993b) also examined the coordination of the water molecules in water clusters found in the crystal structure of Vitamin B_{12} coenzyme at 15 K by Bouquiere et al. (1993) and found a number where the C—H····OW bonds completed the tetrahedral coordination of the water, as shown in Figure 5.8.

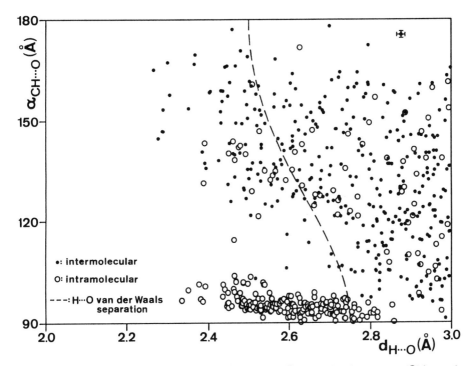

Figure 5.7.   Scatterplot of the angle C—H---O against $d_{H \cdots O}$. (●) intermolecular contacts; (○) intramolecular contacts; (----) H· · ·O van der Waals separation based on a spherical O atom with r ≈ 1.50 Å and a spheroidal H atom with a side-on radius r_S ≈ 1.25 Å and a head-on radius r_h ≈ 1.0 Å. The "classical" van der Waals separation would be a vertical line at $d_{H \cdots O}$ ≈ 2.6 Å. The error bars of the top right point represent the typical experimental uncertainty ± 0.01 Å and ± 1° (from Steiner and Saenger, 1992c, reprinted with permission).

The role of C—H---O hydrogen bonds in determining the molecular packing and conformation in a number of crystal structures was explored by Berkovitch-Yellin and Leiserowitz (1984) using atom-atom potential energy calculations.

A number of crystal structures were selected by Bernstein, Etter, and Leiserowtiz (1994) as examples where and ≡C—H--- $\overset{C}{\underset{C}{|||}}$ and =C—H---O=C play a significant role in determining the crystal packing are shown in Figure 5.9. In the crystal structure of benzoquinone, quinhydrone, and the complex of benzoquinone with p-chlorophenol, the C—(H)——O=C distances range from 3.33 to 3.52 Å. In the crystal structure of β-trans-fumaric acid by Bednowitz and Post (1966), the hydrogen bonded chains of the dicarboxylic molecules are separated by C—(H)----O contacts of 3.04 and 3.17 Å.

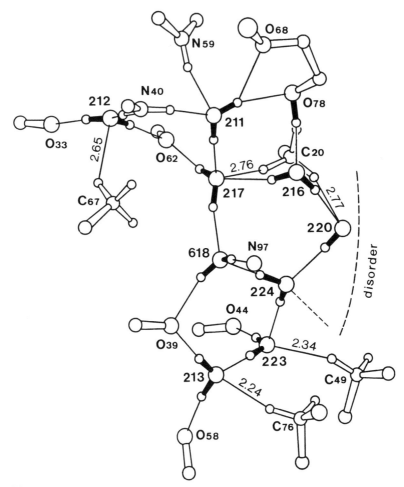

Figure 5.8. Hydrogen-bond network in the ordered water pocket region of vitamin B_{12} coenzyme. The lines with C—H$\cdots$$O_W$ distances are hydrogen bonds (from Steiner and Saenger, 1993b, reprinted with permission).

The eclipsed methyl group observed in the crystal structure of a tricyclic orthoamide by Seiler et al. (1987) is interpreted as due to the formation of three C—H$\cdots$$O_W$ hydrogen bonds, prompting an ab-initio calculation for the CH_3—OH_2 dimer which gave a binding energy of 1.7 kcal/mol^{-1}.

Recently Lutz, van der Mass, and Kanters (1994) measured the C—H stretching frequencies in the gas, liquid, and solid phases and in solution in CCl_4, of some molecules containing the $\equiv$C—H group, including some steroids. The free C—H stretching frequencies, which were in the relatively narrow range of 3330 ± 5 cm^{-1} in the gas phase, shifted by 18–62 cm^{-1} on hydrogen bonding in the liquid or solid phases, depending upon the compound examined.

In a series of studies by Desiraju (1989, 1990), the C—H$\cdots$O distances in crystal structures of chlorinated alkanes, alkyl, and alkene structures with C—H$\cdots$O interactions were examined. Using a database of 1075 C—H$\cdots$O geometries in crystal struc-

Figure 5.8. *(Continued)*

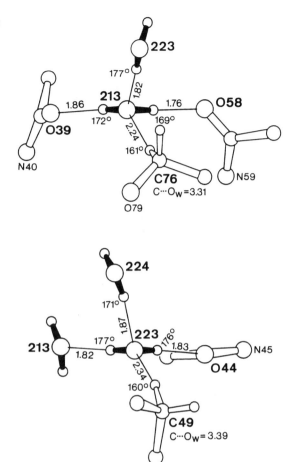

tures of containing $(C\ell_{3-n}C_n)$ C—H donors, Desiraju obtained the distribution shown in Figure 5.10. The shortest and mean C----O distances progressively increase in the expected sequence $C\ell_3CH > C\ell C\ell_2CH > C\ell_2CCH > C\ell_3CH$. That for $C\ell_3CH$ shows two maxima at 3.2 and 3.4 Å, suggesting two- and three-center bonds. Since the hydrogen positions can be calculated with reasonable reliability assuming tetrahedral carbon bonds and a C—H distance of 1.10 Å, a plot of the H----O hydrogen bond lengths could have been constructed and might have revealed more information by distinguishing between two- and three-center bonds. These results led to the suggestion by Pedireddi and Desiraju (1992) of a new scale of carbon acidity by using C—(H)----O distances in crystal structures. The latest pronouncement on the existence of the C—H----O bond by Desiraju (1991) is that "it certainly is."

In a neutron diffraction study of deuteronitromethane at 15 K by Jeffrey et al. (1985), seven C—D----O distances between 2.465(2) and 3.090(2) Å were observed, one of which was chelated. C—H----Pt and C—H----Cℓ hydrogen bonding is involved between the cationic and anionic species in a neutron diffraction study of $[NPr^n_4][PtC\ell_4]$ cis $[PtC\ell_2(NH_2Me)_2]$ at 10 K by Brammer et al. (1991).

In complexes of $(CH_3)_3\overset{+}{N}H$ with polyethers studied by Meot-Ner and Deakyne (1985) by mass-spectrometry methods, it is suggested that the —C—H----O interactions contribute significantly to the binding energy in complexes such as **(I)** and **(II)**.

$$H_2C-H---O$$

$$H_3C-\overset{+}{N}-H---O$$

$$H_2C-H---O$$

(I)

$$H_3C-\overset{\underset{|}{CH_3}}{\underset{CH_3}{\overset{|}{N}}}-C\overset{H--}{\underset{H-}{<}}H----OH_2$$

(II)

Multiple Bonds and Aromatic Rings as Weak Hydrogen Bond Acceptors in Crystals

There are several early references to aromatic rings as bases in Pimental and McClellan (1960). Green (1974) includes benzene and a number of methyl benzene and halogenated benzenes in a list of acceptors. The gas-hase dimer studies discussed earlier in this chapter provide ample evidence that π electrons can function as hydrogen bond bases. Oki and Iwamura (1967) postulated O—H----π bonds in 2-hydroxybiphenyl from infrared and ultraviolet spectral data. Short C—H----N≡C contacts with H----N ≈ 2.2 Å were observed in the early crystal structure analysis of N≡C—C≡C—H by Shallcross and Carpenter (1958).

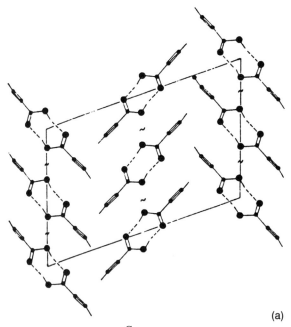

(a)

Figure 5.9.   (a) C≡C—H--- $\overset{C}{\underset{C}{|||}}$ hydrogen bonding in propiolic acid;

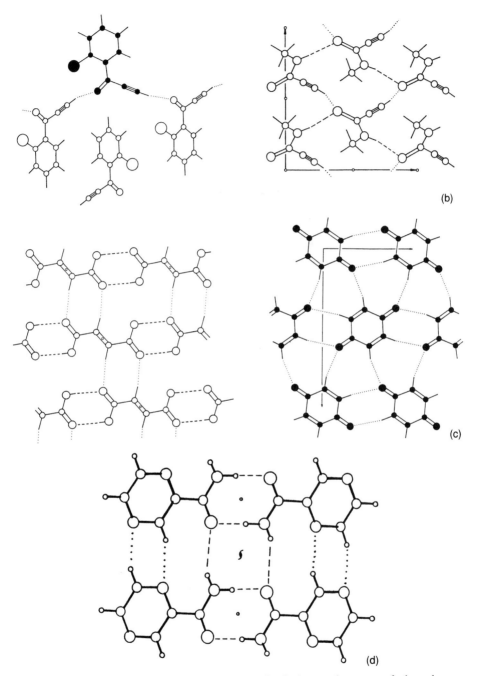

Figure 5.9. (*Continued*) (b) C≡C—H····O hydrogen bonding in the crystal structures of o-bromoben-zoylacetylene and N-methyl propiolamide; (c) C≡C—H····O hydrogen bonding in the crystal structures of fumaric acid and benzoquinone; (d) C≡C—H····N hydrogen bonding in β-pyrazine carboxamide. C—H····N bonds are dotted, N—H····O bonds are dashed (from Bernstein, Etter, and MacDonald, 1990, reprinted with permission).

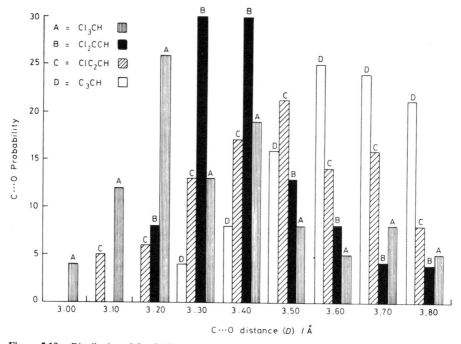

Figure 5.10. Distribution of C····O distances in crystal structures containing $(C\ell_3)_{3-n}$—C_n—C—H····O contacts (from Desiraju, 1989, reprinted with permission).

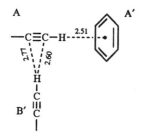

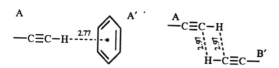

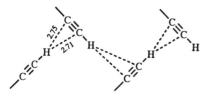

Figure 5.11. C≡C—H····π interactions in crystal structures of I, II, and III (from Steiner et al., 1995, reprinted with permission).

Crystal structure analyses of compounds (**III**), (**IV**) and (**V**) by Steiner et al. (1995) showed the interactions shown in Figure 5.11 with H----π (center) distances as short as 2.5 Å.

HC≡C—C—CH$_3$
         |
        OH

(**III**)   (**IV**)   (**V**)

In the crystalline state C—H----π bonds are reported in crystal structures by Hardy and MacNicol (1976), Atwood et al. (1991, 1992), and Ferguson et al. (1994) with H----π distances ranging from 2.1 to 2.7 Å.

Steinwender et al. (1993) identified O—H----π and ≡C—H----π bonds from infrared data. They also identified 45≡C—H----O hydrogen bond interactions in crystal structures with H----O bond lengths less than 2.5 Å. ⟩N—H----π hydrogen bonding in proteins has been suggested by Burley and Petsko (1985) and Levitt and Perutz (1988).

Mass spectrometry studies by Deakyne and Meot Ner (1985) indicated strong hydrogen bonding between $\overset{+}{N}H_4$ ions and ethylene, benzene, and fluorobenzene.

Compelling crystallographic evidence for C—H----π bonding comes from the low temperature X-ray crystal structure analyses of 2-butyne-HCℓ and 2-butyne-2HCℓ by Mootz and Deeg (1992) and the crystal structure analysis of toluene-2HCℓ and mesitylene-HCℓ by Deeg and Mootz (1993), shown in Figure 5.12. The hydrogen bond lengths

are H---- $\underset{C}{\overset{C}{\|\|}}$ = 2.30 Å and H---- ◎ = 2.32 and 2.53 Å.

C—F *and* C—Cℓ as Hydrogen Bond Acceptors

The gas-phase and theoretical calculations on dimers of HF and HCℓ with the chloromethanes discussed earlier in this chapter show that covalently bonded halogens can be hydrogen bond acceptors. Evidence for the C—F bond as a hydrogen bond ac-

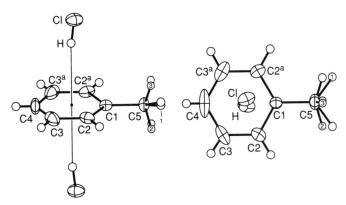

Figure 5.12. Cℓ—H---- ◯ hydrogen bonding in a toluene-2HCℓ adduct (from Deeg and Mootz, 1993).

ceptor in crystal structures was sought by Murray-Rust et al. (1983) with special attention to C—OH---F—C bonds in carboxylic acids. The results were not conclusive. It is relevant to note that in the highly hydrogen bonded crystal structures of 4-fluoro-β-fructopyranose by Podlahová and Loub (1984) and 2-fluoro-β-mannopyranoside by Choong et al. (1975), there is no evidence of O—H---F—C hydrogen bonds. While halide ions are strong hydrogen bond acceptors, there is no evidence from crystal structures supporting hydrogen bonds to halogens. This striking difference between the halogen ions and the covalently bonded halogen atoms must be a consequence of the absence of dipoles or monopoles on the uncharged halogen atom.

5.3 | FORCED C—H---O AND C—H---N CONTACTS

In the carbohydrate and nucleoside and nucleotide crystal structures, where there are many hydrogen bonding functional groups, the molecules are brought closer together by hydrogen bonding than if the intermolecular forces were van der Waals interactions. As a consequence, the C—H groups may be *forced* into closer contact to oxygen or nitrogen atoms than their normal equilibrium distances.

Notable examples are the $\diagdown$C—H---O=C bonds in the Watson-Crick A—U and Hoogsteen A—T base pairing in the oligonucleotide recently discussed by Leonard et al. (1995), shown in Figure 5.13, and in a r(UUCGCG) oligonucleotide with a novel U—U base pairing reported by Wahl, Rao, and Sundaralingam (1996). In the latter

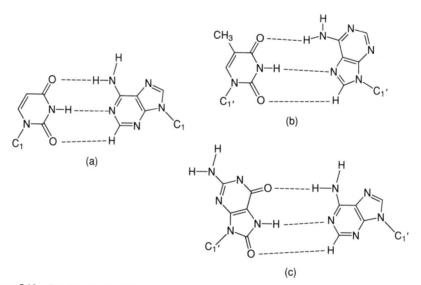

Figure 5.13.   Possible C—H---O hydrogen bonding in base pairing. (a) Watson-Crick; (b) Hoogsteen; (c) the A (anti), 0.8G (syn) of d[CGCAAATT(08G)GCG] (from Leonard et al., 1995).

case, this may not be a forced bond since there is only one NH----O=C bond linking the base pairs and any C—H----N repulsion could be avoided by twisting the bases out of coplanarity. In these configurations, the C—H----O geometry is determined by the planarity of the bases and the stronger N—H----O=C and N—H----N hydrogen bonds.

The question is, are these forced C—H----O contacts attractive or repulsive, and, do they add to or subtract from the total hydrogen bonding energy? As discussed in Chapter 2, this depends upon the balance between the two main terms; the repulsive exchange repulsion and the attractive electrostatic. If the other stronger hydrogen bonds force closer contact than the free equilibrium distances, it is possible that the shorter forced C—H----O/N contacts will be repulsive, while the longer C—H----O/N contacts will be neutral and only long C—H----O/N contacts will be weakly attractive. A rather unconventional thought for most structure analysts, but worth considering when evaluating the role of C—H----O/N interactions in biological structures.

CHAPTER 6

Cooperativity, Patterns, Graph Set Theory, Liquid Crystals

6.1 | COOPERATIVITY

The concept that a particular configuration of single and multiple covalent bonds has a total energy greater than the sum of the energies of the individual bonds is familiar in chemistry. In valence bond theory, this extra energy is called resonance energy. In molecular orbital theory, it is called delocalization energy.

The same principle applies to certain patterns of hydrogen bonds, where it is known generally as *nonadditivity* or *cooperativity*. In the general sense cooperativity applies to all intermolecular interactions and represents the difference between calculating energies using atom-pair potentials and many-atom potentials. It is particularly important in hydrogen bonding because of the diffuse nature and high polarizability of the hydrogen and lone-pair electron densities.

There are two aspects of hydrogen-bond cooperativity. One involves hydrogen-bonding between molecules with conjugated multiple π-bonds. This is described as *Resonance-Assisted Hydrogen Bonding* (RAHB) by Gilli et al. (1989). This descriptor was first applied to hydrogen bonding in β-diketone moieties and has since been extended and made more general by Gilli and Bertolasi (1990), Bertolasi et al. (1991), Gilli et al. (1993), Ferretti et al. (1995), and Gilli et al. (1995). It is referred to as π-cooperativity in the biological crystal structures discussed by Jeffrey and Saenger (1991). The concept is represented in Figure 6.1.

The second type of cooperativity occurs when hydrogen-bonding functional groups with both donor and acceptor properties form continuous chains or cycles. This occurs principally with hydroxyl groups and was first recognized in the crystal structures of small carbohydrate molecules by Jeffrey, Gress, and Takagi (1977), Jeffrey and Lewis (1978), Jeffrey and Takagi (1978), and Jeffrey (1992b), and from the crystal structures

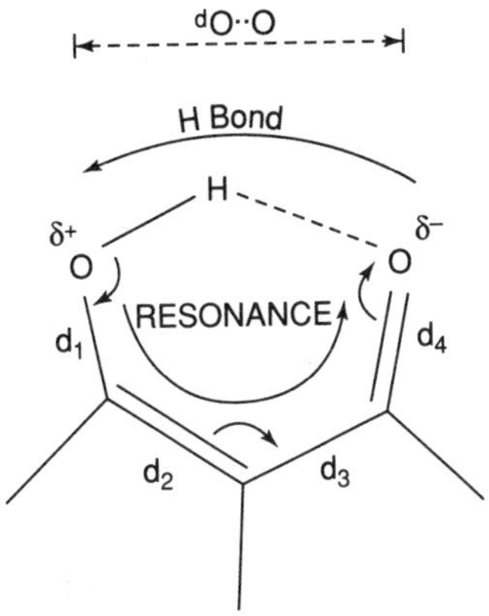

Figure 6.1. The concept of Resonance Assisted Hydrogen Bonding. As a result of the O—H---O hydrogen bond, d_1 and d_3 are shortened and d_2 and d_4 are lengthened (from Gilli et al. 1989).

of the cyclodextrins by Saenger (1979), by Berzel et al. (1984), and Steiner, Mason, and Saenger (1990). The theoretical aspects of these observations were discussed by Tse and Newton (1977). It has been described as σ-bond cooperativity, in contrast to π-bond cooperativity, by Jeffrey and Saenger (1991), since there are no multiple bonds involved.

6.2 | RESONANCE ASSISTED BONDING (RAHB)

The observations relating to this concept are not new. Huggins (1936b) recognized the formation of chains of O—H---O=C bonds in carboxylic acids and carboxylate hydrates, as shown in Figure 1.3. In Coulson's (1952) *Valence*, we find the comment that "in certain systems such as β-oxalic acid in which hydrogen bonding stretches continuously over many molecules, the O---O distance is decreased from 2.8 Å to about 2.5 Å and the bond energy is nearly doubled" (p. 305). Bailinger et al. (1964) pointed out that the intramolecular hydrogen bond in o-nitrophenol was a *special case*. Its unusual strength is a result of resonance contribution (**I**).

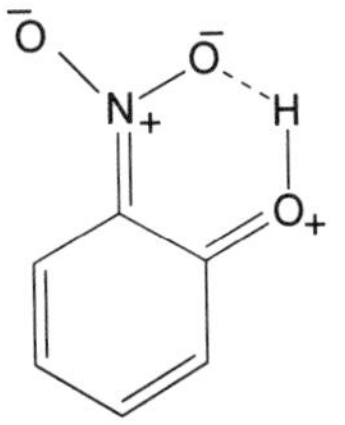

(**I**)

The gas diffraction studies by Almenninger, Bastiansen, and Motzfeld (1969) and Derissen (1971) of the monomers and hydrogen-bonded dimers of formic, acetic, and proprionic acids showed an increase in the C=O bond length and a decrease in the

C—OH bond lengths corresponding to an increase in resonance energy as a result of the formation of the hydrogen-bonded dimers, (**II**). The effect could equally well be described as *hydrogen bonding assisted resonance*.

The relative importance of the dimer and chain (caternary) hydrogen bonding in determining the molecular packing in the crystal structures of the carboxylic acids is discussed, with many examples, by Leiserowitz (1976). The most commonly observed motif includes the hydrogen-bonded dimer (**II**).

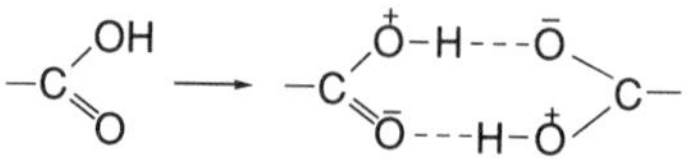

(**VI**)

A similar result was observed by Stevens (1978) and calculated by Ottersen (1975) and by Jeffrey et al. (1981) for formamide, where the C=O bond length increased by 0.018 Å and the C—N *peptide* bond length decreased by 0.023 Å, due to formation of the closed (**III**) or open (**IV**) dimers,

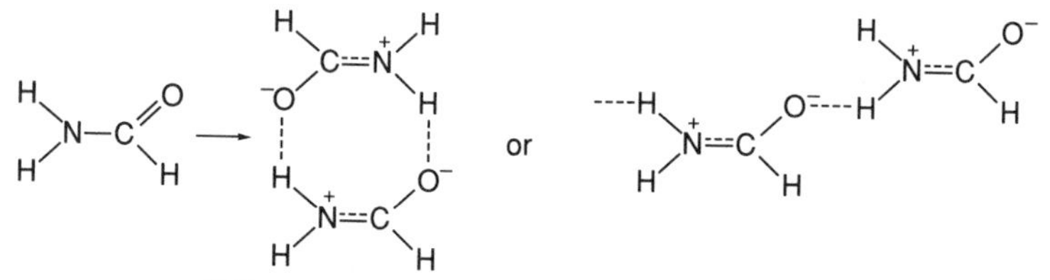

(**III**)   (**IV**)

An analysis of the molecular packing of N-methylamides by Leiserowitz and Tuval (1978) illustrates the role of N—H----O=C in a series of five crystal structures where the types of RAHB or π-cooperative bonding shown in Figure 6.2 are observed.

As pointed out in Chapter 3, there are two factors contributing to strong hydrogen bonds; the charges on the donor and acceptor groups which deshield the proton or increase the electron density of the acceptor, and a configurational constraint that forces a close contact between donor and acceptor. Added to this is the RAHB effect, as in the strong intramolecular hydrogen bonds in the 1,3-diaryl-1-hydroxy and the proton sponges (Tables 3.9 and 3.10). The relative contributions are difficult to assess with intramolecular O—H----O or N—H----N bonds because the hydrogen bonding is an intrinsic characteristic of the molecular structure in which there is a configurationally forced contraction of O—(H)----O or N—(H)----N distances.

The β-diketone enols, (**V**), provide examples of resonance-assisted intermolecular O—H----O=C hydrogen bonding.

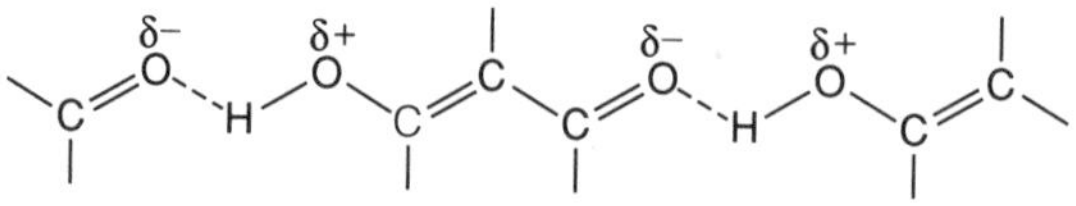

(**V**)

A survey by Gilli et al. (1993) of 13 X-ray and one neutron diffraction analyses of 2-en-2,3-diol-1-one derivatives gave hydrogen-bond O----O distances from 2.465 to 2.685

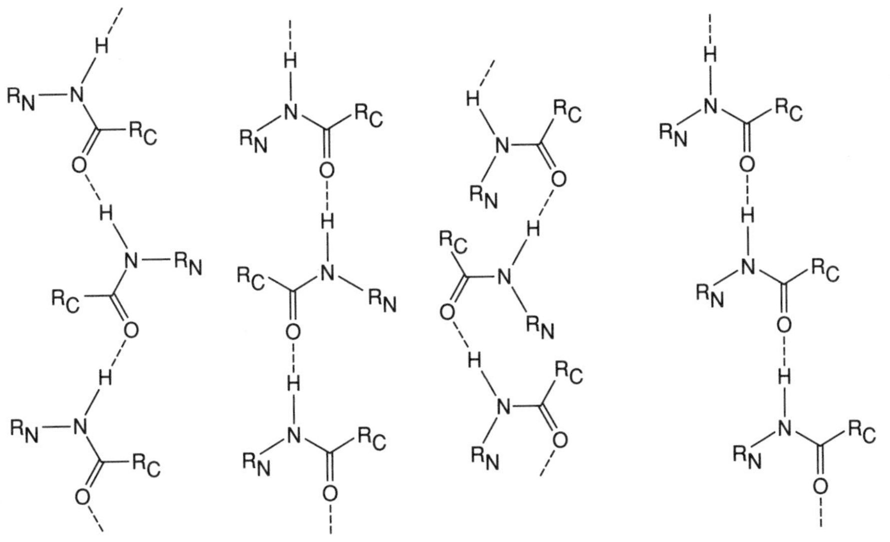

Glide, or 2_1-axis, motif   Translation motif

Figure 6.2. Cooperative chain RAHB patterns observed in the crystal structures of some N-methylamides (from Leiserowitz and Tuval, 1978).

Å. This corresponds to hydrogen bond lengths of 1.5–1.7 Å, which are comparable to those observed for O—H---O=C in the carboxylic acid crystal structures. Using a bond length / π delocalization index, Q, where $Q = d_1 - d_4 + d_3 - d_2$ in the sequence of single and double bonds, i.e.,

$$HO—C{=}C—C{=}O.$$
$$d_1 \quad d_2 \quad d_3 \quad d_4$$

Gilli and Bertolasi (1990) and Gilli et al. (1993) obtained the relationship between O---O distances and Q, the measure of conjugation, shown in Figure 6.3.

Examples of RAHB in the crystal structures of enols, di- and triketones and enediols are given in the review by Gilli and Bertolasi (1990), some examples of which are shown in Figure 6.4, and in some ketohydrazones by Bertolasi et al. (1995). RAHB is invoked to explain the structural differences between two aliphatic keto-carboyxlic acid dimers by Maurin, Les, and Winnicka-Maurin (1995). The role of RAHB in strengthening intermolecular N—H---O=C bonds in a number of small molecule crystal structures with conjugated rings is discussed by Bertolasi et al. (1995).

Gilli et al. (1994) distinguish between negative (O—H---O) and positive (=O—$\overset{+}{H}$---O=) charged and neutral O—H---O resonance assisted bonds and argue that with O---(H)---O distances less than 2.7 Å, the covalent component exceeds the electrostatic component to the bond energy. A comprehensive account of RAHB is presented by Gilli et al. (1995).

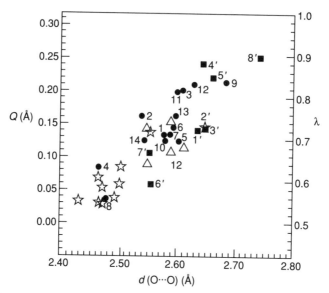

Figure 6.3. Scatter plot of π-delocalization parameter Q *vs* the hydrogen-bonded O—(H)····O distances for enolones and enediolones, showing decrease of hydrogen-bond distances with increase of π-delocalization. (complete delocalization, Q = 0; no delocalization, Q = 0.32) From Gilli et al. (1993). (●) enolones; (■) enediolones; (△) enolone chains; and (☆) intramolecular bonds.

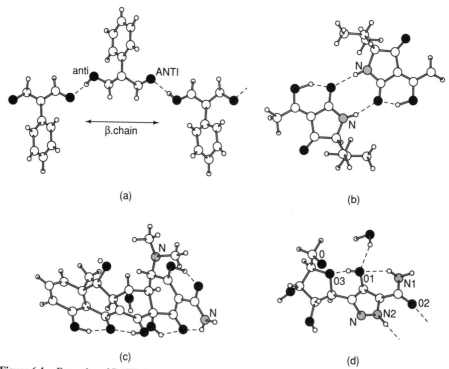

(a)

(b)

(c)

(d)

Figure 6.4. Examples of RAHB in some crystal structures of di- and triketones (from Gilli and Bertolasi, 1990). (a) 2-phenyl-2-propen-3-ol-1-one; (b) 5-acetyl-5-isopropylpyrrolidine-2,4-dione; (c) oxytetracycline; and (d) pyrazofuran monohydrate.

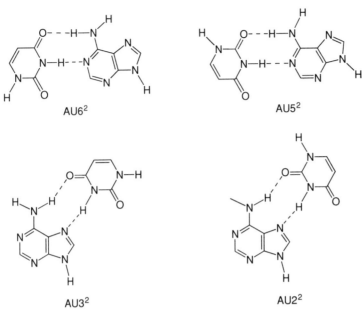

Figure 6.5. Some interesting examples of hetero hydrogen-bond base-pairing which is stabilized by RAHB. AU6^2 is Watson-Crick base-pairing, AU5^2 is reversed Watson-Crick, AU3^2 is Hoogsteen base-pairing, and AU2^2 is reversed Hoogsteen (from Jeffrey and Saenger, 1991).

Resonance Assisted Hydrogen Bonding is Important in Many Biological Structures

There are many ways of homo and hetero base-pairing observed in the crystal structures of purines, pyrimidines, and their complexes, described by Voet and Rich (1970) and Jeffrey and Saenger (1991). In these crystal structures, the hydrogen bonding extends beyond the base pairs to other molecules, and infinite hydrogen bond—π bond networks link the molecules throughout the crystal structures.

In the nucleic acids, base-pairing between purines and pyrimidines involves the hydrogen bonds which link two conjugated ring systems. RAHB will therefore play an important role both in strengthening the hydrogen bonding and increasing the resonance or delocalization energy of the molecules involved, as illustrated in Figure 6.5.

In proteins, the main-chain is not an extended conjugated system, since the two peptide units are separated by a single C—C bond. There will be RAHB, however, in the pleated-sheet hydrogen-bond structures running laterally across the main chains, as described in Chapter 10.

6.3 | POLARIZATION ENHANCED HYDROGEN BONDING

This is also not a new concept. It was perhaps first verbalized by Frank (1958) with his *flickering clusters* theory of water and aqueous solution.

The first definitive evidence came from the early ab-initio molecular orbital calculations of Del Bene and Pople (1970) on cyclic and chain water polymers and from Hankins, Moskowitz, and Stillinger (1970) on linear water trimers. More recent calculations of linear and cyclic water and methanol trimers are reported by Mo, Yañez, and Elguero (1992, 1994).

The calculations of the cyclic sequential $(H_2O)_n$, later referred to as *homodromic* by Saenger (1979), showed an increase of hydrogen bond energy per bond from 5.6 kcal/mol for $n = 3$ to 10.6 kcal/mol for $n = 5$ and 10.8 kcal/mol^{-1} for $n = 6$. There was a corresponding shortening of the calculated H---O bond lengths from 1.57 to 1.45 Å. Trimer calculations which distinguished between sequential, double donor, and double acceptor gave the calculated energy vs hydrogen-bond length shown in Figure 6.6. This is the same as the distinction between homodromic, antidromic, and heterodromic cycles in the Saenger (1979) notation.

The quadrilateral homodromic cycle of (O—H)$_4$ hydrogen bonds appears to be particularly stable. It was observed in the early crystal structure analyses of pentaerythritol by Llewellyn, Cox, and Goodwin (1937) and the neutron diffraction analysis of α-resorcinol by Bacon and Curry (1962) and in many crystal structures since then. It is responsible for the remarkable insolubility of galactaric acid (mucic acid), I, in cold water, where, according to Jeffrey and Wood (1982), the crystal lattice energy exceeds the solvation energy and entropy due to the combination of a homodromic quadrilat-

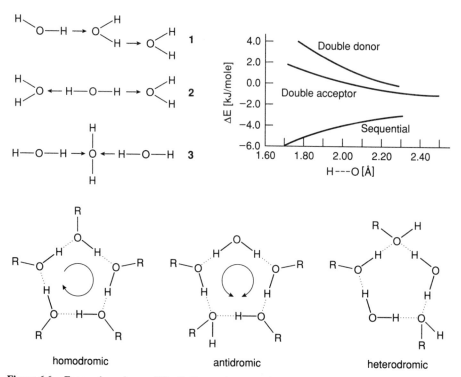

Figure 6.6. Energy dependence of H---O distance in water trimers (from Hankens, Moskowitz, and Stillinger, 1970) and distinction between homodromic, antidromic, and heterodromic (ROH)$_6$ cycles (from Saenger, 1979).

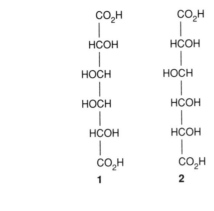

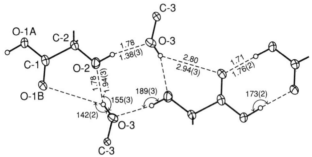

Figure 6.7.   The strong system of hydrogen bonds in the crystal structure of mucic acid (1) responsible for the low solubility and high melting point, 206°C, of polyhydroxy carboxylic acid in cold water and the high density of the crystals ($d_{25} = 1.80$ g/cm^{-3}). In contrast, the crystals of saccharic acid (2) are very soluble, in fact deliquescent, and melt at 125°C (from Jeffrey and Wood, 1982).

eral and a carboxylic acid dimer of hydrogen bonds, shown in Figure 6.7. It was identified in the cyclodextrin hydrates by Saenger and Lindner (1980). It is also responsible for the unusual *head-to-tail* crystal structures of a number of amphiphilic alkyl gluconamides and related compounds, as discussed later.

Recent ab-initio molecular orbital calculations by Mhin et al. (1991) suggest the rather surprising result that the book-shaped configuration with two fused four-membered rings for six hydrogen-bonded water molecules is close in binding energy to the hexagonal ring conformation.

σ-Bond Cooperativity in Carbohydrate Crystal Structures

The most obvious structural manifestation of σ-bond cooperativity is in the predominance of linear chains of ----O—H---O—H---O—H---- bonds in the crystal structures of the monosaccharides and the cyclic hydrogen bond structures in those of oligosaccharides and cyclodextrins.

In the polyols, $CH_2OH(CHOH)_nCH_2OH$, the hydrogen bonds form infinite chains as shown in Table 6.1. These link the molecules laterally so that they stack like pencils in a box. In the pyranoses and pyranosides, there are ring and glycosidic ether oxy-

Table 6.1.

The Infinite-chain Type Hydrogen Bonding in the Crystal Structures of the Alditols.

Alditol	Hydrogen bond donor sequence
meso-Erythritol	→ O(1)H → O(1) →
	→ O(4) → O(4) →
	→ O(2) → O(3) → O(2) → O(3) → O(2) →
D-Threitol	→ O(4)H → O(1)H →
	→ O(2)H → O(3)H → O(2)H →
	← HO(2) ← HO(3) ← HO(2) ←
D,L-Arabinitol	→ O(1) → O(5) → O(4) → O(3) → O(2) → O(1) →
Xylitol	→ O(2) → O(2) → O(2) →
	→ O(1) → O(3) → O(5) → O(1) →
Ribitol	→ O(1) → O(5) → O(2) → O(3) → O(1) →
	→ O(4) → O(4) → O(4) →
D-Glucitol, A-form[2]	→ O(1) → O(5) → O(3) → O(1) →
	→ O(2) → O(5) → O(4) → O(2) →
Galactitol	→ O(2) → O(6) → O(2) →
	→ O(5) → O(3) → O(4) → O(1) → O(5) →
K-D-Mannitol	→ O(1) → O(2) → O(1) →
	→ O(6) → O(5) → O(4) → O(3) → O(6) →
β-D-Mannitol	→ O(1) → O(2) → O(1) →
	→ O(3) → O(4) → O(5) → O(6) → O(3) →
Allitol	→ O(1) → O(2) → O(3) → O(1) →
	→ O(6) → O(5) → O(4) → O(6) →
D-Iditol	→ O(3) → O(5) → O(1) → O(3) →
	↑
	→ O(4) → O(2) → O(6) →

gens which can only be hydrogen-bond acceptors. They act as *chain stoppers*. In consequence, four different hydrogen-bond patterns are observed in the crystal structures of carbohydrates:

1. Infinite chains, with ring (O_r) and glycosidic (O_g) oxygens excluded.
2. Infinite chains, with separate finite chains terminating with O_r or O_g.
3. Finite chains, terminating with O_r or O_g.
4. Infinite chains with O_r and O_g included as minor components of three-center bonds.

The energetics of all four patterns must be comparable, since they were found by Jeffrey and Mitra (1983) to be equally populated. Examples are shown in Figure 6.8.

An analysis by Ceccarelli, Jeffrey and Taylor (1981) of the effect of σ-bond cooperativity on hydrogen bond lengths in the neutron diffraction crystal structures of monosaccharides showed a small, but definite, bond-shortening from a mean value of 1.869 Å for 19 values of noncooperative bonds to 1.805 Å for 59 cooperative bonds. This analysis also showed that the anomeric hydroxyl tended to be a better donor than acceptor and formed slightly shorter hydrogen bonds.

In carbohydrate hydrates, the water molecules use their double-donor double-acceptor properties to link the chains of hydroxyl bonds into three-dimensional nets. Alternately, they act as single acceptors and double donors to form branched chains,

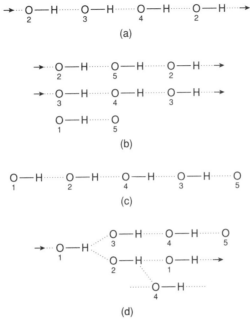

Figure 6.8.   Four types of hydrogen bonding patterns observed in anhydrous carbohydrate crystal structures. (a) infinite chains in methyl β-arabinoside; (b) infinite chains with separate O—H····O in methyl α-glucopyranoside; (c) finite chain termination at O in methyl α-xylopyranoside; and (d) infinite chain branching with three-center bonds in β-lyxose.

as shown schematically in Figure 6.9. As the carbohydrates become more complex as in the di-, tri-, and tetrasaccharides and the cyclodextrins, the cyclic patterns appear and predominate, as illustrated in Figure 6.10. It was the predominance of these cyclic patterns in the cyclodextrin hydrates that led Saenger (1979) to distinguish between the homodromic, antidromic and heterodromic cycles. For structures beyond the tetrasaccharides, the alternate hydrogen bonding diagram shown in Figure 6.11 is preferable.

6.4 | BOND PATTERNS IN CRYSTAL STRUCTURES

The role of hydrogen bonding in determining the packing motifs of molecules in crystals requires the recognition and understanding of the cooperative systems of hydrogen bonding, i.e., *the hydrogen bond structure*. As with molecules, any description of structure requires *connectivity* first, followed by *geometry*. The connectivity is referred to as the *hydrogen-bonding pattern*.

In describing the results of crystal structure analysis, this aspect of the crystal structure is often ignored, particularly if the purpose of the analysis is directed elsewhere.

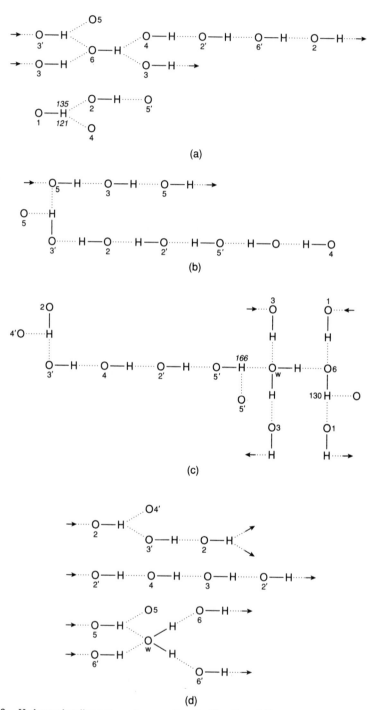

Figure 6.9. Hydrogen bonding patterns in some disaccharides (from Jeffrey and Saenger, 1991). (a) cellulose; (b) methyl cellobioside; (c) β-maltose monohydrate; and (d) methyl β-maltoside monohydrate.

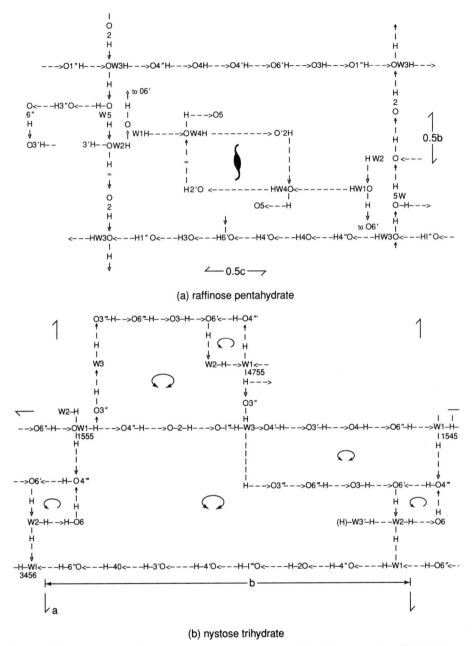

(a) raffinose pentahydrate

(b) nystose trihydrate

Figure 6.10. Hydrogen-bonding patterns in the crystal structures of the (a) trisaccharide raffinose pentahydrate and the (b) tetrasaccharide nystose trihydrate.

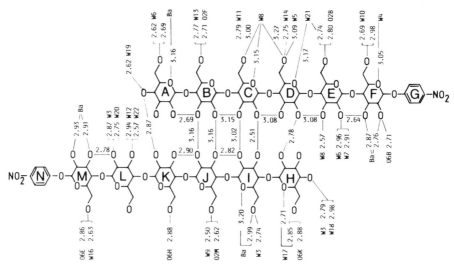

Figure 6.11.   Another type of schematic hydrogen-bonding diagram: p-nitrophenyl α-maltohexaoside, Ba(I)$_3$·27H$_2$O (from Hindricks and Saenger, 1990, reprinted with permission).

The common practice is to provide a table of hydrogen-bond O----O distances or H----O lengths and angles to a central molecule (that at 1.x,y,z). This is generally accompanied by a molecular packing diagram in which the hydrogen bonds are indicated. For complex structures, it becomes very difficult, if not impossible, to follow the connectivity of the chain or cycle of hydrogen bonds in these figures, even when presented in stereo.

To better observe the hydrogen-bond connectivity, the sequence of relevant bonds must be followed from each molecule to its symmetry-related neighbor. To simplify the pattern, a graphic display can be used in which all atoms are removed except those involved in the hydrogen bonding, as shown in Figure 6.12. With more complex crystal structures, this becomes difficult and the schematic methods displaying only the connectivity of the hydrogen-bond structure illustrated in Figures 6.9–6.11 can be used. A similar two-dimensional topology diagram is used by Ravishanker, Vijayakumar, and Beveridge (1994) for representing the hydrogen bonding in protein crystals. These diagrams are the equivalent of the familiar chemical configurational formulae. Like those formulae, these diagrams lack conformational information, but are useful for relating patterns in similar families of structures.

6.5 | USE OF GRAPH-SET THEORY

The need for some systematic way in which to relate and correlate hydrogen bonding patterns led Etter (1990) to propose a *graph-set approach*. This method has been further described and developed by Etter, MacDonald, and Bernstein (1990), Etter (1991), Bernstein (1991), and Bernstein et al. (1995) with numerous examples of its applica-

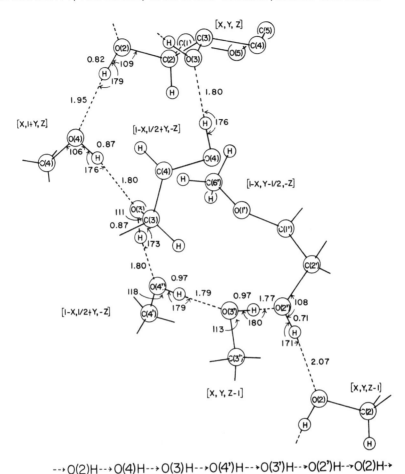

$--> O(2)H--> O(4)H--> O(3)H--> O(4')H--> O(3')H--> O(2')H--> O(2)H-->$

Figure 6.12. Constructing a connectivity diagram from a graphic display of the OH groups in the crystal structure of methyl α-D-xylopyranoside.

tion to individual crystal structures and to groups of structures, e.g., the nitroanilines, by Panunto et al. (1987) and the polymorphs of iminodiacetic acid by Bernstein, Etter, and MacDonald (1990). This method permits the identification of a particular type of hydrogen-bond structure with a symbol consisting of letters and numbers. If incorporated in a crystal structure database, it would permit the retrieval of like or related hydrogen-bond patterns irrespective of the nature of the chemical compounds involved. The basic principles and definitions are simple.

1. All hydrogen bond patterns can be described in terms of chains (C), rings (R), finite complexes (D), and intramolecular hydrogen bonds (S).

2. The number of hydrogen bond donors (d) and acceptors (a) can be identified, and the number of symmetry-independent hydrogen bonds in the pattern can be counted (n).

3. These symbols are combined to form a graph-set designation, e.g., $C_d^a(n)$.

The application of this simple approach to other than very simple hydrogen- bond patterns revealed some ambiguities and complexities. It became apparent that in addition to the basic, or first-level, designators, higher levels are necessary to describe completely the pattern or to avoid ambiguity. This problem arises because of a choice of different pathways to describe the connectivity of hydrogen bonds in a particular crystal structure. The levels are described as $N_n(a,b,c,\cdots)$ where n is the level and a, b, c,$\cdots$ denote the different types of hydrogen bonds. For α-glycine, Figure 6.13, for exam ple, in addition to the pathway $R_4^4(16)$ $\overline{abab}$, there are $R_6^6(26)$ $aa\overline{baab}$, $R_6^6(26)$ $ab\overline{babb}$, $R_8^8(36)$ $abba\overline{abad}$, $R_8^8(36)$ with $aabb\overline{aabb}$. Even so, advocates of three-center bonds might object that the three-center bond from a $\overset{+}{N}$—H to two $\overset{-}{C}$=O was not included. Advocates of resonance-assisted hydrogen bonding might object because the O=CH—CH$_2$—$\overset{+}{N}$H$_3$ is not a π-bonded structure, and the only π-cooperative chain is the $C_2^2(6)$. Advocates of C—H---O hydrogen bonds might object to the omission of a

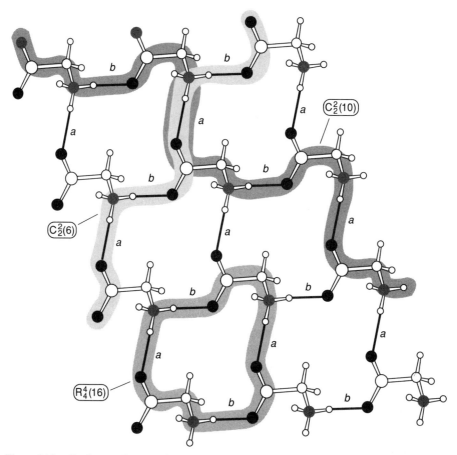

Figure 6.13. Graph-set assignments for the binary level of hydrogen bonding in the crystal structure of α-glycine (from Bernstein et al., 1995).

three-center C—H---O=C bond (2.38 and 2.45 Å). Clearly this potentially useful method requires further refinement and a system of priorities before it can fully address the complexity of all types of hydrogen-bonded patterns.

One of the objections to both this type of symbolism and the configurational hydrogen-bond patterns is that they contain less information content than that of a display of the actual crystal structure which used to be made with ball and stick models and is now possible using the rotational features of computer graphics. No doubt the time will come when the description of complex hydrogen-bonded structures can be published in the form of a virtual image video tape, or through a computer network system.

6.6 | USE OF BOND PATTERNS TO SYNTHESIZE NEW COMPOUNDS

As pointed out by Panunto et al. (1987) Etter & Frankenbach (1989) and Etter (1991), a knowledge of commonly occurring hydrogen-bond patterns associated with particular donor and acceptor function groups can be used to synthesize new organic complexes. Crystallographers have long been familiar with the different ways in which molecules associate in crystals, through such publications as *An Introduction to Crystal Chemistry* by Evans (1946), *Organic Chemical Crystallography* by Kitagorodskij (1955), and *Structural Inorganic Chemistry* by Wells (1962), but the concept that this information can be used in the design of new solid-state compounds appears as a novelty in intermolecular synthetic organic chemistry. It introduced the concept of engineering supramolecular chemistry in which hydrogen-bonding plays an important role, as discussed by Rebek (1990), Garcia-Tellado et al. (1991), Bernstein, Etter, and Leiserowitz (1994), and Philp and Fraser Stoddart (1996), for example. One of the objectives of the graph-set descriptions is to assist in utilizing hydrogen bonding to obtain controlled supramolecular assemblies of molecules.

6.7 | BONDING IN LIQUID CRYSTALS

Liquid crystals are supramolecular assemblies in one and two dimensions. As the name implies, they are a state of matter midway between crystals and liquids. They can equally well be described as *ordered liquids.* Instead of the three-dimensional order of the molecules in a crystal, the molecules in a liquid crystal have one- or two-dimensional partial order. Characteristically the molecules, (*mesogens*) are rod-like (*calamatic*) or disc-like (*discotic*). When the molecules are hydrophilic at one end and hy-

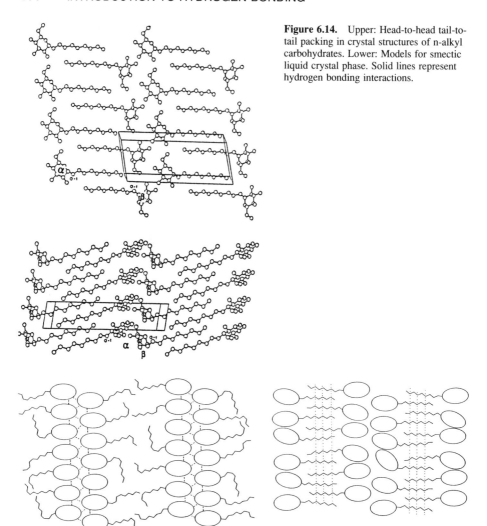

Figure 6.14. Upper: Head-to-head tail-to-tail packing in crystal structures of n-alkyl carbohydrates. Lower: Models for smectic liquid crystal phase. Solid lines represent hydrogen bonding interactions.

drophobic at the other, they are *amphiphilic*. Intermolecular hydrogen bonding then plays an important role in the way the molecules assemble. Mesogens which form liquid crystals due to increase of thermal motion are *thermotropic*. Those which form liquid crystals by contact with a polar liquid, usually water, are *lyotropic*.

In thermotropic calamatic liquid crystals, the role of the hydrogen bonding is to increase the length of the rods. Examples where hydrogen bonding increases the length of the rod are p-n-alkoxybenzoic acid, (**VI**), trans-p-methoxy and ethoxy cinnamic acids, and others described in an excellent review by Paleos and Tsiourvas (1995).

$$H_3C(CH_2)_2-\bigcirc-C\overset{OH-----O}{\underset{O-----HO}{\diagdown}}C-\bigcirc-(CH_2)CH_3$$

(**VI**)

Hydrogen bonded complexes can also form liquid crystals, such as that of 4-butoxybenzoic acid with trans-4-[(4-ethoxybenzoyl)oxy]-4-stilbzole (**VII**) from Kato and Fréchet (1989).

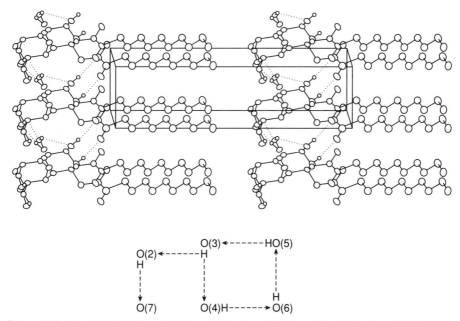

(**VII**)

Ferroelectric liquid crystals have been made through the hydrogen bonding of the 2-methylbutyl ester of terephthalic acid and 4-alkoxy-4′-stilbazoles by Yu (1993). Non-amphiphilic molecules with no hydrogen bonding functionality, such as the biphenyl cyanides used in liquid crystal technology form *nematic* thermotropic liquid crystals in which the molecules have partial one-dimensional order, with random alignment in the second dimension.

In the crystal structures of amphiphilic mesogens, such as the n-alkyl substituted carbohydrate molecules reviewed by Jeffrey and Wingert (1992), and the glycolipids reviewed by Pascher et al. (1992), the hydrophilic head groups are hydrogen bonded laterally and cooperatively to form the familiar head-to-head tail-to-tail bilayer patterns, shown in Figure 6.14. This lateral hydrogen bonding persists in the thermotropic liquid crystal phases to form the two-dimensional order associated with the *smectic* phases, described by Gray and Goodby (1984).

Figure 6.15.   Top: Head-to-tail packing in crystal structure of 1-deoxy-(N-methyl undecamido)glucitol. Bottom: Associated hydrogen bonding of the head groups.

The more soluble of these molecules also form *lyotropic* liquid crystal phases in which the hydrophilic components form hydrogen bonds at the water interface to form a variety of phases, known as *laminar*, *hexagonal* and *cubic*, as shown, for example, by the studies of the n-alkyl β-glucopyranosides by Chung and Jeffrey (1989) and Sakya, Seddon, and Templer (1994).

An unusual head-to-tail packing is observed in the crystal structures of certain N-(n-alkyl)-gluconamides, (**VIII**), and 1-deoxy(N-methyl alkananido) glucitols, (**IX**), which have strong hydrogen bonding which includes the homodromic quadrilateral shown in Figure 6.15. On heating, the crystals undergo an unusual phase transition to normally more stable head-to-head packing prior to forming the thermotropic smectic A phase, as described by Jeffrey and Maluszynska (1990b).

The influence of hydrogen bonding on the conformation of the glycosphringolipids, (**X**), has been studied by comparing crystal structures and by molecular mechanics calculations by Nyholm, Pascher, and Sundell (1990).

$$CH_3(CH_2)_nNH.CO(CHOH)_4CH_2OH$$

(VIII)

$$CH_3(CH_2)_nCO.NH(CH_3).CH_2(CHOH)_4CH_2OH$$

(IX)

$$CH_3(CH_2)_n.CO.NH$$
$$CH_3(CH_2)_nHC=CH.HOC \diagup \diagdown CH.CH_2O.(C_6H_2O_4)$$

(X)

Disorder, Proton Transfer, Isotope Effect, Ferroelectrics, Transitions

7.1 | HYDROGEN BOND DISORDER

Disorder in hydrogen bonds occurs when the donor and acceptor groups wholly or partially switch their functions. It is clearly demonstrated in crystals by the fractional occupancies of hydrogen or deuterium sites in hydrogen bonds, i.e., A—H(1—x)····H(x)—B. It is most common in structures with strong hydrogen bonds with low-barrier potentials or those with extended networks.

The classical example is Ice I_h where $x = 0.5$, which provides the well-known residual entropy first described by Pauling (1935) on the basis that the two hydrogens of each water molecule are equally distributed between the four tetrahedrally disposed hydrogen bonds. This calculation was later elaborated by Nagle (1986) to give a more precise number.[1]

All the high pressure ices described in the following chapter are reported to have symmetrically disordered O—H····O bonds except Ice II. However, values of 0.49 and 0.51 have recently been reported by Londono, Kuhs, and Finney (1993) for Ices III and IV. This raises an interesting question of how this slightly unsymmetrical disorder is related to structure and suggests that a closer examination of disorder in the other high pressure Ices is necessary.

Two mechanisms are available for proton disorder in hydrogen bonds. One is *conformational*, since it only involves bond reorientation. The other is *configurational*, since it involves hydrogen bond breaking and making with a reorganization of the adjacent single and double covalent bonds.

---

[1]There are six possible arrangements for the hydrogen positions in Ice I. The probability that a site is vacant is 0.25. Therefore the total number of possible configurations for n water molecules $N = (6/4)^W$. The residual entropy is

$$\delta = k\log N = R\log(1.5) = 0.805 \text{ cal mol}^{-1}\text{deg}^{-1}.$$

The Nagle extension gave a value of 0.8145 ± 0.0002 cal mol^{-1}deg^{-1}, in agreement with the experimental value of 0.82 ± 0.05 cal mol^{-1}deg^{-1}. A good discussion of this is to be found in Fletcher (1970).

Conformational disorder is possible when there is orientational freedom of the A—H bond, as in C—O—H, N—O—H, P—O—H. The protons can then switch bonds by means of a rotation about the single covalent bond:

$$--\overset{\frown}{C}O-H--\overset{\frown}{C}O-H \longrightarrow H-O---H-O$$

This was called the *flip-flop* mechanism by Saenger et al. (1982) in relation to the solid-state disorder observed in the crystal structures of the cyclodextrin hydrates as described by Steiner, Mason, and Saenger (1991). No covalent A—H bonds are broken and there is no necessity for reorganizing the electron distribution in the adjacent C—O or C—C bonds. This is a relatively low-energy transformation which may be further reduced by the intermediate formation of three-center bonds, as originally suggested by Savage (1986a) in the crystal structure of vitamin $B_{12} \cdot 12H_2O$.

Steiner, Saenger, and Lechner (1991) used quasi-elastic neutron scattering to study the dynamics of the flip-flop mechanism in β-cyclodextrin$\cdot 11H_2O$. They found that the experimental data could be interpreted in terms of a simple two-site jump model with two different jump distances. One is orientational jumps with distances about 1.5 Å. The other is diffusive motions of the water molecules in the cyclodextrin cavity with H---H distances of about 3.0 Å. At room temperature the jump rates were 2×10^{16} and 2×10^{11} jumps per second.

Configurational disorder takes place when there is no orientational freedom of the covalent bonds in the donor and acceptor groups and proton transfer takes place, as in an enol $\rightleftharpoons$ keto transformation,

$$C—OH---O=C \rightleftharpoons C=O---HO—C$$

or an amino $\rightleftharpoons$ imino transformation,

$$\text{>N—H---N<} \rightleftharpoons \text{>N---H—N<.}$$

Then the observation of proton disorder requires not only the breaking of the covalent A—H bond but also an electronic redistribution to change multiple C=O or C=N bonds from double to single and vice versa. This would appear to require more energy than the conformational change, except with strong $\overset{+}{O}$—H---O or O—H---$\overset{-}{O}$ bonds where the potential energy surface is flat, i.e., the low barrrier hydrogen bonds discussed in Chapter 3.

When the double minimum potential is symmetrical, as in homo-nuclear hydrogen bonds, quantum mechanical tunnelling can significantly reduce the energy required. This does, however, require a symmetrical environment, as pointed out by De LaVega (1982). Effective tunnelling can take place in **I**, when $R_1 = R_2$ but not when $R_1 \neq R_2$.

(I)

Neither keto $\rightleftharpoons$ enol nor amino $\rightleftharpoons$ imino disorder is observed in crystals,[2] but is postulated to occur in solutions and in some reactions. Recent inelastic neutron scattering experiments on partially deuterated N-methyl acetamide and polyglycine by Kearley et al. (1994) suggest that proton transfer does take place in the solid state.

7.2 | PROTON TRANSFER

Proton transfer is the basis of acid-base chemistry, of which that involving hydrogen bonds is only one component. As pointed out by Arnett and Mitchell (1971) proton transfer and hydrogen bonding are very different concepts, although the former may facilitate the latter. An interesting discussion of hydrogen bonding with and without proton transfer in the ammonium and methyl ammonium halides is presented by Legon (1993). A good example of the difference is provided by the pyridine-hydrogen fluoride complexes discussed in Chapter 3. Pyridine forms a 1:1 complex with HF, which has a F—H$\cdots$N$\Big\langle$ hydrogen bond. There is no proton transfer although the F—H distance is long, 1.13 Å, and the H$\cdots$N$\Big\langle$ distances is short, 1.32 Å. In the 1:2 and 1:3 complexes, there is proton transfer forming a pyridinium cation and HF_2^- and $H_2F_3^-$ anions. In this ionic complex, the $\overset{+}{N}$—H bond is normal, 0.8–0.9 Å, and the $\overset{+}{N}$—H$\cdots$F hydrogen bonds are 1.3–1.5 Å.

The fact that hydrogen bonding facilitates, or restricts, proton transfer is considered by some to be the most important chemical property of the hydrogen bond. The facility of hydrogen bonds to transmit H^+ (or H_3O^+) and OH^- ions in water or an aqueous media provides a catalysis mechanism for many reactions, examples of which are shown in Table 7.1. In the field of molecular biology proton transfer has come to be recognized as a significant component of enzyme catalysis and the transmission of ions through membranes.

Zundel (1976) characterizes the chains of O—H$\cdots$O—H$\cdots$O—H hydrogen bonds described in Chapter 6 as *homoconjugated* or proton *polarizable*. He associates them with the appearance of two intense continua in the 3000 to 1600 cm^{-1} region of infrared spectra, and with anomalous proton conductivity. This theme has been applied to a number of hydrogen bonded systems and to proton pathways in biological molecules by Eckert and Zundel (1988a,b), Zundel and Brzezinski (1992), and Zundel (1992, 1994).

The anomalously high electrical conductivity of ice and water attracted attention and was associated with the transport of H^+ and O^- ions before the concept of hydrogen bonding was fully developed. Both Bernal and Fowler (1933) and Huggins (1936a) discussed the role of the hydrogen bonds in this process. An excellent account of the early history and the relevant experimental measurements, techniques and theories is given by Eigen and De Maeyer (1958). The concept most commonly accepted is that introduced by Bjerrum (1952) in which the conductivity is due to small num-

---

[2]When $R_1 = R_2$ in (**I**), it cannot be detected in a crystal structure analysis.

Table 7.1.
H⁺ or OH⁻ Catalyzed Reactions Involving Water.

H_2O	Approximate $-\Delta G^{0'}$ pH 7.0 kcal/mole	Catalyst
$+RC\overset{O}{\underset{NR'H}{\diagdown}} \rightleftharpoons RC\overset{O}{\underset{OH}{\diagdown}} + H_2NR'$	3	H^+, OH^-
$+RC\overset{O}{\underset{OR'}{\diagdown}} \rightleftharpoons RC\overset{O}{\underset{OH}{\diagdown}} + HOR'$	3	H^+, OH^-
$+RC\overset{O}{\underset{SR'}{\diagdown}} \rightleftharpoons RC\overset{O}{\underset{OH}{\diagdown}} + HSR'$	7	H^+, OH^-
$+RO\overset{O}{\underset{OH}{\overset{\|}{P}}}OH \rightleftharpoons ROH + HO\overset{O}{\underset{OH}{\overset{\|}{P}}}OH$	3	H^+
$+RO\overset{O\ O}{\underset{HO\ OH}{\overset{\|\ \|}{P}P}}OH \rightleftharpoons RO\overset{O}{\underset{OH}{\overset{\|}{P}}}OH + HO\overset{O}{\underset{OH}{\overset{\|}{P}}}OH$	7	H^+
$+ \overset{-O}{\diagdown} \diagup \overset{}{\underset{OR'}{\diagdown}} \rightleftharpoons \overset{-O}{\diagdown} \diagup \overset{}{\underset{OH}{\diagdown}} + HOR'$	3	H^+
$+ \rangle{=}O \rightleftharpoons + \diagup\overset{OH}{\underset{OH}{\diagdown}}$	3	H^+, OH^-
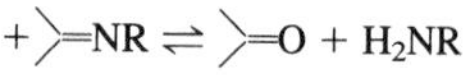 $+ \rangle{=}NR \rightleftharpoons \rangle{=}O + H_2NR$	3	$H+$

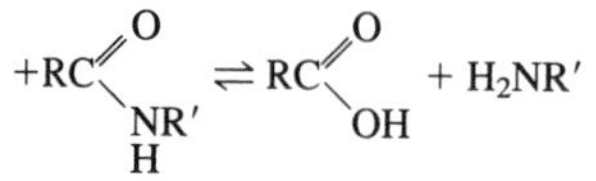

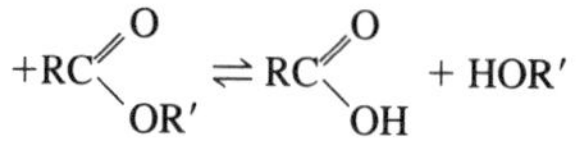

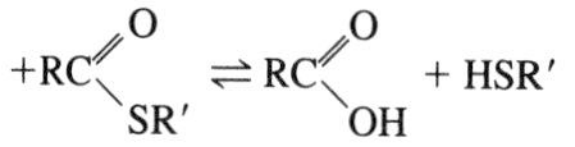

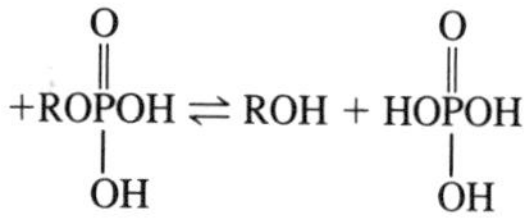

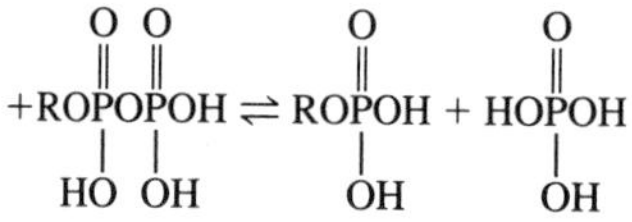

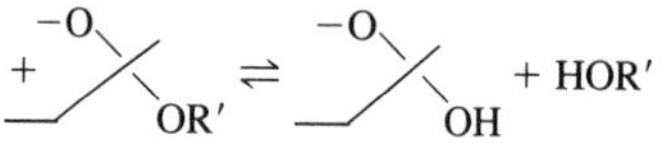

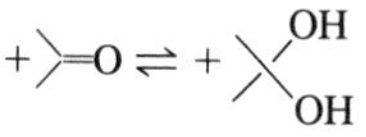

bers of the D and L defects. In this model, a water molecule rotates through 120°, so that a O—H---H—O bond is formed, leaving an O----O bond. The former is a D-defect, *doppelt besetze*, while the latter is an L-defect, *leire bindung*. These defects then diffuse through the network of hydrogen bonds in opposite directions in a two-stage process illustrated in Figure 7.1. The formation of these defects has been discussed by Cohan et al. (1962) and Cohan and Weissman (1964).

Two processes for forming a D-defect can be envisioned; one is a rotation of the water molecule, the other is a proton jump, both of which require the breaking of two hydrogen bonds.

Since these defects cannot be observed directly and are inferred from conductivity measurements, a number of conflicting models were presented from 1962–1966. Dunitz (1963), for example, thought that having two protons between a pair of oxygens was a very high energy process. He proposed an X-defect, in which the rotation of the water molecule is only 60°. This results in an intermediate defect configuration

Figure 7.1. (a) Formation and movement of Bjerrum defects. (i) unperturbed chain; (ii) formation of D and L defects; and (iii) separation of D and L defects (from Jaccard, 1972). (b) Model for proton transfer along a hydrogen-bonded chain of water molecules (from Nagle, 1992). (c) Model for a Bjerrum L defect, showing 120° rotations (from Newton, 1983).

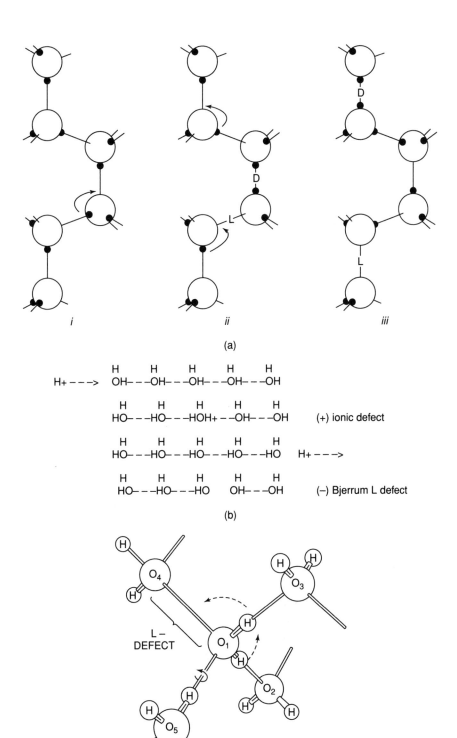

(a)

$$
\begin{array}{ccccc}
\text{H} & \text{H} & \text{H} & \text{H} & \text{H} \\
\text{H+ -- ->} \quad \text{OH--} & \text{--OH--} & \text{--OH--} & \text{--OH--} & \text{--OH}
\end{array}
$$

$$
\begin{array}{ccccc}
\text{H} & \text{H} & \text{H} & \text{H} & \text{H} \\
\text{HO--} & \text{--HO--} & \text{--HOH+--} & \text{--OH--} & \text{--OH}
\end{array}
\quad \text{(+) ionic defect}
$$

$$
\begin{array}{ccccc}
\text{H} & \text{H} & \text{H} & \text{H} & \text{H} \\
\text{HO--} & \text{--HO--} & \text{--HO--} & \text{--HO--} & \text{--HO}
\end{array}
\quad \text{H+ -- ->}
$$

$$
\begin{array}{ccccc}
\text{H} & \text{H} & \text{H} & \text{H} & \text{H} \\
\text{HO--} & \text{--HO--} & \text{--HO} & \text{OH--} & \text{--OH}
\end{array}
\quad \text{(--) Bjerrum L defect}
$$

(b)

(c)

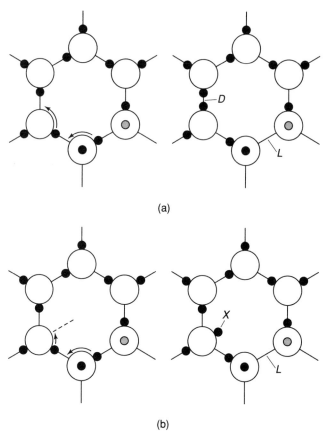

Figure 7.2. (a) Bjerrum D,L defects; (b) Dunitz X,L defects.

shown in Figure 7.2. This is, in fact, a three-center bonding configuration, such as proposed by Giguère (1984, 1986, 1987) and Dannenburg (1988) to exist as an intermediate state in water. An ab-initio molecular orbital calculation by Newton (1983) used the concept of a three-centered bond for the activated complex for the migration of the Bjerrum L defects. In view of the emphasis on three-center bonding by water molecules in recent years, this concept might now be viewed more favorably than perhaps it was thirty years ago. Other investigators, notably Gränicher (1958) and Onsager and Runnels (1969), considered the addition of water vacancies (Schottky) and interstitial (Frenkel) defects[3] similar to those used as models for the conductivity of metals and salts, shown in Figure 7.3. Finally the measurements described by Maidique, Von Hippel, and Westphal (1971) raised the question whether the conductivity of ice was a surface or bulk effect. Nagle (1992) in a recent NATO conference on *Proton Transport in Condensed Matter* is vague on the details of the mechanism.

---

[3]Schottky and Frenkel defects are models used to explain conductivity in crystals. A Schottky defect is an irregularity in the crystal lattice when a site occupied by an atom or ion becomes unoccupied. In a Frenkel defect, the atom or ion moves to an interstitial site (Uvarov, Chapman, and Isaacs, 1971).

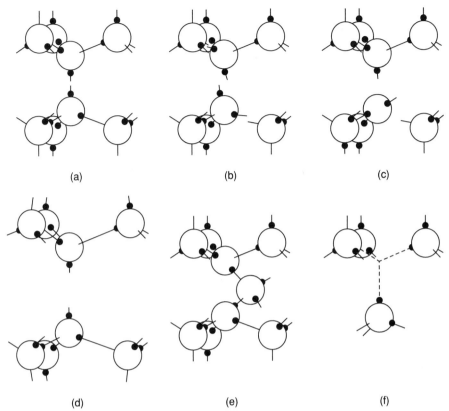

Figure 7.3.   Models for D defects. (a) rigid lattice; (b) Cohan and Weissman (1964); (c) Dunitz (1963); (d) Eisenberg and Coulson (1963); (e) interstitial, Haas (1962); (f) vacancy, Onsager and Runnels (1963) (from Jaccard, 1972).

This subject appears to be one of water's more enduring mysteries. A catalogue of alternatives is presented in Table 7.2.

In acidic solutions, migration of $H_5O_2^+$ and $H_9O_4^+$ complexes has been proposed by Weiderman and Zundel (1970), as shown in Figure 7.4.

7.3 | THE ISOTOPE EFFECT

Early experiments by Robertson and Ubbelohde (1939) and Ubbelohde and Woodward (1945) showed that substituting deuterium for hydrogen in O—H----O hydrogen bonded crystals resulted in changes in lattice dimensions.[4] Although these were small (<0.05 Å), they were experimentally significant since lattice parameters could be measured very precisely by powder diffraction. From these differences, very small changes in

[4]Ubbelohde was one of the first to explore the availability of heavy water for structural research.

Table 7.2.

Catalog of Proposed Defects in the Lattice of Pure Ice (Translated From Siegle and Weithase, 1969).

Author	Defect	Originally proposed as responsible for	Remarks
Bjerrum (1952)	simultaneous processes		Simultaneous translation of several protons or rotation of molecules following the BF-rules. Usually because of high activation energy.
	ion states OH^-, H_3O^+	DC conductivity	Movement of protons along the O—O interconnecting lines.
	L,D-defects (Bjerrum defects)	dielectric relaxation	Empty or doubly occupied O—O interconnecting lines. Movement: molecular rotation around O—H axis causes proton jump.
Gränicher (1958)	Schottky and Frenkel defects	self diffusion	H_2O-lattice vacancies (movement through neighbouring molecules jumping into lattice vacancies) and H_2O on interstitial positions
Haas (1962)	combined defects (Frenkel and Bjerrum defects)	dielectric relaxation and self diffusion	Combined movement; usually termed Haas defect
Onsager and Runnels (1963)	combined defects (Schottky and D-defects	mechanical, dielectrical and nuclear magnetical relaxation, self diffusion	Combined movement, other terms: D-fault diluted by a vacancy, singly invested vacancy.
	free interstitial	self diffusion and nuclear magnetic relaxation	Molecule on interstitial position exchanging with molecule on regular lattice position after passing several lattice vector distances

O—(H)----O hydrogen bond distances on deuterium replacement were estimated as shown in Table 7.3. Subsequent single crystal structural studies confirmed that the changes in hydrogen bond geometries on deuterium substitution were indeed very small and not consistent.

An X-ray study of $KHCO_3$ and $KDCO_3$ at 298 K, 219 K, 95 K by Thomas, Tellgren, and Olovsson (1974) showed similar changes in O----O distances of <0.03 Å. X-ray studies of deuterium substitution in the strong single minimum $[F—H—F]^-$ by Ibers (1964a, 1964b) and McGaw and Ibers (1963) showed only very small contractions of the F----F distances of 0.0024 Å in KHF_2 and 0.0046 Å in $NaHF_2$. A more detailed neutron diffraction study by Hussain and Schlemper (1980) gave the results shown in Table 7.4. Ichikawa (1978) reported the data shown in Figure 7.5a, which indicates that deuterium expands bonds for O----O distances between 2.4 and 2.65 Å and again beyond 2.85 Å.

Novak (1974) reported the $\nu_s OH/\nu_s OD$ isotopic frequency ratios vs $\nu_s OH$, $\nu_y OH$ and O----O distances for a number of hydrogen-bonded crystals ranging from ice to

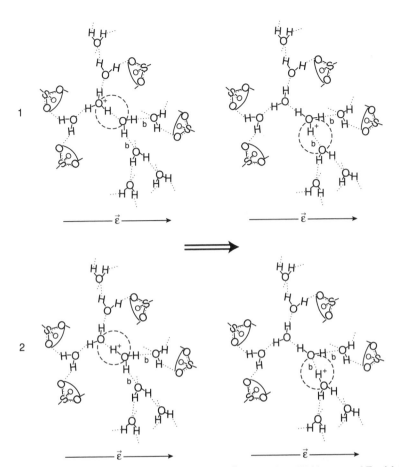

Figure 7.4. Proton transfer by structure diffusion of $(H_5O_2)^+$ groups (from Weiderman and Zundel, 1970).

Table 7.3.
Equivalent[a] Bond Length Changes on Replacing H by D (Measured Values). (From Ubbelohde and Gallagher, 1955).

Crystal	r_{AB} (Å)	$\Delta r\ (r_D - r_H)$ (Å $\times 10^{-3}$)
Oxalic acid dihydrate	2.491	+40.5
Acetylene dicarboxylic acid dihydrate	2.56	+34.0
Succinic acid	2.64	+18[b] approx.
Ice	2.76	+1.5
β-Resorcinol	2.75	No measurable effect
	2.70	
Urea	2.99	Small contraction
NH$_4$H$_2$PO$_4$	2.49	+10.0
KH$_2$PO$_4$	2.52	+9.7
KH$_2$AsO$_4$	2.54	+8.0
CuSO$_4\cdot$5H$_2$O	—	Small contraction
KHF$_2$	2.26	Small contraction

[a]These are calculated from expansions of the unit cell on the basis of assumptions discussed in the various papers.
[b]This figure was estimated from measurements in a crystallographic zone which is now known to be the least favorable for accurate determination.

Table 7.4.

Changes in O—H⋯O Hydrogen-Bond Geometries Upon Deuterium Substitution.

Compounds	H⋯O(Å)	D⋯O(Å)	Δ O—H(D)(Å)	Δ O⋯O(Å)	Δ O—H(D)⋯O(D)
			Δ = D—H		
ImH$^+$·HMal$^-$	1.197(5)	1.214(5)	−0.010(5)	+0.006(4)	+0.5(5)
Quinolinic acid	1.238(5)	1.257(2)	−0.025(4)	−0.005(3)	−0.8(5)
Ni(ao-H$_2$)Cℓ·H$_2$O	1.242(5)	1.391(10)	−0.129(()	+0.019(8)	−0.2(7)
(COOH)$_2$·2H$_2$O	1.480(7)	1.493(2)	+0.005(5)	+0.018(3)	−1.9(6)

$KH(CH_2COO)_2$. The plot of ν_sOH/ν_sOD in Figure 7.5b shows a minimum in the range of 2.45 to 2.6 Å, corresponding to the maximum in Figure 7.5a. A comparison of ab-initio, NMR and diffraction data for the O—H⋯F$^-$ bonds in $(HCO_2H)^+F^-$, $K^+(HC_O2H)_2F^-$ and $(Cs^+)_2(CHO_2H)_2F^-$ by Mortimer et al. (1992) also showed minima in plots of theoretical energies vs H⋯F hydrogen bond distances, at H⋯F = 1.40 and 1.45 Å.

Altman et al. (1978) studied the effects of deuterium and tritium substitution on the NMR chemical shifts of a number of hydrogen bonding compounds in $CD_2Cℓ_2$ solution. The differences were very small (<1.0 ppm) and positive except for the hydrogen maleate and phthalate anions. Similar results were obtained by Gunnarsson et al. (1976).

An interesting structural change on deuteration is that reported by Mootz and Schilling (1992) where deuterium substitution changes the hydrogen bonding in trifluoracetic acid tetrahydrate from a cationic layer structure of hydrogen-bonded $(H_3O^+·3H_2O)_n$ enclosing $F_3C·COO^-$ anions to molecular hydrogen-bonded $(D_2O)_n$ layers enclosing F_3CCOOH molecules.

In principle, isotope substitution could provide valuable experimental evidence relating to the hydrogen-bond potential energy surfaces. The problem lies in the small effects involved and possibly in the difficulty of ensuring the complete deuteration in crystals by recrystallizing from D_2O.

The relative stability of O—H⋯O and O—D⋯O bonds has been studied by Scheiner and Čuma (1996). They conclude that the energy differences are due to differences in the zero-point energies. This is reversed in ionic hydrogen bonds. Raising the temperature tends to favor H-bonds.

The difference in ^{13}C NMR chemical shifts on substituting deuterium or hydrogen, known as the *two-bond isotope effect* ($^2\Delta$) has been applied to intramolecular O—H⋯O bonds in phenols and enols by Reuben (1986). The $\ell n.^2\Delta$ values correlated well with the hydrogen bond energies derived from the chemical shifts of the hydroxyl protons by Schaefer (1975), as shown in Figure 7.6. This method is suggested as a fingerprint for distinguishing between the *cis* and *trans* orientation of vicinal hydroxyl groups in 1,3-diols and monosaccharides by Reuben (1984).

One interesting application of the isotope effect is its use to locate hydrogen atoms in protein crystal structures by making use of the large neutron scattering difference between hydrogen and deuterium. The problems associated with this method have been described by Harrison, Wlodawer, and Sjölin (1988).

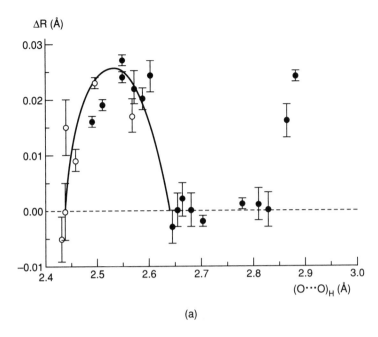

(a)

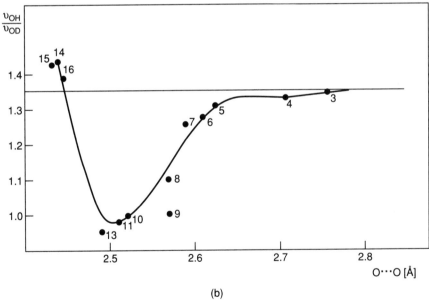

(b)

Figure 7.5. (a) the geometric isotrope effect, ΔR, as a function of $(O\text{----}O)_H$ length (from Ichikawa, 1978); (b) the ν_{OH}/ν_{OD} frequency ratio as a function of $O\text{----}O$ distance (from Novak, 1974).

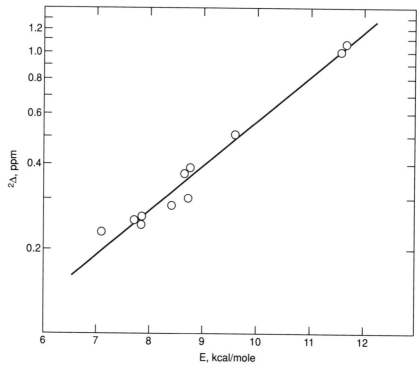

Figure 7.6. Plot of two-bond deuterium isotope effect on logarithmic scale vs hydrogen bond energies for hydroxyl portion in orthosubstituted phenols. $\ell n(^2\Delta) = 2.783 + 0.354\ E$ (from Reuben, 1986).

7.4 | TRANSITIONS IN FERROELECTRICS

In contrast to these small structural effects, deuterium substitution can have a very large effect on the transition temperatures of ferroelectrics. This suggests that hydrogen bonding plays a significant role in those ferroelectrics with hydrogen-bonded crystal structures. A *para* to *ferroelectric* phase transition results from changes in crystal structure accompanied by spontaneous polarization which can be reversed by the application of an electric field. The phenomenon is related to that of *pyroelectricity* where the crystals become polar on heating; *piezoelectricity* when they become polar on application of pressure; and *ferroelasticity* when they become polar under mechanical stress. The structure relationships between these properties is discussed by Abrahams (1994).

The ferroelectric, pyroelectric or piezoelectric crystal must be polar; i.e., have a polar space group. This should not be confused with a noncentric space group. All polar space groups are noncentric, but the reverse is not true. For example, $P2_12_12_1$ is a common noncentric space group which is not polar, whereas $P2_1$ is both noncentric and polar.

A good discussion of this topic is in Hamilton and Ibers (1968), but there seems to have been relatively little basic structural research since then and the role of the hydrogen bonds is generally not well understood.

In recent years, the field of ferroelectricity has become much broader with many more ferroelectrics being discovered, without a correspondingly better understanding of the structural basis for the phenomenon. A recent prediction of the number of potential ferroelectrics on the basis of structure by Abrahams (1990) examined 2943 polar space groups in the Inorganic Crystal Structure Data Base and suggested about 50 new ferroelectrics.

Transitions at the *Curie temperature* from nonpolar to polar crystal structures involve small nonhydrogen atomic displacements greater than 0.1 Å and less than 1 Å. The quantity measured is the low-frequency dielectric susceptibility, η, along the polar axis which is approximately

$$\eta \cong C/(T - T_o).$$

The Curie constant is of the order of 10^2 K for crystals containing hydrogen bonds and much larger and 10^4 K for those that do not. The atomic displacements can be one-dimensional along the polar axis, two-dimensional in a plane containing the polar axis, or three-dimensional.

About a quarter of the known ferroelectrics are hydrogen-bonded crystal structures. In these crystals, substitution of hydrogen by deuterium produces an isotope effect which can be large, over 100°C, as in potassium hydrogen phosphate, or a few degrees, as in Rochelle salt. The occurrence of an isotope effect, whether large or small, is regarded as evidence that some changes occur in the hydrogen bonds which are involved in the para to ferroelectric transition. Much of the crystal structural work relating to the nature of these changes was carried out prior to 1970 and the full power of modern X-ray and neutron diffraction methods have only been utilized in a few cases.

The nonhydrogen containing ferroelectrics, such as the perovskites, have received more attention, possibly because the ferroelectric effect is larger and the crystals more stable. A review from the structural point of view is provided by Abrahams and Keve (1971), in addition to the more general account by Jona and Shirane (1962) and Lines and Glass (1977). A discussion of hydrogen-bonded ferroelectrics from the theoretical point of view is by Blinc (1980). Ferroelectric compounds in which there are hydrogen-bonding functional groups are given in Table 7.5.

The transition to a ferroelectric phase can take place by a translation or reorientation of the ions or water molecules in the crystals, or by a proton transfer across a homonuclear hydrogen bond, i.e., O—H---O $\rightleftharpoons$ O---H—O or N—H---N $\rightleftharpoons$ N---H—N. These are classified as *order-disorder ferroelectrics*. In either case, an analysis of the role of hydrogen bonding in the ferroelectric transitions involving those compounds which contain hydrogen-bonding groups requires *a priori* knowledge of the positions of the hydrogen or deuterium atoms involved in the hydrogen bonds in the para and ferroelectric phases.[5] This requires three-dimensional single crystal neutron diffraction analysis of both phases. With modern X-ray technology, accurate X-ray analysis might suffice if the phase transitions take place below room temperature permitting minimization of the thermal motion of the hydrogen atoms.

---

[5]As Coulson (1972) pointed out, it is wise to find out what the atoms are doing before developing mathematical theories to account for phase transitions.

Table 7.5.

Examples of Hydrogen-Bonded Ferroelectric Crystals.

Compound	Curie temp. (K)		Space group		Reference
	(H)	(D)	Para	Ferro	
Sodium trideuterium selenate	—	270	$P2_1/n$	Pn	McMullan, Thomas, and Nagle (1982)
Ammonium sulphate	223	308	Pnam	$Pna2_1$	Schlemper and Hamilton (1966), O'Reilly and Tsang (1967)
Triglycine sulphate	323		P2/m	$P2_1$	Kay and Kleinberg (1973), Itoh and Mitsui (1974), Hoshina, Okaya, and Pepinsky (1959), Bjorkstam (1967) Toyoda et al. (1959), Shibuya and Mitsui (1961)
Triglycine fluoroberylate	343		P2/m	$P2_1$	Hoshina et al. (1957)
Potassium dihydrogen phosphate (KHP)	122	213	F4d2	Fdd2	Frazer and Pepinsky (1953), Bacon and Pease (1953, 1955)
Sodium potassium tartrate·4H$_2$O (Rochelle salt)	255, 297	251, 308	$P2_12_12_1$	$P2_1$	Frazer, McKeown, and Pepinsky (1954), Mazzi, Jona, and Pepinsky (1957)
Potassium ferro-cyanate trihydrate (KFCT)	251		$I4_1/a$ or C2/c	Cc	Waku et al. (1959), Rush, Leung, and Taylor (1966), Tsang and O'Reilly (1965), Taylor, Mueller, and Hitterman (1967)
Colemanite Ca[B$_3$O$_4$OH$_3$]H$_2$O	266		$P2_1/a$	$P2_1$	Hamilton and Ibers (1968)

Only for ammonium sulphate and sodium trideuterium selenite is there adequate crystal structural information available. *Ammonium sulphate* was studied in both phases by single crystal neutron diffraction by Schlemper and Hamilton (1966). When the space group changes from the paraelectric Pmma to the ferroelectric and polar $Pna2_1$, there are only small shifts in the positions of the central N and S atoms of the NH_4^+ and $SO_4^=$ tetrahedra. For the S atoms, it is less than 0.02 Å and for the N atoms it is less than 0.09 Å. Within the experimental errors, the $SO_4^=$ ions are tetrahedral, but the NH_4^+ ions show some deviations from ideal tetrahedra with H—N—H angles ranging from 100 to 116° in the para phase and 105 to 114° in the ferro phase. The transition occurs with a reorientation of the ions leading to a quite different distribution of the $\overset{+}{N}$—H---O$=\overset{=}{S}$ closest H---O contacts as shown in Table 7.6.

In the paraelectric phase, one NH_4^+ ion has 11 oxygen neighbors with H---O distances ranging from 1.96 to 2.48 Å. The other has 12 oxygen neighbors ranging from 1.85 to 2.43 Å. These resemble the packing of an ionic structure, modified by two three-centered $\overset{+}{N}$—H---O bonds; the two components of each are across a mirror plane of symmetry, N(1)H(1)---O(1)a and O(1)b, N(2)H(6)---O(2)a and O(2)b.

Table 7.6.
Hydrogen Bond (H---O) Geometries of Para and Ferroelectric Ammonium Sulphate.

$\overset{+}{N}$—H	Para (RT)				Ferro (LT)			
	O=$\bar{\text{S}}$ (Å)	N—H (Å)	H---O (Å)	N—H---O (°)	O=$\bar{\text{S}}$ (Å)	N—H (Å)	H---O (Å)	N—H---O (°)
N(1)—H(1)	O(1)[a]	0.99	1.96	156	O(1)	0.99	2.38	119
	O(1)[b]		1.96	156	O(2)		2.13	148
N(1)—H(2)	O(1)[c]	0.94	2.48	139	O(3)	1.02	1.92	161
	O(1)[d]		2.48	125				
	O(3)[a]		2.48	125				
N(1)—H(3)	O(3)[e]	0.97	2.14	155	O(4)	1.04	1.82	174
	O(2)[e]		2.27	136				
	O(2)[f]		2.27	136				
N(1)—H(4)	O(3)[e]	0.97	2.14	155	O(2)	1.04	1.92	166
	O(2)[e]		2.27	136				
	O(2)[f]		2.27	136				
N(2)—H(5)	O(2)[c]	0.94	2.36	122	O(2)	1.01	1.85	166
	O(2)[d]		2.36	122				
	O(3)[g]		2.43	116				
	O(3)[h]		2.43	116				
N(2)—H(6)	O(2)[a]	0.99	1.85	175	O(2)	0.99	2.28	129
	O(2)[b]		1.85		O(3)		2.39	109
					O(4)		2.59	112
N(2)—H(7)	O(3)[e]	0.98	2.05	160	O(4)	1.00	1.87	168
	O(1)[e]		2.39	135				
	O(1)[f]		2.39	135				
N(2)—H(8)	O(3)[e]	0.98	2.05	160	O(1)	1.01	1.96	165
	O(1)[e]		2.39	135	O(3)		2.46	130
	O(1)[f]		2.39	135				

[a] = x, y, z; [b] x, y, $1/2-z$; [c] = $1/2+x$, $1/2-y$, z; [d] = $1/2+x$, $1/2-y$, $-z$; [e] = $-x$, $-y$, $-z$; [f] = $-x$, $-y$, $1/2-z$; [g] = $1/2+x$, $1/2-y$, z; [h] = $1/2+x$, $1/2-y$, $1/2-z$.

In the ferroelectric phases, the N(1)H(1) forms longer three-center bonds to O(1) and O(3), while N(2)H(6) forms a weak four-center bond, and N(2)H(8) an unsymmetrical three-center bond. The major change is the formation of five strong two-center bonds with H----O = 1.82 to 1.92 Å and N—H---O angles of 161–173°. The coordination around the NH$_4^+$ ions is reduced to five H----O distances for N(1) and eight for N(2). The deuteron magnetic resonances and proton relaxation times corresponding to this transition are reported by O'Reilly and Tsang (1967). Further discussion of this interesting structure is found in Hamilton and Ibers (1968).

Sodium trideuterium selenite, Na(H$_{0.06}$D$_{0.94}$)$_3$(SeO$_3$)$_2$, has been studied in both para and ferroelectric phase by neutron diffraction by McMullan, Thomas, and Nagel (1982). Both structures contain Na$^+$, DSeO$_3^-$ ions and D$_2$SeO$_3$ molecules. The space group changes from P2$_1$/n to Pn at the Curie temperature of 270 K. The two symmetry independent disordered O—($1/2$D)----($1/2$D)—O hydrogen bonds in P2$_1$/n, change to six ordered O—D----O hydrogen bonds with the reduction of symmetry. In the ferroelectric phase, the ordered D----O hydrogen bond lengths range from 1.480 Å to 1.671 Å; in the paraelectric phase, the disordered ($1/2$D)—O distances are 1.563 Å and 1.607 Å.

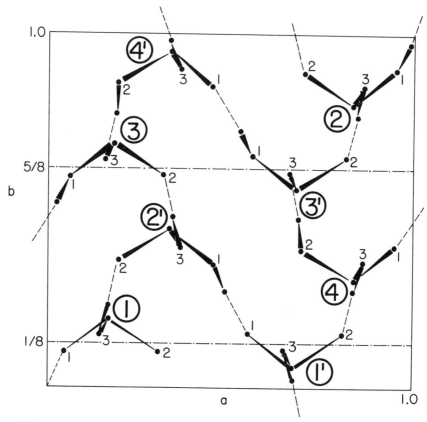

Figure 7.7. Hydrogen bonding of $DSeO_3^-$ ions and D_2SeO_3 molecules in the ferroelectric phase of sodium trideuterium selenite (from McMullan, Thomas, and Nagle, 1982, reprinted with permission).

The anions and molecules are hydrogen bonded into layers as shown in Figure 7.7. In the ferroelectric state, the ions 1–4 are related to $1'$ and $4'$ by the n glide symmetry in Pn. In the paraelectric $P2_1/n$ phase, the even and odd numbered species are related by the $\bar{1}$ and the even primed and odd unprimed species by the 2_1 axis.

For *triglycine sulphate* there is an X-ray crystal structure analysis of the paraelectric phase at 57°C by Hoshino, Okaya and Pepinsky (1959), and both paraelectric and ferroelectric phases at two temperatures, 19°C and 37°C, by Itoh and Mitsui (1974). A neutron diffraction analysis of the ferroelectric phase at 19°C by Kay and Kleinberg (1973) located all the hydrogens. The structure contains two glycinium cations $NH_4^+ \cdot CH_2 \cdot COOH$ (glycines I and III) and a glycine zwitterion $NH_4^+ \cdot CH_2COO^-$ (glycine II), with two sulphate anions.

An unusual feature of both analyses is that in the glycinium cation I, the nitrogen atom is 0.45 Å out of the plane of an otherwise planar (nonhydrogen) configuration, while the glycinium cation III and the zwitterion II are almost planar with the N atoms 0.06 Å out of the plane. There are two short hydrogen bonds involving the carboxylic acid C—OH in the glycinium ions; in I, C—OH----O=S of 1.52 Å, and in III C—O—H to O=C in II of 1.41 Å. All other N—H----O=S bonds are 1.72 and 2.09 Å.

If this transition involved order-disorder of the hydrogen bonds, the strong bond linking a glycinium ion to a glycine zwitterion is the candidate:

The out-of-plane nitrogen in one of the two glycinium ions is unusual, but occurs in both para and ferroelectric phases. There has been much research by other methods which is described in Lines and Glass (1977). This is a good illustration of the necessity for determining where the hydrogen atoms are and how they change their positions in the two phases before attempting to interpret other less direct observations.

There are two isostructural ferroelectric crystals, triglycine selenate, and triglycine-fluoroberylate. Ferroelectric mixed crystals have been made, none of which have been studied by X-ray or neutron diffraction.

Potassium dihydrogen phosphate, KH_2PO_4 (KDP) is one of the best understood of the ferroelectrics from the structural point of view. The space group change is from tetragonal noncentric $\overline{F}4d2$ to polar orthorhombic Fddd. Two-dimensional neutron diffraction analyses of both phases by Bacon and Pease (1953, 1955) showed that both the structures consist of two interpenetrating lattices of $H_3PO_4^-$ ions and of K^+ ions. The $H_3PO_4^-$ ions are linked by $P-OH\cdots O=P^-$ bonds which are disordered above the Curie points and become ordered in the ferroelectric phase, and can reverse with the application of an electric field. There are small displacements of the P and S atoms of <0.1 Å and larger reorientations of the PO_4 tetrahedra. However, the hydrogen bonds are not aligned with the polar axes, and are in fact nearly perpendicular to it. It is therefore questioned by Jona and Shirane (1962) whether the ferroelectric properties should really be ascribed to changes in the hydrogen bonding or to small displacements and reorientation of the ions. Bjorkstam (1967) observed an abrupt change in the deuterium NMR spectrum at the Curie temperature. Above the Curie temperature the deuterons jump across the hydrogen bond at times $<<10^{-5}$ sec; at the Curie temperature and below, the deuterons are ordered. The theory associated with transition and other experimental data is discussed by Lines and Glass (1977). A modern three-dimensional neutron diffraction analysis of both phases would be a worthwhile contribution to this problem.

Rochelle salt, sodium potassium tartrate tetrahydrate, is the oldest known ferroelectric. The role of hydrogen bonding in its structure and thermal properties was discussed by Ubbelohde and Woodward (1945). It has one of the smallest isotope effects and exhibits both upper and lower Curie points (at $-18°$ and $24°C$). Its crystal structure was quite complex by 1950 standards and qualifies for one of the least well-understood transitions. There is a two-dimensional neutron diffraction analysis by Frazer and Pepinsky (1953) and Frazer, McKeown, and Pepinsky (1954), and a two-dimensional X-ray analysis of both the para and ferroelectric phases by Mazzi, Jonas and Pepinsky (1957). Beyond showing that there are only small shifts in the nonhydrogen atoms, both analyses are clearly inadequate to investigate the role of the hydrogen bonding. There is a discussion of this transition in Jona and Shirane (1962).

Potassium ferrocyanide trihydrate (KFCT) is ferroelectric in space group Pc below $-22°C$, changing to C2/c above the Curie point. X-ray and NMR studies by

Kiriyama et al. (1964) indicated that there are no significant changes in the nonhydrogen positions and the transition is believed to involve the hydrogen-bond configuration of the water molecules. The quadrupole splittings and line widths of the deuterium magnetic resonance of single crystals observed by Tsang and O'Reilly (1965) were also interpreted in terms of reorientation of the water molecules. Inelastic neutron scattering experiments by Rush, Leung, and Taylor (1966) gave the interesting result that the water molecules in the crystals had a greater degree of rotational freedom than those in either water or ice. The crucial neutron crystal structure analyses have not been made.

The *alums* are a very large family of isomorphous cubic crystals with general formula $M^{+1}M^{3+}(RO_4)_2 \cdot 12H_2O$ with M^{+1} monovalent cations such as K, Rb, Cs, $T\ell$, NH_4, CH_3NH_3, NH_3OH and trivalent cations such as $A\ell$, Fe, Cr, V, In, Ga, and R is S, Se and Te. They are very easily prepared as large single crystals, as every school child should know. The ferroelectric transition is unusual in that it is from cubic to monoclinic with little change in unit cell dimensions. The transition temperatures range from 71 K to 261 K. When there is deuterium substitution, the isotope effect is -20 K for $ND_4Fe(SO_4)_2 \cdot 12D_2O$ and -39 K for $CH_3ND \cdot A\ell(SO_4)_2 \cdot 12D_2O$. There have been no definitive crystal structure analyses.

Colemanite is a mineral, $CaB_3O_4(OH)_3 \cdot H_2O$, which is ferroelectric below 25°C. The ferroelectric phase is $P2_1$ with two independent molecules in the unit cell, changing to $P2_1/a$ above the Curie point. According to Hamilton and Ibers (1968), the phase transition involves changes in the hydrogen bonding between the water molecules and the hydroxyl groups. As with the ammonium sulfate phase transition, transitions involving three-center hydrogen bonds may be involved. While the presence of an isotope effect indicates that hydrogen bonding is involved, it is by no means clear how important it is, or whether it plays a primary or secondary role with other structural changes in providing the driving force for the transition. An interesting discussion of the geometric and quantum aspects of these phase transitions is given by Ichikawa (1995) in which the linear relationship between transition temperatures and O----O distances is reported. In this paper the hydrogen bonded ferroelectrics are classified according to 0-, 1-, 2-, and 3-dimensional bonding networks.

This is a field where the current advances in neutron diffraction technology could clearly play the crucial role in providing a sound structural basis for interpreting this important phenomenon.

Water, Water Dimers, Ices, Hydrates 8

8.1 | WATER: THE MYSTERIOUS MOLECULE

Water has probably received more scientific and technological attention than any other substance. Anyone disbelieving this statement should start with a monograph entitled *Properties of Ordinary Water Substance*, by Dorsey (1940), and follow by reading *Water. A Comprehensive Treatise*, volumes 1–7, edited by Franks (1972–1980). This is understandable, since water is the most common molecule on the earth's surface and constitutes about 70% of the human body and the food it consumes. It is also the smallest molecule with the greatest potential for hydrogen bonding. This chapter is limited to those aspects of water structure and properties that are most directly concerned with hydrogen bonding.

Water maintains some scientific mysteries which are concerned with this hydrogen-bond functionality.[1] There are two apparently conflicting theories for the hydrogen-bond structure of liquid water. The finer details of the hydrogen-bonding in ordinary ice (I_h) have evaded resolution by one of the most powerful diffraction instruments in the world. The anomalous density maximum of liquid water at 4°C, which is so important for sea life, evades a thoroughly comprehensible explanation.[2] Water has a high dielectric constant associated with the distortion or breaking of hydrogen bonds. It also has the anomalous conductivity associated with the transfer of H_3O^+ and OH^- ions through the hydrogen-bonded structure discussed in Chapter 7. Other properties peculiar to water are the degree to which it can be supercooled and the increase of fluidity under pressure. The complexity of understanding these phenomena is discussed by

---

[1]Water exercised its hypnotic quality during the 1960s in an expensive scientific folly known as Polywater, as described by Franks (1981).

[2]It is said to be explained by C. H. Cho et al. (1996). They used a model having competition in the second neighbor structure between the open 4.5 Å O----O coordination of I_h and a denser structure having bent hydrogen bonds such as are observed in some of the high pressure forms of ice.

Angell (1988). Water is recognized as being essential for life processes, but its actual role in these processes is relatively poorly understood, as discussed in Chapter 10.

The Theories of Liquid Water Structure

Water has short-range order with a peak in the O----O X-ray diffraction radial distribution curve shown in Figure 8.1 at 2.85 Å. Notably this is 0.1 Å shorter than the gas-phase value and 0.1 Å longer than that observed in the ices. These differences must reflect similar differences in mean hydrogen bond lengths. The peak for the radial distribution curve corresponds to about 4.4 nearest oxygen neighbors which is larger than that from a hydrogen-bonded tetrahedron.

Considerable effort has gone into recording the X-ray and neutron diffraction data from liquid water at various temperatures. The X-ray data give O----O pair correlations, while the neutron data provide both O----D and O----O correlations. Beyond confirming the expected, the nearest neighbor O—D, D----O, and O----O distances, the interpretation of these results is not simple. A good review of the present status is given by Bellissent-Funel (1991).

Historically there are two apparently conflicting models for water structure: the uniform continuum model and the cluster or mixture model. The *continuum model* of Pople (1951) or the closely related *random network model* of Bernal (1964) envision a continuous network structure with first neighbor tetrahedral coordination. Both envision flexible hydrogen bond valence and torsion angles which can deform such that second and third neighbor O----O distances are averaged to give a continuous distribution of O----O distances beyond the first neighbor. Variations of the continuum model is that by Orentlicher and Vogelhut (1966) in which the hydrogen bonds break rather than bend. Wenes and Rice (1972) developed a thermodynamic lattice gas model based on a ($H_{10}O_5$) body-centered cubic cell.

Mixture models have been proposed with more variety. An early version of the mixture model is the *flickering clusters* of Frank and Wen (1957) and Frank and Quist (1961) shown in Figure 8.2(a). A variation of this by Nemethy and Scheraga (1962) has tetrahedrally bonded clusters with molecules making two or three bonds at the periphery, but no monomers and dimers. From the infrared spectroscopy properties, Walrafen (1972) proposed a mixture of species having two-, three-, and four-coordinated species. There have also been suggestions by Giguère (1984, 1986, 1987) and Dannenberg (1988) that there is a significant population of an intermediate state in which there are three-centered hydrogen bonds. Three-center bonding involving *defect* water molecules is necessary to account for the observation that the diffusion and rotation rates of water molecules are comparable to those of nonassociated liquids, according to Sciortino, Geiger and Stanley (1991). This defect structure appears to be similar to the X-defect proposed by Dunitz (1963) to account for the high electrical conductivity of ice discussed in Chapter 7.

Water rings and clusters are becoming more fashionable. In the model supported by Symons (1972), the properties of water are interpreted in terms of the relative concentrations of broken and unbroken hydrogen bonds. A statistical thermodynamic treatment by Hagler, Scheraga, and Nemethy (1972) is based on a distribution of water clusters with a medium size of 11 waters. Symons (1981, 1989) discusses the more

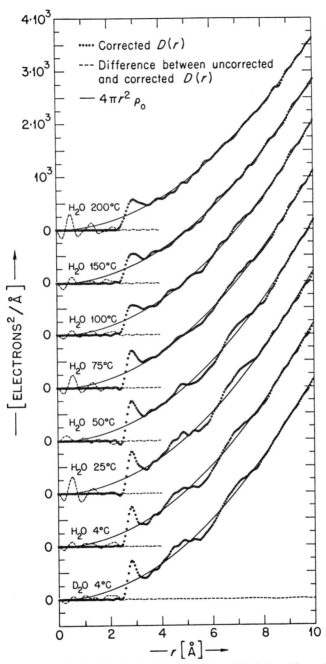

Figure 8.1. X-ray radial distribution function for H_2O (from Narten, Danford, and Levy, 1966, reprinted with permission). The functions below 75°C show a second minor peak at O---O=3.8 Å, which is not reproduced by most models. Some investigators believe it, others do not.

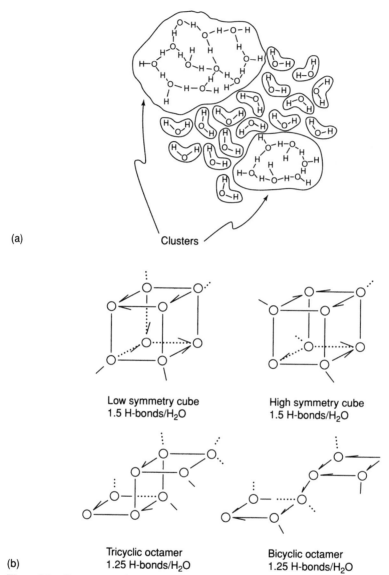

(a)   Clusters

Low symmetry cube
1.5 H-bonds/H_2O

High symmetry cube
1.5 H-bonds/H_2O

(b)
Tricyclic octamer
1.25 H-bonds/H_2O

Bicyclic octamer
1.25 H-bonds/H_2O

Figure 8.2. Some proposed water clusters. The old and the new. (a) the flickering cluster model of Frank and Wen (1957) (from Kavenau, 1964); (b) the octamer ⇌ tetra model of Benson and Siebert (1992). (c) ab-initio molecular orbital minimum energy configurations for $(H_2O)_6$ (from Mhin et al., 1991); and (d) predicted $(H_2O)_{20}$ clusters (from Kirschner and Shields, 1994).

chemical aspects of water and aqueous solutions, from the spectroscopists point of view, in terms of three- and four-coordinated water; the former having free hydroxyls or nonbonded lone-pairs. He envisions a three-dimensional network containing short-term four-, five-, six- and seven-membered rings. In an early molecular dynamics simulation of liquid water, Rahman and Stillinger (1973) found a nontrivial contribution from polygons greater than octagons. Newton (1983) discusses small water clusters

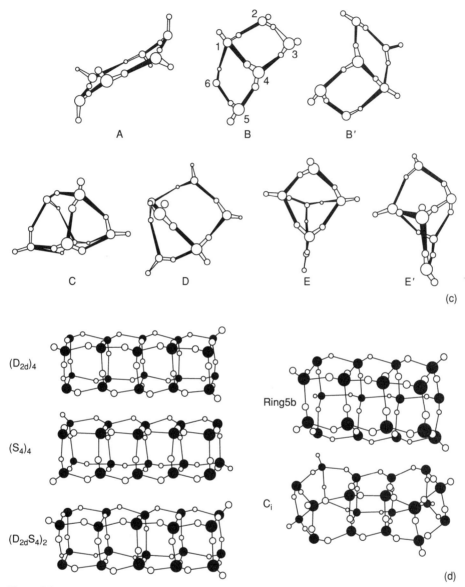

Figure 8.2. (*continued*).

from the ab-initio point of view. In a more recent addition, Benson and Siebert (1992) argue that the heat capacity data for water can only be modelled with discrete water clusters. They propose hydrogen-bonded octamers, shown in Figure 8.2(b). This is supported by the infrared data of Libnau et al. (1994). Recent ab-initio molecular orbital calculations of $(H_2O)_6$ by Mhin et al. (1991) suggest that a book-shaped configuration with two fused quadrilaterals has a binding energy comparable to that of the cyclic hexamers, shown in Figure 8.2(c). Small water clusters, $(H_2O)_n$ with n = 2, 3, 5, 5, are observed by infrared molecular beam depletion and fragment spectroscopy in an expanded helium-water gas at 100°C by Huisken, Kaloudis, and Kulcke (1995). The theoretical studies of Kirschner and Shields (1994) examined the energetics of various types of water clusters, shown in Figure 8.2(d), including those as large as $(H_2O)_{20}$ and $(H_2O)_{40}$. These large clusters, such as $(H_2O)_{21}H^+$, can be formed in the gas phase, as shown by Castleman (1994), but even a transient existence in the liquid seems unlikely. These are discussed in Chapter 9.

The earlier crystallographically inspired clathrate hydrate model of Pauling (1959), the filled-ice model of Danford and Levy (1962) and the multiple ice-like structures of Kamb (1968) are also cluster models. At the time, they were considered *too crystalline*. Since water clusters appear in the gas-phase, this criticism may now seem to be unjustified.[3]

Thermodynamists, and most diffractionists favor the continuum model. Spectroscopists had mixed opinions, with Buijs and Choppen (1963) and Marchi and Eyring (1964) favoring the mixture model, while Stevenson (1965) and Wall and Hornig (1965) favored the continuum model. Since the former describes structure on the basis of a long-time space average, while the latter is a snapshot of structure with an exposure of less than 10^{-4} seconds, both concepts might be correct, as in the wave vs particle theories of radiation. These are the two views of structure, referred to as the long-time/space average *D(diffusionally averaged)-structure* and the short-time snapshot as a *V(vibrationally averaged)-structure* of Eisenberg and Kauzman (1969). Crystallography is neutral in this argument, since the ices and clathrate hydrates are based on four-connected nets with tetrahedral coordinate, while the hydrates of small organic molecules provide examples of water molecules with between one and five hydrogen bonds, as shown later. A general view of the role of hydrogen bonding in the structure of ice and water by a principal theoretician in the field is provided by Stillinger (1980).

8.2 | THE WATER DIMER: A THEORETICAL GUINEA PIG

Water forms a hydrogen-bonded dimer, (**I**), the structure of which was determined by microwave spectroscopy by Dyke, Mack, and Meunter (1977). The derived structural parameters are $O\text{----}O = 2.96(\pm0.04)$ Å, $\theta = 55°$.

(**I**)

Since the water dimer is one of the simpler dimers, its structure and binding energy is recalculated by semi-empirical and ab-initio quantum mechanics each time that

---

[3]Understanding the structure of liquid water is certainly not getting any simpler. There seem to be as many different models as authors who have written on the subject.

an advance in computer technology makes more sophistication possible. A table of references to 38 $H_2O\cdots H_2O$ semi-empirical and minimal basis set ab-initio calculations before 1974 is in Joesten and Schaad (1974). The most revealing calculation from the chemist's viewpoint was that of Singh and Kollman (1985), shown in Figure 2.3, where the energy components were decomposed by the Morokuma method with varying $O\cdots O$ distances and $O\text{—}H\cdots O$ angles, as discussed in Chapter 2. Modern calculations using a variety of basis sets and correlations give binding energies ranging from 3.61 to 5.75 $kcal/mol^{-1}$, as reported by Chakravorty and Davidson (1993). Most of the calculated values were significantly smaller than the experimental value of 5.4 ± 0.7 $kcal/mol^{-1}$ of Dyke and Muenter (1974).[4] A detailed history and discussion of ab-initio studies of the water dimer found in Scheiner (1994).

Water is a Weak Hydrogen Bond Donor in the Gas-Phase

The adducts of H_2O with HF, HCℓ, HBr, NCN, HC≡CH, NH_3, and N_2 have also been studied by gas-phase microwave spectroscopy. They are of interest because of the question, which is donor and which is acceptor? Water is known to be a stronger hydrogen bond acceptor (i.e., base) than donor (i.e., acid), so it is not surprising that it is an acceptor in FH—OH_2. Less expected is that it is also an acceptor with HCℓ, HBr, HCN, and HC≡CH. Only for NH_3 is water the donor. Legon and Millen (1987a, 1992) have quantified these observations in terms of the limited gas-phase nucleophilicities (N) and electrophilicities (E) as described in Chapter 4.

Since the hydrates of HF and HCℓ studied by Mootz, Ohms, and Poll (1981) contain the H_3O^+ ion, the question arises, is there any $H_3\overset{+}{O}$ in the dimers in the gas phase? From thermodynamic arguments, Legon and Millen (1992) conclude that the amount is very small, but may be a major component in the liquid phase.

In addition to the water dimer discussed earlier, many gas-phase dimers involving water as an acceptor or donor have been studied by theoretical methods. References to those studied prior to 1974 are in Joesten and Schaad's book. The results of ab-initio calculations at the MP2/6-21+G (d,f) level reported by Del Bene (1988a, 1988b) for dimers formed by H_2O with a variety of ions and molecules are given in Table 8.1. The calculations agree with the observation that water is a strong acceptor but weak donor. Only for NH_3 is water the hydrogen bond donor.

8.3 | POLYMORPHISM OF SOLID H_2O[5]

Solid H_2O has 12 known crystalline forms, shown in Table 8.2. It also has an amorphous form. In its diversity it resembles SiO_2 which has six natural polymorphic crystalline forms, an amorphous form, and a large number of synthetic forms, the zeolites.

---

[4]Theory has yet to catch up with Pauling (1939) who estimated the energy required to break a water hydrogen bond at 5.1 $kcal/mol^{-1}$.

[5]An excellent text on the *Chemical Physics of Ice* is by Fletcher (1970).

Table 8.1.

Calculated Hydrogen Bond Distances and Energies For H_2O Complexes, Based On Ab-initio Molecular Orbital Calculations at MP4/6$-$31+G (2d,2p) Level by Del Bene (1988b).

Complex	$-\Delta H^a$	$-\Delta H_{exp}$	r_{O-H} (Å)	r_{O-B} (Å)	O—H···B (°)
HOH—F$^-$	27.3	23.3	1.09	2.417	180
HOH—OH$^-$	26.4	26.8	1.04	2.523	176
HOH—Cℓ^-	13.5	13.1	1.01	3.266	161
NH$_4^+$—OH$_2$	18.9	19.9, 17.3		2.772	180
OH$_3^+$—OH$_2$	32.7	31.8	1.05	2.445	178
FH—OH$_2$	7.4	6.2		2.720	175
CℓH—OH$_2$	4.4			3.266	161
HOH—NH$_3$	4.7		0.96	3.039	178
HOH—OH$_2$	3.6	3.7	0.96	2.971	178

$^a-\Delta H$ is the standard enthalpy at 298 K in kcal mol^{-1} for the reaction H_2O + B → B···H_2O.

There are, in fact, analogies between the ices and the silicates, with O—H···O replacing Si—O—Si. The clathrate hydrate host structures described in Chapter 9 have their analogues in the *clathrasils*.

Ordinary ice (I_h) and cubic ice (I_c) are formed at normal pressures, the I_c being stable below $-120°C$. The hydrogen bonding in I_h and I_c is very similar. Only the arrangement of the oxygen atoms is different. In I_c, the oxygens are arranged on a cubic diamond lattice, whereas in I_h the arrangement corresponds to the less common hexagonal *Lonsdalite* form of carbon.

The forms of ice other than Ice I are formed under pressure and their structures were determined by Kamb (1968) and his collaborators.[6] The crystal data and some hydrogen bonding characteristics are given in Table 8.3. All these ice polymorphs contain four-, five-, and six-membered cycles of hydrogen bonds, shown in Figure 8.3.

In both I_h and I_c, the hydrogen bond patterns contain cyclohexane-type buckled hexagonal motifs, chairs in I_c and boats in I_h. I_h is isostructural with the hexagonal ZnS wurtzite structure and I_c is isostructural with the cubic ZnS zincblende with O—($^1/_2$H)—($^1/_2$H)—O replacing Zn—S—Zn. I_c is made by compressing and cooling ice under pressure. It can also be made by condensing water vapor on a cold copper rod. It was first recognized as a polymorph of ice from X-ray diffraction patterns by Burton and Oliver (1935). Below 200 K, amorphous ice is formed which transforms to I_c at $\sim150°C$. An important distinction is that in I_c all four hydrogen bonds are equivalent by symmetry. In I_h, the bond in the direction of the hexagonal axis can be different from the other three, depending upon the c/a axial ratio.

In the high pressure ices the hydrogen-bonding patterns become increasingly complex with pressure increasing until the very high pressure ices VII and VIII are formed. Ice II contains nearly planar hexagonal rings of hydrogen bonds arranged in columns. The O···O···O angles are greatly distorted from tetrahedral, with some close nonbonded

---

[6]An informal and more interesting account of this work by Kamb is to be found in *Crystallography in North America*, edited by D. McLachlan and J. Glusker (1983).

Table 8.2.
The Structures of the Ices.

Ice	Method[a]	H or D atom sites	Cyclic motifs	Density g cm⁻³	Space group	P (kb)	Geometry		
							H···O (Å)	O—H···O (°)	O···O (Å)
I_h	N 60 K 123 K 223 K	symmetrically disordered	hexagons	0.931	P6$_3$/mmc	0.0	1.746	180	2.750
I_c		symmetrically disordered	hexagons	0.93 (calc)	Fd3m	0			
II	N 210 K	ordered	hexagons and octagons	1.18	R$\bar{3}$	2.1	1.81 to 1.86	166 to 178	2.77 to 2.84
III	N (TOF) (powder)	disordered	pentagons, heptagons, and octagons	1.16	P4$_1$2$_1$2	2.1	1.70 to 1.96	163 to 166	2.700 to 2.850
IV	X 110 K	disordered	hexagons and octagons	1.27		5.4			2.785 to 2.918
V	X 100 K	disordered	quadrilaterals, pentagons, hexagons, and octagons	1.23	A2/a	3.4			2.766 to 2.867
VI	N 225 K (powder)	disordered	quadrilaterals and octagons	1.31	P4$_2$/nmc	6.2	1.80	157 to 175	2.73 to 2.77
VII	N (powder)	disordered (highly ionized)	hexagons	1.50	Pn3m	21.5	1.958	180	2.901
VIII	N 10 K (powder)	ordered (antiferroelectric)	hexagons	1.50		21.5			
IX	N 110 K (TOF) (powder)	almost ordered Ice III	pentagons, heptagons, and octagons				1.789 to 1.821	167 to 175	2.714 to 2.805
X	IR	centered(?)		2.5 (extrapol)		4.40			

[a]N = neutron diffraction, X = X-ray diffraction, IR = infrared spectroscopy, TOF = time-of-flight analysis.

Table 8.3.

(a) Effect of the Cationic Environment of Water Molecules on the O—H⋯O Hydrogen Bond Lengths.[a]

Class I

R, R'	H,H (E)	H,M$^+$ (G)	H,M^{2+} (H)	H,M^{n+} (H″)	M$^+$,M$^+$ (A)	M$^+$,M^{2+} (H')	M^{2+},M^{2+} (B)
No. of structures	13	14	13	2	37	18	5
No. of O—H⋯O in survey	24	26	21	4	46	32	8
H⋯O min (Å)	1.747	1.742	1.650	1.668	1.699	1.679	1.520
max	2.099	2.258	2.034	1.742	2.242	1.952	1.827
mean	1.877	1.924	1.835	1.694	1.919	1.782	1.748

Class IIA (planar)   Class IIB (pyramidal)

R	H (F)	M$^+$	M^{2+} (D)	M^{n+} (M)	H (K)	M$^+$	M^{2+} (J)	M^{n+} (N)
No. of structures	8	0	26	7	10	0	27	2
No. of O—H⋯O in survey	13	0	43	14	17	0	44	4
H⋯O min	1.817	—	1.656	1.636	1.765	—	1.680	1.790
max	2.059	—	2.176	2.132	2.137	—	2.069	1.982
mean	1.897	—	1.897	1.811	1.894	—	1.777	1.898

(b) Relationship to Pauling's (1929) bond strength.[b]

	D,J,A,E,G	B,H	F,K	M,N,H,H'
No. of O—H⋯O in survey	183	12	30	71
H⋯O mean (Å)	1.875	1.730	1.895	1.810
W⋯O mean	2.820	2.702	2.829	2.770
S (valence units)	0.33	0.67	0.24	0.50

[a]Adapted from Chiari and Ferraris (1982). Their notation in parentheses.
[b]Adapted from Ferraris, Fuess and Joswig (1986).

O⋯O separations of 3.24 Å. Ice III is composed mainly from hydrogen-bonded pentagons with some heptagons, but with no hexagonal rings. Ice IV, which is metastable, is even more distorted with O⋯O⋯O angles between 88 and 128° and a large number of nonhydrogen-bonded O⋯O contacts between 3.14 and 3.29 Å. Some oxygens lie at the center of hexagonal rings.

Ice V contains quadrilaterals, pentagons, and hexagons, in which respect it resembles some of the clathrate hydrate host lattices discussed in Chapter 9. Ice VI is the first of the structures with two independent penetrating hydrogen-bond lattices (e.g., *self-clathrates*). The nonbonded distances between the oxygens on the two lattices of 3.4 Å is shorter than the hydrogen-bonded O⋯O distances.

Ices VII and VIII have much more regular interpenetrating lattices, each with the

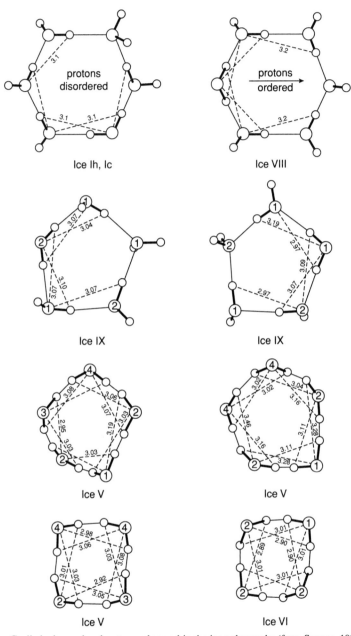

Figure 8.3. Cyclic hydrogen bond patterns observed in the ice polymorphs (from Savage, 1986b).

I_c structure with little distortion from tetrahedral bonded coordination. Each oxygen has eight nearest neighbors, four hydrogen-bonded at O----O = 2.88 Å, and four non-bonded at a shorter distance of 2.74 Å, demonstrating that it is easier to compress a van der Waals O----O distance than a O—H----O distance. Until the formation of interpenetrating lattices, the principal effect of increasing pressure is to produce more complex hydrogen-bond patterns with large departures from tetrahedral coordination around the oxygen atoms; with the interpenetrating hydrogen bond lattices, the patterns become much simpler.

Ice IX has the same structure as ice III, except that the hydrogen atoms are ordered in IX and disordered in III. The transformation III → IX is between 208 and 165 K. The positions of the hydrogen atoms are only well-determined in the neutron single crystal analyses of ices I, II, and IX. The neutron powder data of VI, VII, and VIII by Kuhs et al. (1984) are less definitive. The newer pulsed neutron powder diffraction methods using time-of-flight methodology were applied to ice VIII by Jorgensen et al. (1984) and to ice III and IX by Londono, Kuhs, and Finney (1993). They found that the hydrogen atoms are not fully ordered in III and not quite symmetrically disordered in IX. So the hydrogen-bonding patterns in these high pressure ices may be more complex than originally envisioned.

A very high pressure ice X, which was predicted to be *symmetrical ice*, with the hydrogen atom midway between the oxygen atoms, was identified by infrared spectroscopy with the transition at 440 kbar by Polian and Grimsditch (1984).

In addition to the ices, there are a large number of ice-like structures with four-coordinated water molecules which contain voids which are only stable when occupied by other molecules. These hydrate inclusion compounds are discussed in Chapter 9.

The Structure of I$_h$ Defeats Modern Technology

The early single crystal neutron diffraction study by Peterson and Levy (1957) determined that in D$_2$O I$_h$ the O—D bond lengths were 1.000(9) and 1.007(7) Å, and the D—O—D angles were 109.5(2)° with linear O—H----O bonds, giving the oxygen atoms an almost ideal tetrahedral coordination. However, the gas-phase spectroscopy values for the water molecule are O—H = 0.958–0.965 Å, H—O—H = 104.5–104.8° from Harmony et al. (1979). Similarly, distances of 0.95–0.97 Å or shorter are commonly observed for O—H bonds in the neutron diffraction analyses of many hydrates, including the clathrate hydrates discussed later. Among the inconsistencies in I$_h$ noted by Whalley (1974) were differences between observed and calculated O—H stretching frequencies.

Various models were suggested to explain this discrepancy, including one by Chidambaram (1961) which bent the O—H----O angles and disordered the hydrogen positions off the hexagonal axis. To resolve this question, a neutron diffraction analysis of H$_2$O and D$_2$O at 60, 123, and 223 K was carried out by Kuhs and Lehman (1983, 1985) at the highest level of precision possible at that time. It showed that as well as having disordered hydrogens, the oxygen atoms were also disordered about sites 0.06 Å off the hexagonal axis. While this resolved the bond length/bond angle discrepancy, it did not completely determine the structure. There were three models, illustrated in Figure 8.4, that fit the experimental data. Although the agreement between observed and calculated structure factors was extremely good, it was not possible to distinguish between any of these models or a disordered combination of these models, which seems most likely.

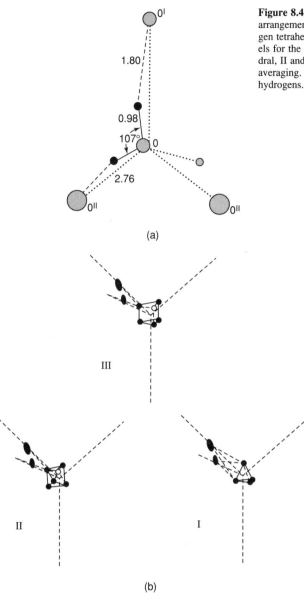

Figure 8.4. The structure of I_h. (a) atomic arrangement in one of the disordered oxygen tetrahedra; and (b) three possible models for the oxygen disorder, with I tetrahedral, II and III trigonal. Trigonal space averaging. Circles = oxygens; ellipses = hydrogens. From Kuhs and Lehman (1985).

8.4 WATER COORDINATION IN HYDRATES

As pointed out by Bernal (1952), inorganic hydrates, like the silicates, can be classified into those containing isolated water molecules, small clusters of water molecules, one-dimensional chains or ribbons, two-dimensional layers or three-dimensional lattices. The same prediction was made for organic hydrates by Jeffrey and Mak (1965). Only for the

Figure 8.5. Coordination of water molecules in inorganic hydrates (from Falk and Knop, 1973, reprinted with permission), according to Chidambaram, Sequira, and Sikka (1964). The numbers refer to the number of examples reported.

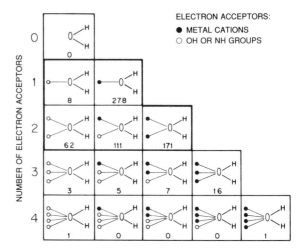

hydrate inclusion compounds, discussed in Chapter 9, have there been any recent attempts at a major classification of the inorganic or organic hydrates on this basis. What has been done is to examine the donor and acceptor geometries of individual water molecules.

The classification of water acceptor geometries which included both hydrogen bonds and cationic lone-pair interactions was made by Chidambaram, Sequeira, and Sikka (1964) and by Falk and Knop (1973), as shown in Figure 8.5. The majority were three-coordinated with two donors and one acceptor bond or one cation or four-coordinated with two donors and two acceptors or cations. There were a few examples, less than 5%, of five-coordinated, and less than 1% of six-coordinated.

When more neutron diffraction data became available, the water coordination was reviewed again for mostly inorganic salt hydrates by Ferraris and Franchini-Angela (1972) and Chiari and Ferraris (1982). Ninety-seven crystal structures were used, providing 296 O_W—H----O hydrogen bonds. In these structures, the water molecules were also three- and four-coordinated. In the three-coordinated waters the bisector of the oxygen lone-pair directions was directed towards the cation with the O-H bonds either planar or pyramidal with respect to that direction, as in (**II**) and (**III**). The four-coordinated waters were approximately tetrahedral with both oxygen lone-pairs directed toward cations, as in (**IV**).

$$M\text{---}O\begin{matrix}\diagup H\text{---}O\\\diagdown H\text{---}O\end{matrix} \qquad M\blacktriangleleft O\begin{matrix}\diagup H\text{---}O\\\diagdown H\text{---}O\end{matrix} \qquad \begin{matrix}M\\M\end{matrix}\diagdown O\begin{matrix}\diagup H\text{---}O\\\diagdown H\text{---}O\end{matrix}$$

(**II**)   (**III**)   (**IV**)

Within this rather small sample of crystal structures, the distributions were equally probable. The length of the O_W—H----O hydrogen bonds is clearly influenced by the charge on the cations, as shown in Table 8.3. The greater the cationic charge, the greater is the deshielding of the water protons and the shorter the hydrogen bonds. In Table 8.3b, the relationship to Pauling's (1929) bond strength ($\bar{S}$) is given. Pauling's bond strength is the valence of the cation divided by its coordination.[7]

---

[7]The concept of bond strength has been applied to the discussion of bonding in oxides by Brown and Shannon (1973).

A correlation between the activation energy of the 180° *flip-motion* of the water molecules in a number of inorganic hydrates with hydrogen bond distances and deuterium coupling constants by Larsson, Lindgren, and Tegenfieldt (1991) distinguished between waters having configuration (**II**) or (**III**) vs those with configuration (**IV**). The activation energies, measured by NMR spectroscopy, covered a range of ~1.5 kcal/mol^{-1} for the stronger bonds with H----O = 1.7 to 1.8 Å, to ~3 kcal/mol^{-1} for the weaker bonds with H----O 1.8–2.0 Å. Not surprisingly, the activation energies for the four-coordinated waters were larger.

Falk and Knop (1973) reported the infrared stretching frequencies for uncoupled HDO in 33 inorganic salt hydrates and the ices II, III, V, and VI. The ν(OH) values lie in the range 3170–3608 cm^{-1} and the ν(OD) from 2421 to 2657 cm^{-1}. The highest frequencies observed are for $Na_2[Fe(CN)_6NO]2H_2O$, where the O_WH nearest neighbor is a nitrogen atom corresponding at best to a weak O_W—H----N bond of 2.6 Å. In their review, Falk and Knop (1973) examined 663 inorganic hydrates for the site symmetry of the water molecules. Only 11 had the symmetry C_{2v} of the isolated water molecule. There were 34 examples of C_2 symmetry, but the majority, 570, were asymmetric with C_1. In two structures, $K_3MnC\ell_4·2H_2O$ and $Mo_6Br_{12}·2H_2O$, the waters were at sites with four-fold symmetry, indicating spatial or dynamic disorder. Examples are reported of inorganic hydrates in which the waters are hydrogen-bonded $(H_2O)_n$ with units n = 2–8, and in infinite chains and sheets. The review also included tables of hydrogen bond distances and angles for two-centered and three-centered O_W—H----O_W hydrogen bonds. The two-centered H----O_W bond lengths ranged from 1.65 Å in $KCr(SO_4)_2·12H_2O$, to 2.18 Å in $UO_3(NO_3)_2·6H_2O$. Many of the three-center bonds were almost symmetrical with H----O distances 2.1–2.7 Å and angles ~130°. The sequences of O_W----B mean distances were acceptors for B: F^- = 2.68 ± 0.04 Å, O^- or O^{2-} = 2.79 ± 0.24 Å, N^+ = 3.04 ± 0.10 Å, $C\ell^-$ = 3.19 ± 0.17 Å, Br^- = 3.42 ± 0.20 Å, and I = 3.63 ± 0.04 Å.

Infrared spectroscopy suggests existence of $(H_2O·X^-)_2$ clusters in some low hydrates. Recent research using infrared powder spectroscopy on a number of mono and dihydrates by Harman and co-workers suggests the formation of $(H_2O·X^-)_2$ clusters where X is halide or hydroxide ions. The compounds studied are given in Table 8.4. Some of these compounds form higher hydrates which are thought to be new clathrate or inclusion compounds. Hydrogen bond energies from equilibrium vapor pressure measurements are in the 10 kcal/mol^{-1} range, suggesting strong hydrogen bonds. The spectroscopic evidence suggests that the clusters are planar with C_{2h} symmetry rather than tetrahedral. This is supported by the crystal structural analysis of tetramethyl ammonium monohydrate by Loehlin and Kvick (1978) which contains the cluster (**V**) in which O----$C\ell^-$ = 3.204 and 3.247 Å and $C\ell$—O—$C\ell^-$ = 104°, and O—$C\ell^-$—O = 76°.

(V)

Table 8.4.

Crystal Structures Containing $(H_2O—X^-)_2$ Clusters Deduced From Infrared Spectroscopy.

Tetrapropylammonium chloride monohydrate, by Harmon, Toccalino, and Janos (1989).

Tetramethylammonium fluoride monohydrate, tetraethylammonium fluoride, and chloride monohydrates, N,N-dimethyl pyrrolidinium fluoride monohydrate, by Harmon, Gabriele, and Harmon (1989).

N,N,N-trimethyl-1-adamantyl ammonium fluoride hemihydrate (the trihydrate is believed to form an inclusion hydrate), by Harmon, Mounts, and Wilson (1991).

N-methylhexamethylene tetramine chloride, bromide, and iodide monohydrates, by Harmon and Keefer (1992).

N-methylquinoilidinium halide and hydroxide monohydrates, by Harmon and Southworth (1993).

N,N'-dimethyltriethylene diammonium halide dihydrates, by Harmon and Brooks (1993).

Hexamethonium $(CH_3)_3N^+—(CH_2)_6—N^+(CH_3)_3$ halide mono and dihydrates, by Harmon, Brooks, and Keefer (1994).

N,N,N-trimethyl-1-adamantyl ammonium hydroxide monohydrate (the tetrahydrate is believed to be a hydrate inclusion compound), by Harmon, Southworth, and Mounts (1993).

Strong acids and bases form many hydrates. The strong acids such as the hydrogen halides, H_2SO_4, $HC\ell O_4$, HNO_3, etc. form a large number of hydrates in which the water is protonated to form oxonium cations with strong O—H----O bonds, as discussed in Chapter 3 and shown in Table 3.7. The lower hydrates form oxonium clusters, the higher hydrates form clathrate or layer inclusion compounds discussed in Chapter 9.

The strong bases, such as the hydroxides, also form many hydrates. $NaOH \cdot xH_2O$, for example, occurs with x = 1, 2, 3.5, 4a, 4b, 5, and 7. The structure of $\beta NaOH \cdot 4H_2O$ determined by Mootz and Seidel (1990a) contains $Na(H_2O)_6^-$ octahedra in a hydrogen bonded $OH^-H_2O)_n$ network. The hydrates $(CH_3)_4N \cdot OH \cdot xH_2O$ were studied by Mootz and Stäben (1992), Stäben and Mootz (1993), and Mootz and Stäben (1993). The lower hydrates form strong O—H----$\bar{O}$ bonds, discussed in Chapter 3. The higher hydrates with hydration numbers from 4.6 to 10 form the inclusion hydrates discussed in Chapter 9.

The Stereochemistry of Water Molecules in the Hydrates of Small Biological Molecules is Surprisingly Varied

Hydrogen-bonded water plays an important and not too well understood role in biological structure and function. The nature of the hydrogen bonding in biological structures is obscured by the difficulty of locating hydrogen atoms in macromolecular crystal structure analyses. For this reason, the hydrogen bonding in the crystal structures of the small molecule components, where hydrogens can be located, has been analyzed in some detail, as discussed in Chapter 4.

The stereochemistry of the water molecules in the crystal structures of 311 hydrates of amino acids, peptides, carbohydrates, purines, pyrimidines, nucleosides, and nucleotides, giving 621 O_WH----A hydrogen bonds, was examined by Jeffrey and Maluszynska (1990a). Only neutron diffraction and good X-ray analyses were used where there was no disorder or ambiguity about the positions of the hydrogen-bonding hydrogen atoms. In the X-ray analyses, the covalent O—H and N—H bond lengths were normalized to 0.97 and 1.00 Å, respectively, to make them more consistent with the neutron diffraction data. Data where H—O_W—H angles were greater than 130° and less than 90° were discarded. The results of this survey are shown in Table 8.5. The

Table 8.5.
The Stereochemistry of the Water Molecules in the Hydrates of Small Biological Molecules (from Jeffrey and Maluszynska, 1990a).

Structure	Count	Structure	Count	Structure	Count
$X—H\cdots O$ with $H\cdots A_1$, $H\cdots A_2$	107	X^1H, X^2H on O with $H\cdots A_1$, $H\cdots A_2$	168	X^1H, X^2H, X^3H on O with $H\cdots A_1$, $H\cdots A_2$	2
$X—H\cdots O$ with A_1, $H\cdots A_2$, $H\cdots A_3$	38	X^1H, X^2H on O with A_1, $H\cdots A_2$, $H\cdots A_3$	69	X^1H, X^2H, X^3H on O with A_1, $H\cdots A_2$, $H\cdots A_3$	2
$X—H\cdots O$ with A_1, $H\cdots A_2$, $H\cdots A_3$, A_4	11	X^1H, X^2H on O with A_1, $H\cdots A_2$, $H\cdots A_3$, A_4	29	X^1H, X^2H, X^3H on O with A_1, $H\cdots A_2$, $H\cdots A_3$, A_4	4
$X—H\cdots O$ with A_1, H, A_2, H, A_3	7	X^1H, X^2H on O with $H\cdots A_1$, $H\cdots A_2$, A_3, A_4	18	O with $H\cdots A_1$, $H\cdots A_2$	9
$X—H\cdots O$ with $H\cdots A_1$, $H\cdots A_2$, A_3, A_4	3	X^1H, X^2H on O with A_1, $H\cdots A_2$, $H\cdots A_3$	3	O with $H\cdots A_1$, $H\cdots A_2$, A_3	5
$X—H\cdots O$ with A_1, $H\cdots A_2$, $H\cdots A_3$, A_4	2			O with $H\cdots A_1$, H	1

results corresponded with those observed with the inorganic salt hydrates in three respects: (1) both three-coordinated waters, with one acceptor and two donor bonds, and four-coordinated waters are observed; (2) three-coordinated water can have a planar or pyramidal configuration of hydrogen bonds, with a continuum from strictly planar to pyramidal with 110° angles; and (3) there were a few examples of other coordinations.

About 25% of the bonds are three-centered, which is about the same proportion as observed with O—H---O bonds generally, as described in Chapter 4. There are 10 examples of bifurcated/three-center bonds, 15 examples of water molecules which accept no hydrogen bonds, especially in the amino acids and peptides. There are eight examples, only in the carbohydrates and amino acids, where the water molecules accept three hydrogen bonds. Steiner and Saenger (1993a), in their study of C—H---O hydrogen bonds in carbohydrates and cyclodextrins, reported in Chapter 5, have suggested that C—H---O_W bonds complete the four-coordination shell in the case of three-coordinated pyramidal waters and sometimes increase the four-coordination to five.

The role of water in the hydration of biological macromolecules such as proteins and nucleic acids is discussed in Chapter 10.

8.5 | WATER IN MOLECULAR RECOGNITION

Crystallization must involve molecular recognition. As the solvent content is reduced, the collision between molecules increases until it becomes energetically more favorable for them to nucleate rather than solvate. Thereafter the molecules deposit on the crystal surfaces in the correct orientation. The crystallization of proteins is more surprising however, since more than half the volume of the crystals is often still occupied by water. The water molecules near the surface of the protein crystal show evidence of some order, so-called *bound water*, but the water further away, the *unbound water*, shows no evidence of order. The question is, how do these large and sometimes irregularly shaped molecules get oriented into a three-dimensional lattice which is sufficiently ordered to permit X-ray diffraction when they are separated by large volumes of water? What is the means of communication between their electrostatic potentials?

As pointed out in Chapter 6, water molecules in carbohydrates and cyclodextrin hydrates link the chains of O—H$\cdots$O—H$\cdots$O—H bonds in two ways: either to form branched chains or three-dimensional nets. Irrespective of whether the continuum or mixture model of liquid water is envisioned, there is agreement that on a short-time transient basis, there are chains of —$O_WH\cdots O_WH$— hydrogen bonds. In fact, such chains are implicit in the defect models necessary to explain the anomalous conductivity of water, discussed in Chapter 7.

One process of molecular recognition which is believed to specifically involve hydrogen-bonding, is the phenomenon of *sweet taste*. The current, and only, hypothesis that has persisted with extraordinary resilience for more than a quarter of a century is due to Shallenberger and Acree (1967).[8] More recent infrared and Raman spectroscopic work by Mathiouthi and Portman (1990) supports this hypothesis. It requires three contact points of the sweet-tasting molecule that interact with the receptor sites on the tongue. How do the sweet molecules reach the receptor sites? Is it by Brownian motion? This is equivalent to a drunken guest in a dark corridor placing his key in the correct key-hole in the correct orientation (compliments to Emil Fisher's Schlüsse-Schloss-Prinzip, as described by Lichtenhalter (1994). It is suggested by Jeffrey (1994) that nature is more efficient and uses the cooperativity of water for the purpose of molecular recognition.

The electrostatic potential of the sweet molecule, irrespective of whether it has hydrogen bonds, will affect the polarizability of the adjacent water molecules. This change

---

[8]This may be because of the absence of competing hypotheses. Sweet and sour taste is a major commercial factor in the food industry. The biology of the human taste buds is complex, as elegantly described by McLaughlin and Margolskee (1994). It will be some years before the tastee receptor site proteins are extracted, purified and crystallized. Thereafter, crystal structure analyses will follow and the sites where various tastants adhere will be located. What they do when they get there is a more difficult chemical problem.

in polarizability can be transmitted along a cooperative chain of hydrogen bonds while the chain persists, to a complementary electrostatic potential at the receptor site, which includes hydrogen bond functional groups, since it is a protein. Since the role of water in molecular recognition is a field wide open for speculation, several mechanisms can be considered.

1. The polarization of the chain of water molecules directs the tastant to the receptor site.

2. The polarization induced by the electrostatic potential of the tastant and transmitted through the hydrogen-bonded chain is sufficient to activate a neural response at the receptors. A concept inspired by the experiments through artificial membranes by Kurikara et al. (1991).

3. The cooperative effect of the polarized water chain permits proton transfer along the chain to the receptor site, as suggested by the experiments of Zundel and collaborators, described in Chapter 7, and by the inelastic neutron scattering experiments of Kearley et al. (1994).

This subject is discussed from a different view point by Lemieux (1996).

Inclusion Compounds 9

9.1 | THE CONCEPT OF INCLUSION

Inclusion compounds are a class of molecular complexes in which macromolecules or a group of molecules enclose other molecules without the formation of covalent or ionic bonds between them. The molecules which do the enclosing are referred to as the *hosts*, while the includees are the *guests*. The hosts form a lattice type structure with voids which are large enough to accommodate the guests. Appropriately, the chemistry of the guests must not be such as to prevent the formation of the walls which enclose them. The hosts may be macromolecules, such as the zeolites, or large molecules, such as crown ethers, calix[r]arenes or cyclodextrins, or assemblies of small molecules forming a host lattice with ionic bonds, hydrogen bonds or by intercalation. In some cases, the hosts are covalently bonded molecules with holes big enough to accommodate guests, as in the cyclodextrin or the [12]-collarene shown in Figure 9.1. In others, inclusion compounds are formed simply because the host molecules cannot pack efficiently through van der Waals interactions without the presence of guest species. The many examples of the porphyrin-based lattice clathrates reported by Byrn et al. (1993), shown in Figure 9.2, are such an example.

Molecular inclusion is a large and fast-growing field of supramolecular chemistry in which crystal structure analysis has played, and will continue to play, the key role. This chapter is concerned with those compounds where hydrogen bonding is an essential component of the host structure.

Many of these inclusion compounds were known as *molecular complexes* long before the reasons for their formation were revealed by crystal structure analysis. Reviews of the earlier work are provided by Hagan (1962) and Powell (1964). More comprehensive and contemporary reviews are to be found in *Inclusion Compounds*, volumes 1 and 2, edited by Atwood, Davies, and MacNicol (1984). The inclusion

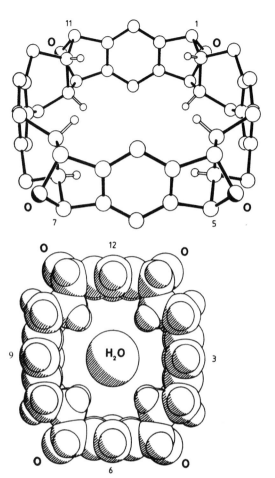

Figure 9.1. A covalent bond has lattice formed by a [12]-collarene; an inverted clathrate hydrate with the water as guest (from Ashton et al., 1988, reprinted with permission).

compounds where hydrogen bonding provides the structure of the host lattices are the hydrates (reviewed by Jeffrey, vol. 1, chap. 5); the hydroquinones, phenols, and Dianin's compound (reviewed by MacNicol, vol. 2, chap. 1); the ureas, thioureas, and selenoureas (reviewed by Takemoto and Sonado, vol. 2, chap. 2); the choleic acids (reviewed by Giglio, vol. 2, chap. 7); and the cyclodextrins (reviewed by Saenger, vol. 2, chap. 8). The hydrate inclusion compounds and the cyclodextrins are described from the hydrogen-bonding point of view in *Hydrogen Bonding in Biological Structures* by Jeffrey and Saenger (1991). Recent, well-illustrated reviews of the field are provided in chapter 7 of *Crystallography in Modern Chemistry*, by Mak and Cong-Du-Zhou (1992), and in volume 6 of MacNichol, Toda, and Bishop (1996).

The clue to the raison d'etre of these compounds came from the crystal structure analysis of a hydroquinone —SO_2 complex by Palin and Powell (1947). The crystal structure contains a host lattice formed by two interpenetrating lattices of hydrogen-bonded hydroquinone molecules. The inclusion process, which is illustrated

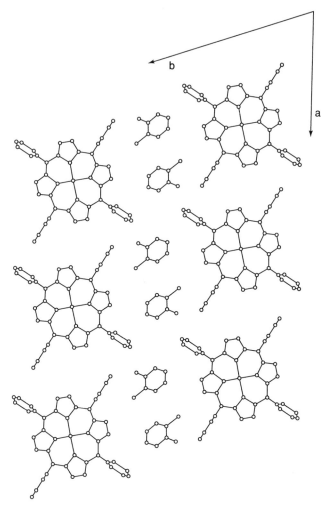

Figure 9.2.   Porphyrin-based van der Waals inclusion lattice (from Byrn et al., 1993).

in Figure 9.3, was called *clathration*, from the Latin word *clathratus* meaning enclosure by a grating. A few years later, Schlenk (1949, 1951) and Smith (1950, 1952)
determined the channel-type structures of the urea and thiourea complexes and used
the word *inclusion* to describe the phenomenon. Inclusion is the more general term,
of which the cage-like clathrates and the channel and sandwich type structures, are
special cases. It is not necessary that all the voids in the host lattices are occupied
to form a stable crystalline compound, therefore *nonstoichiometry* is a characteristic
of these compounds. This is one of the reasons that these compounds defied interpretation through chemical analysis, since their structures could not be deduced from
their stoichiometry.

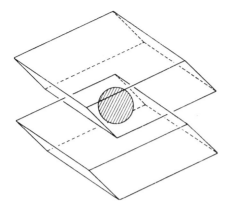

Figure 9.3   The concept of clathration. Top: enclosure by interpenetrating host lattice. Bottom: one hydrogen-bonded hydroquinone host lattice. Long lines represent hydroquinone molecules. Short lines are O—H----O—H bonds (from Evans, 1946, reprinted with permission).

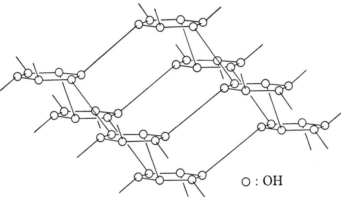

O : OH

9.2 | CLATHRATES

The *hydroquinone-SO₂ complex* studied by Palin and Powell (1947) is a member of the clathrate class of inclusion compounds that has been known since the 19th century, with H_2S as guest in 1849 and SO_2 as guest in 1859. Hydroquinone itself has three crystalline modifications, α, β, and γ. All are hydrogen-bonded structures which can form inclusion compounds with the general formula $3C_6H_4(OH)_2xG$. The β form, which is that studied by Palin and Powell, forms three types of host structures: Type I with symmetry $R\bar{3}$, Type II with symmetry R3; and Type III with symmetry P3. In all three host lattices, the hydroxyl groups of the hydroquinone molecules are hydrogen-bonded to form $(OH)_6$ homodromic hexagons which are nearly planar with O----O distances from 2.65 to 2.80 Å. The benzyl moieties are directed above and below the $(OH)_6$ plane at alternate hydroxyls. In Type I, the host lattice is formed by two interpenetrating lattices. The Type II host lattice is similar, with a different arrangement of the hydroquinone molecules. In the CH_3OH and HCℓ complexes studied by Mak (1982), the

O----O distances around the hexagon are unequal, which is taken as evidence that there can be a host-guest hydrogen-bonding interaction.[1] Only one Type III structure is known, with CH_3CN as guest. It is a very complex structure, with three symmetry-independent hydroquinone molecules forming 54 molecules in a $R\overline{3}$ symmetry unit cell. Two of the three molecules form interpenetrating lattices similar to those in the Type II hosts. The third molecule forms double helices of hydrogen-bonded chains of molecules which extend along a three-fold screw axis of symmetry. The interpenetrating lattices and the helical chains are linked by hydrogen bonds. The calculated density of the α-hydroquinone is high, 1.38 g/cm^{-1}, and the local density of the helical region is even higher, 1.43 g/cm^{-1}. This structure has three cages which can accommodate guests.

Phenol Forms Clathrate Host Lattices Which are also Stabilized by Hydrogen Bonds

Since phenol has a hydrogen-bonding functional OH group on one side of the molecule only, half of the clathration is provided by the packing of the benzene rings and half by the hydrogen bonding. As in the hydroquinone host lattices, the benzene moieties point alternatively above and below the [OH]$_6$ rings. The packing of the sextet of phenol molecules then forms host lattices with two cavities, one small and one large. The general formula is $12C_6H_5OH \cdot nX \cdot mY$, where X refers to occupancy of the small cage, Y to the large cages. Structures are known with n = 1, m = 4; n = 0, m = 4; n + m = 4.

Hexagons of (—OH)$_6$ Hydrogen Bonds Stabilize Another Clathrate Host Lattice

This is the host lattice formed by Dainin's compound, **I**; 4-p-hydroxyphenyl-2,2,4-trimethylchroman,[2] and a number of related thiochroman derivatives.

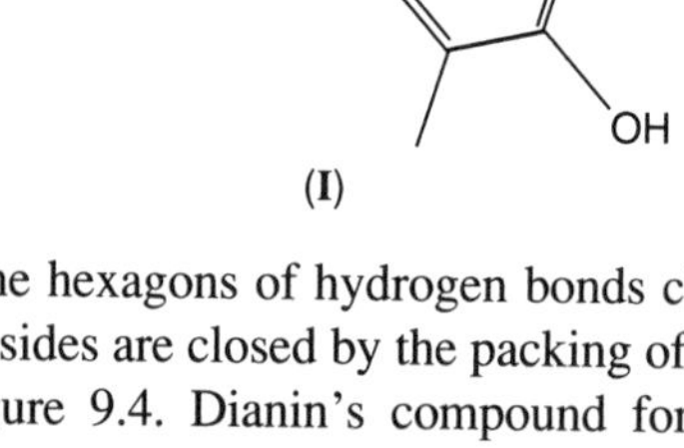

(I)

As with the phenols, the hexagons of hydrogen bonds close only the top and bottom of the voids. The other sides are closed by the packing of the remainder of the molecule, as illustrated in Figure 9.4. Dianin's compound forms inclusion compounds

---

[1]This makes the compound a *semi-clathrate*, as discussed later in the chapter.

[2]This compound was reported in 1914 from the condensation of phenol and mesityloxide by Dainin, who was a Russian chemist and a student of the chemist-composer, Borodin.

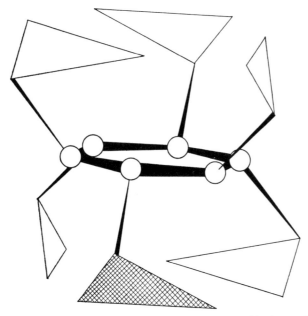

Figure 9.4.   Clathration by half hydrogen-bonding, half van der Waals packing in complexes of Dainin's compound (from Hagan, 1962, reprinted with permission).

$6C_{18}H_{20}O_2 \cdot nM$, where n = 1, 2, or 3. A wide variety of molecules can be included and n depends upon their size. For example, n = 1, bromobenzene, tetrachlorethylene; n = 2, $CC\ell_4$, acetic acid; n = 3, methanol, from a review by Mandelcorn (1959). The molecules are linked by hexagons of —O—H⋯O—H— and point alternately on either side of the $(O—H)_6$ hexagon, forming the "hour-glass" structure shown in Figure 9.5.

The Bile-Acid Inclusion Compounds

Another long-known class of inclusion compounds in which hydrogen-bonding plays a role is that of deoxycholic acid ($3\alpha,12\alpha$-dihydroxy-5β-cholan-24-oic acid, $C_{20}H_{40}O_4$). This molecule, which is a bile-acid, has a per-hydro-1,2-cyclopentenophenanthrene ring system with an acid group at one end and hydroxyl groups at the other end and in the middle. It has five crystalline polymorphs: orthorhombic α and β, tetragonal α and β, and hexagonal, all of which form channel-type inclusion compounds.

Figure 9.5. Hourglass cavity in the Dianin complex with dimethyl acetylene.

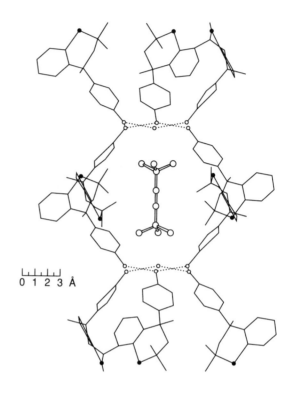

0 1 2 3 Å

In the orthorhombic structures, the molecules form bilayer channels by hydrogen bonding the carboxylic acid group on one molecule to the end hydroxyl and midpoint hydroxyl on another. These channels can include a great variety of guests, including water, ethanol, camphor, and palmitic acid. More than 40 crystal structures have been reported. A recent analysis of a 2:1 inclusion compound with camphor is by DeSantis et al. (1995). The tetragonal structures also contain channels formed by hydrogen-bonded bilayers.

9.3 | THE CLATHRATE HYDRATES

The greatest variety and most intensely studied of the hydrogen-bonded inclusion compounds are the clathrate hydrates. The ability of water molecules to form a wide variety of four-connected nets, which results in the polymorphism of ice, is also apparent in the hydrate inclusion compounds. In the ices, the voids in these nets are small enough that they are not collapsed by the van der Waals attractions across them. In ice I_h, they are large enough to enclose a water molecule, which led to the Danford and Levy (1962) theory for the structure of liquid water. In the hydrate inclusion compounds, the voids are much larger and the nets are unstable unless they are occupied. This is in contrast to the isostructural $(SiO_2)_n$ clathrasils, where the bonds forming the nets are covalent Si—O bonds rather than H----O hydrogen bonds and occupation of the voids is not necessary to form stable crystals.

As with many of the other inclusion compounds, the clathrate hydrates have a long history. Faraday (1823) made a crystalline hydrate of chlorine, believed it to be $C\ell_2 \cdot 10H_2O$, and a few years later a similar hydrate of bromine was reported by Löwig (1829). Throughout the remainder of the 19th century, hydrates of all the well-known gases were prepared, mostly by the French scientists, Forcrand and Viullard. This work was reviewed by Schroeder (1927). It was more than a century later that the structure of the gas hydrates, and chlorine hydrate in particular, was determined from X-ray powder diffraction data by Claussen (1951a,b), von Stackelberg and Muller (1951), and Pauling and Marsh (1952).[3] The structure of the bromine hydrate had to wait much longer for a single crystal neutron diffraction analysis by Brammer and McMullan (1993). During the years when the octet rule and paired electrons were rationalizing constitutional chemistry, the formation of a crystalline compound composed of $C\ell_2$, H_2S, and H_2O without an unpaired electron between them was a mystery.[4] Undoubtedly the clue to understanding this phenomenon came originally from the hydroquinone-SO_2 crystal structure analysis of Palin and Powell (1947).

The voids in the ice-like hydrate inclusion host lattices can be occupied in four ways:

1. By stable atoms or molecules ranging in size from argon to adamantine, C_9H_{14}. These are considered *true clathrate hydrates.*

2. By certain cations, notably the tetramethyl-, tetra-n-butyl, and tetra-iso-amyl ammonium ions. In these structures, the anions are incorporated by hydrogen-bonding into an anionic hydrate framework and are *ionic clathrate hydrates.*

3. By certain anions, notably those of strong acids, where the protons are included in the hydrogen-bonding into a cationic hydrate lattice and are also *ionic clathrate hydrates.*

4. By alkylamines ranging from methylamine to propylamine and beyond. In these structures, the hydrocarbon moiety of the amine occupies the voids while the amine functional groups are incorporated by hydrogen-bonding into the hydrate lattice. These are known as *semi-clathrate hydrates.*

As with the ices, the $(H_2O)_n$ lattices are formed from four-connected hydrogen-bonded nets of water molecules. The predominant, but not exclusive, architectural motif of the gas hydrate structures is the pentagonal dodecahedron, which is a regular polyhedron having 12 pentagonal faces (5^{12}).[5] This is a $H_{40}O_{20}$ unit of water structure,

---

[3]Linus came third. An unaccustomed position, upon which he comments in a footnote to his paper.

[4]When freshman chemists used H_2S in analytical chemistry, it was a mysterious experiment to pass H_2S into cold chlorine water and produce a white precipitate. The same experiment works with a $CC\ell_4$-H_2O mixture.

[5]Polyhedra are denoted by a number indicating the number of edges on a face and a superscript indicating the number of faces. For example, $4^3, 5^6, 6^3$ has three quadrilaterals, six pentagons, and three hexagons. It is interesting to compare the convex polyhedra observed in hydrates with those found in the fullerenes, which also have a preponderance of pentagons, as described by Zhang et al. (1992).

Table 9.1.

Molecules Reported To Form True Clathrate Hydrates—Types I and II.

	Type I	Type II
Rare gases	Xe	Ar, Kr
Diatomic	N_2, O_2, $C\ell_2$, $BrC\ell$	Br_2
Triatomic	CO_2, N_2O, H_2S, H_2Se, SO_2, $C\ell O_2$, COS	CS_2
Inorganic	PH_3, AsH_3	SF_6
Hydrocarbons	CH_4, C_2H_2, C_2H_4, C_2H_6, CH_3SH, cyclopropane	propylene, cyclopropane, C_3H_8, isobutane, cyclopentene, cyclopentane, $C_6H_6{}^a$, cyclohexane^a, isobutylene^a, cis-2-butene^a, allene^a, n-butane^a, norborane^a, bicycloheptadiene^a, neopentane^a
Fluorinated hydrocarbons	CH_3F, CH_2F_2, CHF_3, CF_4, $CH_3C\ell$, CH_3Br, $CHC\ell F_2$, C_2H_3F, C_2H_5, CH_3CHF_2	C_2F_4, $(CH_3)_3CF$, $CC\ell_2F_3$
Chlorated hydrocarbons	$CH_3C\ell$	$CH_2C\ell_2$, $CHC\ell_3$, $C_2H_3C\ell$, $CC\ell_4$, $CH_3CHC\ell_2$, $C_2H_5C\ell$, $CC\ell_3$
Chloro-fluoro-hydrocarbons	$CH_2C\ell F$, $CHC\ell F_2CHC\ell F_2$,	$CHC\ell_2F$, $CC\ell_2F_2$, $CC\ell_3F$, $CH_3CC\ell F_2$
Other alkyl halides	CH_3Br, CH_3I	C_2H_5Br, CH_3I, $CHBrF_2$, $CBrF_3$, CBr_2F_2, $CBrC\ell F_2$, C_3H_8Br
Cyclic ethers	ethylene oxide, trimethylene oxide, dioxalane	trimethylene oxide, dimethyl ether, propylene oxide, 1,3-dioxolane, furan, 2,5-dihydrofuran, tetrahydrofuran, 1,4-dioxane, cyclobutanone, cyclohexene oxide
Ketones	—	acetone, cyclobutanone, methylformate
Cyclic imines and oxyimines		ethylene imine, pyrrolidine, azetidine, pyrroline, propylene-imine, pyrrole, pyrazole, imidazole, isodazole, isothiazole, oxazole, thiazole, pyrrolidone
Miscellaneous	acetonitrile	

^{a}From ^{139}Xe NMR chemical shifts.

shown in Figure 9.6. Of the 40 hydrogen atoms, 30 form the 30 edges of the polyhedron.[6] The remaining 10 donate hydrogen bonds to adjacent polyhedra, which in turn supply the hydrogens for the donor/acceptor equality of the water molecules. However, the pentagonal dodecahedron, unlike the tetrahedron and octahedron, is not a space-filling solid. Other polyhedra have to be involved to form a periodic crystal lattice. In the case of the gas hydrates, these polyhedra are larger, as shown in Figure 9.7. Except for the pentagonal dodecahedron, which is a *regular polyhedron*, these are known as *semi-regular polyhedra*, because they cannot have simultaneous planar faces with equal edges and equal angles.

Molecules known to form gas hydrates are shown in Table 9.1. Those about the size of methane can occupy both the (5^{12}) and the larger $(5^{12}6^2)$ hedra nonstoichiometrically, since the 5^{12} hedra need not all be occupied. The larger polyhedra, however, have diameters of ~ 5.5 Å, and the oxygen-to-oxygen van der Waals forces will collapse them if they are vacant. These compounds are difficult to analyze by chemi-

[6]Convex polyhedra obey Euler's theorem. Faces + vertices = edges + 2 (Lyusternik, 1963). For the pentagonal dodecahedron, 12 + 20 = 30 + 2.

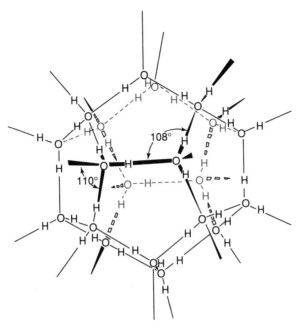

Figure 9.6. The famous $H_{40}O_{20}$ pentagonal dodecahedron of hydrogen-bonded water structure.

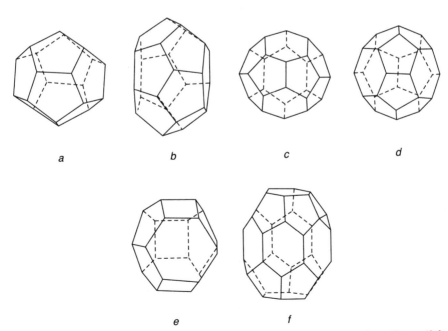

a b c d

e f

Figure 9.7. Hydrogen-bonded $(H_2O)_n$ polyhedra observed in the clathrate hydrates.[4] (a) 5^{12}; (b) $5^{12}6^2$; (c) $5^{12}6^3$; (d) $5^{12}6^4$; (e) $4^35^66^3*$; and (f) $5^{12}6^8*$ (*inferred from clathrasil 1H structure).

Table 9.2.

Large Molecules Forming Hexagonal Type III Clathrate Hydrates On Basis of ^{139}X NMR Chemical Shifts.

methylcyclopentane	adamantine
methylcyclohexane	2-adamantanone
2-methyl butane	bicyclo[2,2,2]oct-3-ene
2,3-dimethyl butane	2,3-dimethyl-2-butene
hexamethylethane	2,3-dimethyl-1-butene
2,2-dimethyl pentane	3,3-dimethyl-1-butene
3,3-dimethyl pentane	3,3-dimethyl-1-butyne
cycloheptene	cis-1,2-dimethylcyclopheane
cyclooctane	tert-butylmethylether
hexachlorethane	tetra-methyl silane
cis-cyclooctane	isoamyl alcohol

cal methods. The crystal structure analyses established the ideal composition of $2X \cdot 6Y \cdot 46H_2O$ for the so-called Type I, 12 Å cubic hydrate, and $16X \cdot 8Y \cdot 136H_2O$ for the Type II, 17 Å cubic hydrate. X are the small guest molecules that occupy or partially occupy the (5^{12}) hedra. Y are the larger molecules which occupy the $(5^{12}6^2)$ and $(5^{12}6^4)$ polyhedra. Large molecules, with van der Waals diameters greater than ~5.5 Å, cannot occupy the (5^{12}) hedra. But occupying them by other small gases raises the melting or decomposition point of the crystals, and hence these were called *help gases*. The most commonly used help gases are Xe or H_2S.

The cut-off between Type I and Type II cubic gas hydrate structures comes with molecules the size of $CHC\ell F_2$, which form both hydrates. The largest molecule observed to form a Type II hydrate is CBr_2F_2 or cyclobutanone. However, the very small guests, Ar and Kr, also form a Type II hydrate, as predicted by Holder and Manganiello (1982) from thermodynamic considerations and confirmed by Davidson et al. (1984).

Gases such as $C\ell_2$ and SO_2 form crystals with water above 0°C at atmospheric pressure, but other gas hydrates have to be prepared under gas pressure. A table of the necessary thermodynamic conditions for hydrate formation is given in a review of the clathrate hydrates by Davidson (1973).

More recently, Ripmeester and Ratcliffe (1990) used ^{129}Xe NMR resonance, to identify a new clathrate hydrate which was shown by X-ray powder patterns to have a host lattice isostructural with a known clathrasil structure. The host lattice had $2(5^{12})$, $1(4^35^66^3)$, and a larger $3(5^{12}6^8)$ polyhedron which could accommodate molecules up to the size of norborene or adamantine, as shown in Table 9.2, with stoichiometry $2X \cdot Y \cdot 3Z \cdot 34H_2O$. Two new gas hydrate structures have been observed under high pressure: a helium hydrate, $He \cdot 6H_2O$ by Londono, Finney, and Kuhs (1992) and a hydrogen hydrate $H_2 \cdot 6H_2O$ by Vos et al. (1993). The host structure is rhombohedral with small voids and similar to that of Ice II. Above 2.3 to 3.0 GPa, a second hydrogen hydrate is observed with a cubic diamond-like host lattice by Vos et al. (1993).

The clathrate hydrates subsequently studied by neutron diffraction are given in Table 9.3. In all of these structures the hydrogen atoms are symmetrically disordered with ranges of O—H, H----O, O----O, and O—H----O distances and angles similar to those in the Ices.

Table 9.3.
Hydrogen Bond Geometries in Clathrate Hydrate Structures Studied by Single Crystal Neutron Diffraction.[a]

Structure	O—D (Å)	D ⇌ O (Å)	O ⇌ O (Å)	O—D ⇌ O (°)
6.C_2H_4O.0.4 air 46.D_2O at 80 K[b] (e.s.d. 0.004 Å)	0.971–0.998	1.726–1.818	2.724–2.784	171.6–180
8CCℓ_4.3.5Xe.136D_2O[c] at 100 K at 13 K (e.s.d. 0.001–0.002 Å)	0.979–0.993 0.986–1.001[d]	1.748–1.813 1.738–1.802	2.735–2.795 2.731–2.785	175.0–180 174.8–180
Polyhedral radii (center to vertices)	5^{12} $5^{12}6^2$ $5^{12}6^4$	3.83–2.96 Å 4.06–4.65 4.64–4.73		
Mean O----O distance	2.90 Å			

[a]A neutron diffraction study of 6Br$_2$48D_2O at 100 K has been briefly reported by Brammer and McMullan (1993) but the complete refinement is not yet published.
[b]Hollander and Jeffrey (1977).
[c]McMullan and Kvick (1990).
[d]The lengthening of the O—D bond lengths with decrease in temperature is due to the bond-shortening effect of riding thermal motion of the D-atoms.

The Pentagonal Dodecahedron Stimulated Ideas

The 5^{12} polyhedron is an *Archimedean solid*, which as its name implies, was known in ancient Greece. Its appearance in chemistry, more than 2,000 years later, stimulated a number of imaginative suggestions, apart from its more mundane use as a novelty calendar; it has 12 faces. Pauling produced a clathrate hydrate model for the structure of liquid water and, with Miller, a clathrate hydrate theory of anesthesia (laughing gas and chloroform form clathrate hydrates).[7] Clathrate hydrates were proposed for the snows of Mars and Venus, the composition of asteroids, permafrost, deep-sea marine deposits, and the lungs of deep sea fish.

Gas hydrates can block natural gas pipelines if they are not buried sufficiently deeply or insulated in cold climates. This led to some unexpected explosions and a considerable amount of technological research both in the United States and the USSR. A sea water desalination process was proposed using the appropriate gas, such as butane or a freon, as both refrigerant and crystal hydrate former. Although pilot plants were built, the method did not work, either because halide ions inhibit clathrate formation or they are hydrogen-bonded into the hydrate lattice, forming an ionic clathrate from which the salt cannot be easily washed. Underground deposits of methane hydrate are found beneath the ocean floor in regions north of the Arctic Circle and off the east coast of the United States and elsewhere. These could be an important energy source in the distant future, as suggested by Figure 9.8.

References to many of these earlier publications, with titles, are to be found in a review of the clathrate hydrates by Jeffrey and McMullan (1967).

---

[7]After reading a paper by McMullan and Jeffrey (1959), Pauling devoted much effort to his clathrate theory of anaesthesia (Goertzel and Goertzel, 1995). It is interesting to note a recent NMR study of the effects of anaesthetics on the hydration of molecules which form clathrate hydrates by Akin and Harmon (1994).

Figure 9.8. Estimated distribution of organic carbon in Earth reservoirs (from *Chemical and Engineering News*, March 6, 1995, p. 40).

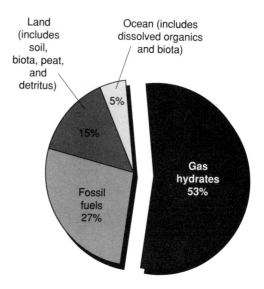

Gas hydrates hold huge reserves of organic carbon

Land (includes soil, biota, peat, and detritus)

Ocean (includes dissolved organics and biota)

5%

15%

Gas hydrates 53%

Fossil fuels 27%

Total organic carbon = 18,800 gigatons

Note: Estimated distribution of organic carbon in Earth reservoirs (excluding dispersed carbon in rocks and sediments, which equals nearly 1,000 times the total amount). Atmospheric carbon estimated to be 3.6 gigatons.
Source: U.S. Geological Survey

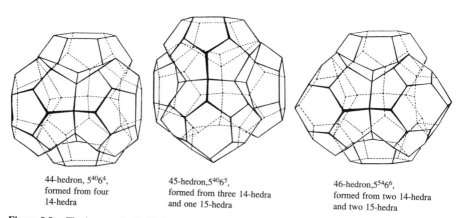

44-hedron, $5^{40}6^4$, formed from four 14-hedra

45-hedron, $5^{40}6^5$, formed from three 14-hedra and one 15-hedra

46-hedron, $5^{54}6^6$, formed from two 14-hedra and two 15-hedra

Figure 9.9. The larger anionic $(H_2O)_n \cdot X^-$ polyhedra found in the tetra-n-butyl and tetra-iso-amyl ammonium salt hydrates.

The structures of some quaternary ammonium salt hydrates was a surprise. Fowler et al. (1940), investigating the thermodynamics of aqueous solutions of alkyl ammonium salts, reported some unusual hydrates of tetra-n-butyl and tetra-iso-amyl ammonium salts. They were indeed unusual in that they had between 30 and 60 *waters of crystallization*. It was not until 19 years later that crystallographic studies by McMullan and Jeffrey (1959) and Feil and Jeffrey (1961) showed that these compounds, with formulae such as $5(n-C_4H_9)_4NF \cdot 164H_2O$ and $2(iso-C_5H_{11})NF \cdot 76H_2O$, were isostructural with the gas hydrates. This result was quite unexpected at that time. It comes about because, by combining the large polyhedra of the gas hydrates, even larger polyhedra, shown in Figure 9.9, are constructed which have four compartments that can accommodate the alkyl chains when a common vertex in the center is occupied by the central nitrogen atom rather than H_2O, as shown in Figure 9.10. As in the gas hydrates, 5^{12} polyhedra complete the packing to form the crystal lattices. The anions are included by hydrogen-bonding into the water host lattice, thereby maintaining a requirement for a four-connected, hydrogen-bonded $(H_2O)_n$ net with twice as many hydrogens as oxygens. The central nitrogen atoms are not hydrogen-bonded, and the anions accept four hydrogen bonds. The host lattice is therefore ionic and distorted, as illustrated in Figure 9.10, to enclose the cations to form ionic clathrates.

A number of these tetra-n-butyl and tetra-iso-amyl ammonium and tri-n-butyl and tri-iso-amyl sulfonium and phosphonium salts were identified by their crystal data by

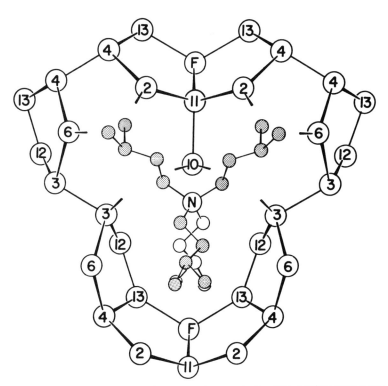

Figure 9.10.   Tetra-iso-amyl ammonium ion water-anion cage in $(iso-C_5H_{11})_4N^+F^- \cdot 38H_2O$. The F^- ions appear to be ordered.

Beurskens, Jeffrey, and McMullan (1963). These included some unknown structural types. By crystal structure analysis, Solodovinikov et al. (1982) identified a new structure for $(iso\text{-}C_5H_{11})P\cdot Br\cdot 32H_2O$ with large $5^{40}6^8$ polyhedra formed by using four $5^{12}6^3$ hedra. Lipkowski et al. (1990) identified a new clathrate hydrate structure for tetraisoamyl ammonium fluoride and Lipkowski et al. (1992) identified a layer clathrate hydrate for tetrapropylammonium fluoride. This suggests that further studies will reveal another series of compounds with larger voids.

This work was reviewed by Jeffrey and McMullan (1967), Davidson (1973), and Jeffrey (1984). In those days of X-ray film methods, only the geometry of nonhydrogen atoms could be determined, with no information available concerning the positions of the hydrogen atoms. The nature of the disorder in the hydrogen bonds introduced by the inclusion of the hydrogen-bonded anions requires careful neutron diffraction analysis and is still unexplored. The guest species are frequently disordered, both in these structures and in the gas hydrates. The disordered and dynamical properties of the guest molecules in the gas hydrates, which were studied by NMR spectroscopy and dielectric relaxation methods, are reviewed by Davidson (1973). As with the gas hydrates, further research may reveal new classes of host structures with larger voids capable of enclosing larger guest species.

The alkylamine *semi-clathrate hydrates* also have a long history. They were first reported by Pickering (1893) in a paper with an interesting title, "The Hydrate Theory of Solutions." Nine of these hydrates were identified crystallographically by McMullan, Jordan, and Jeffrey (1967), of which five single-crystal X-ray structure analyses (shown in Table 9.4) were completed. This work was reviewed by Jeffrey (1969, 1984). One was found to be a true clathrate while the amine groups were hydrogen-bonded into clathrate-like water cages in various different ways in the other four. Both the method of hydrogen bonding of the amine groups and the topography of the cages differed

Table 9.4.
Crystal Structure Characteristics of the Alkylamine Hydrates.[a]

Name, formula per molecule	Amine-water relationship	Hydrogen-bonded n-hedra
t-Butylamine $16(CH_3)_3CNH_2\cdot 156H_2O$	Non-bonded within 17-hedra a true clathrate	$16[4^35^96^27^3]$ $12[4^45^4]$
Trimethylamine $8(CH_3)_3N\cdot 82H_2O$	Hydrogen-bonded to broken 15-hedra and to extra water in broken 14-hedra	$3[5^{12}]$ $2[(4)^25^86^3(6)^2]$ $2[5^6(5)^66^1(6)^1]$
iso-Propylamine $10(CH_3)_2\cdot CH.NH_2\cdot 80H_2$	Hydrogen-bonded in distorted 14- and 16-hedra	$2[4^66^2]$ $2[5^{12}]$ $6[4^25^86^4]$ $4[5^{12}6^4]$
Diethylamine $12(CH_3CH_2)_2NH\cdot 104H_2O$	Hydrogen-bonded in distorted 18-hedra and broken 17-hedra	$4[5^{12}6^6]$ $8[4^35^86^6(7)^2]$
n-Propylamine $16CH_3CH_2CH_2NH_2\cdot 104H_2O$	Hydrogen-bonded in broken 14- and 16-hedra	$4^25^86^1$ $(5^{10}6^1)$ (5^8) (5^86^2) $(4^15^66^2)$

[a]( ) represents partial polyhedra, which can be hypothetically closed to form $5^{12}6^2$, $4^15^{10}6^2$, $4^15^{10}6^5$, and $4^25^86^5$ polyhedra, respectively.

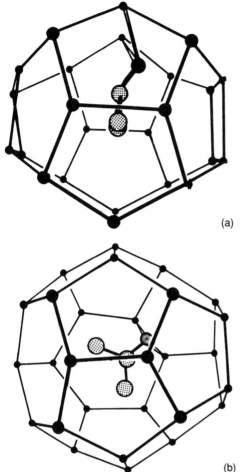

(a)

(b)

Figure 9.11. An example of a semi-clathrate inclusion. Hydrogen-bonded isopropylamine molecules in the crystal structure of $10(CH)_3CH \cdot NH_2 \cdot 80H_2O$. (a) in $4^2 5^8 6^4$ hedron; and (b) in $5^{12} 6^4$ hedron.

with each crystal structure; an example is shown in Figure 9.11. The phase diagram studies of Favier, Rosso, and Carbonnel (1981) reported the existence of many more hydrates of amines and diamines which are expected to be semi-clathrates. Similar studies by Carbonnel and Rosso (1973, 1976), Rosso and Carbonnel (1978), and Grueu, Rosso, and Carbonnel (1980) identified a large number of hydrates of aldehydes and ketones, cyclic imines, and oxyimines which could be either pure clathrates or semi-clathrates. This is a vast and largely unexplored field of hydration chemistry.

Phase diagrams with infrared and NMR studies with crystalline powders form a convenient method for identifying the formation of clathrate hydrates, as illustrated by the research of Harmon et al. (1987) and Harmon and Budrys (1991).[8] From vapor pressure NMR and infrared studies, clathrate hydrate structures are proposed for choline chloride dihydrate by Harmon and Günsel (1984); for calcium phosphoryl choline chlo-

[8]This extensive series of papers by Harmon and his undergraduate students is mainly concerned with the lower hydrates, some of which are discussed in Chapter 8.

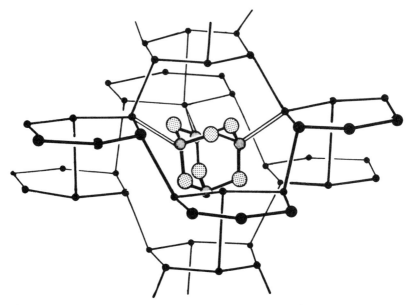

Figure 9.12.   Hexamethyl tetramine hexahydrate. The molecule is held by $\geqslant$N----H—O hydrogen bonds with a $(H_2O)_n$ lattice (from Mak, 1965, reprinted with permission).

ride tetrahydrate by Harmon and Akin (1991); and for N,N,N-trimethyl-1-adamanty-lammonium fluoride trihydrate by Harmon, Mounts, and Wilson (1991).

Prior to crystallization, aqueous solutions of the quaternary ammonium salt hydrates and the alkylamine hydrates become very viscous, suggesting the formation of clathrate-like clusters of water molecules. The question whether these structures exist wholly, or in part, in the melt of $4(CH_3)_4N\cdot41H_2O$ was studied by Folzer, Hendricks, and Narten (1971) by X-ray diffraction, with an inconclusive result. Infrared studies of the solid and liquid phases by Falk (1971) pointed to the absence of any ice-like or clathrate-like structure above the melting point.

Hexamethylene tetramine,[9] $(CH_2)_6N_4$, forms a hexahydrate in which the water molecules form a lattice consisting of hexagonal rings of O—H----O hydrogen bonds linked by other O—H----O hydrogen bonds to form a lattice with voids. The molecules are held within these voids by three O—H----N$\leqslant$ hydrogen bonds, as shown in Figure 9.12.

Clathrate hydrates and channel structures in which the anions and water molecules hydrogen-bond to form the host lattices are represented by the crystal structures of $4(C_2H_5)_4N^+\cdot F^-\cdot11H_2O$ by Mak (1985) and $(C_2H_5)_4N^+\cdot C\ell^-\cdot H_2O$ by Mak, Brunslot, and Beurskens (1986).

Many more $(H_2O)_n$ polyhedra are observed in the hydrates of strong bases and acids. The phase diagram studies with X-ray structure analyses of Mootz and Seidel (1990b) and Mootz and Stäben (1992) on the hydrates of tetramethyl ammonium hydroxide, and by Mootz, Oellers, and Wiebeke (1987) on the higher hydrates of some strong acids, greatly expanded the variety of polyhedra observed. $(CH_3)_4NOH$ forms

[9]Hexamethylenetetramine was the first organic crystal structure analysis by Lonsdale (1929). Its solubility in water is unusual, since it decreases with increase of temperature, that is, with a decrease in hydrogen-bond strength, indicating inclusion takes place in solution.

eight hydrates four of which are dimorphic with low- and high-temperature phases. The $(CH_3)_4N^+$ cations are guests in an anionic $(OH \cdot nH_2O)^-$ host lattice. The acids HPF_6, $HAsF_6$ and $HSbF_6$ form penta- and hexahydrates. The pentahydrates are cubic, based on space-filling $(4^6 6^8)$ polyhedra. The hexahydrates are hexagonal structures which exhibit a variety of polyhedra, including the pentagonal dodecahedron. In these structures, the $(XF_6)^-$ anions are guests in a cationic $(H_3O \cdot nH_2O)^+$ host lattice.

In order to form a four-connected hydrogen-bonded water framework, it is necessary to have $(H_2O)_n$, but the anionic host structure $(OH \cdot xH_2O)_n^-$ occurs in the tetramethyl ammonium hydroxide hydrates, the first of which was studied by McMullan, Mak, and Jeffrey (1966) and later in more detail by Mootz and Seidel (1990b). In the low temperature, phases of the 4.6, 5.0, and 7.5 hydrates, this deficiency of protons results in the formation of a number of incomplete or broken polyhedra, shown in Figure 9.13, with the incomplete face indicated by dashed lines and

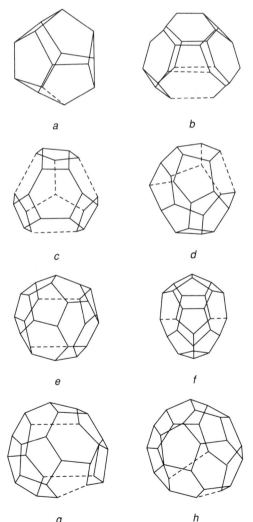

Figure 9.13. Broken polyhedra in the $(CH_3)_4NOH$ hydrates (a) $4^2(4)^1 5^5(5)^1$ in α-7.5H₂O; (b) $4^4(4)^2 6^6(6)^2$ in β-5H₂O; (c) $4^3 5^3(5)^3 6^1(6)^3$ in α-4.6H₂O; (d) $4^1 5^5(5)^5 6^1(6)^1$ in α-4.6H₂O; (e) $4^2 5^7(5)^1 6^3(6)^2$ in 6.67H₂O; (f) $4^1 5^8(5)^2 6^2(6)^2$ in 6.67H₂O; (g) $4^2 5^6(5)^2 6^3(6)^2$ in α-7.5H₂O; and (h) $4^1 5^9(5)^1 6^4(6)^1$ in α-7.5H₂O.

a

b

c

d

e

f

g

h

Table 9.5.
Distribution of O----O Distances (Å) in $(CH_3)_4N^+OH$ Hydrates.

	4.6α (b)[a]	4.6β (c)	5β (b)	6.67 (c)	7.5α (b)	7.5β (c)	8.75 (c)	10 (c)
2.5–2.6								4
2.6–2.7	2			2	12			12
2.7–2.8	11		44	19	44	96		48
2.8–2.9	3	24		3	8	16		24
2.9–3.0				1				
3.0–3.3	1[b]		1[b]			24		
>3.3			1[b]					

[a] b = broken, c = complete.
[b] Not considered hydrogen bonds.

Figure 9.14. Complete hydrogen-bonded $(OH \cdot nH_2O)^-$ and $(H_3O \cdot nH_2O)^+$ polyhedra observed in $(CH_2)_4NOH$ and $HAsF_6$ hydrates.

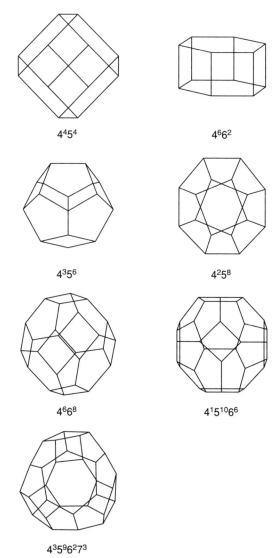

$4^4 5^4$

$4^6 6^2$

$4^3 5^6$

$4^2 5^8$

$4^6 6^8$

$4^1 5^{10} 6^6$

$4^3 5^9 6^2 7^3$

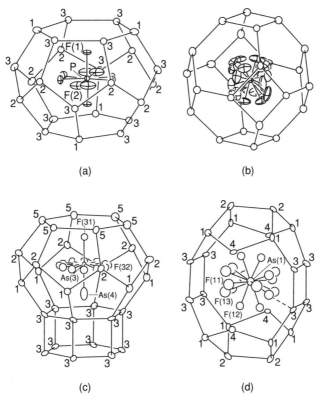

(a)　　　　　　　　　(b)

(c)　　　　　　　　　(d)

Figure 9.15. (a) PF_6 anion disordered in a $5^{12}6^2$-polyhedron in $HPF_6 \cdot 7.67H_2O$; (b) PF_6 anion disordered in a $4^6 6^8$ polyhedron in $HPF_6 \cdot 5H_2O \cdot HF$; (c) AsF_6 anion disordered in $5^{12}6^2$ polyhedron with a vacant $4^6 6^2$ polyhedron in $HAsF_6 \cdot 6H_2O$; and (d) AsF_6 anion disordered in a broken polyhedron in $HAsF_6 \cdot 6H_2O$ (from Mootz, Vellers, and Wiebeke, 1987).

parentheses.[10] But in the high temperature phases of the 4.6, 7.5 hydrates and the 6.67, 8.75, and 10, hydrates, the polyhedra have complete hydrogen-bonded faces, as shown in Table 9.5 and Figure 9.14, despite the *proton deficiency*. The nature of the proton disorder involved is not apparent from the X-ray analyses. There is a wide variety of hydrogen-bonding interactions, as shown by the distribution of O----O distances shown in Table 9.5. Unfortunately, there are no neutron diffraction studies which might clarify the distribution of the hydrogen atoms.

A reverse situation of *excess protons* is found in the hydrates of the strong acids HPF_6, $HAsF_6$, $HSbF_6$, HBF_4 and $HC\ell O_4$ by Mootz, Oellers, and Wiebeke (1987) and Wiebeke and Mootz (1988). These structures are related to the oxonium hydrates discussed in Chapter 4, but again the nature of the hydrogen-bond disorder is not determined from the X-ray studies. These crystal structure analyses are complicated by the disorder of the anionic guests, as shown in Figure 9.15.

---

[10]Incomplete polyhedra of this type are referred to as *stretched polyhedra* by Stäben and Mootz (1995), since they can be related to the complete polyhedra by stretching certain O----O distances in the crystal structure of the semi-clathrate $4(CH_3)_3NH_2 \cdot 29H_2O$. This crystal structure contains eight symmetry-independent amine molecules hydrogen bonded within eight different stretched polyhedra. Similar polyhedra were observed in the crystal structure of $16(CH_3(CH_2)_2)NH_2)105H_2O$ by Brickenkamp and Panke (1973).

Gas-phase Clathrate Hydrates

Charged clusters of water molecules are believed to exist in the earth's outer atmosphere and in outer space. They can be produced in the laboratory and are identified by their infrared spectra. They are studied by Molecular Ion Consequence Spectroscopy, MAMICS, as described by Castleman (1994). Binding energies for $H_3O^+ \cdot n(H_2O)$ for n = 1–7 by gas-phase ion-molecular equilibria gave values of ΔH ranging from 36 kcal/mol^{-1} for $H_3O^+ \cdot H_2O$ down to 10 kcal/mol^{-1} for each hydration step to $H_3O^+ \cdot 7H_2O$.

Of particular interest is the ion $(H_2O)_{20}H_3O^+$, the *magic number entity*, which has been observed under a variety of experimental conditions, such as the expansion of ionized water vapor by Beuhler and Friedman (1982). An obvious suggestion that this was the $H_{40}O_{20}$ pentagonal dodecahedron enclosing a H_3O^+ oxonium ion was made by Kassner and Hagen (1976). More recently, Wei, Shi, and Castleman (1991) have applied time-of-flight mass spectroscopy to ionized proton cluster ions to produce a binary vapor of water and trimethyleneamine (TMA). The most abundant species occurred at $(H_2O)_{21}(TMA)_{10}$, corresponding to the pentagonal dodecahedron enclosing a H_3O^+ with 10 external hydrogen bonding hydroxyls to the amine molecules. The other less abundant peaks are for (21,9), (20,10), (20,11), (21,11), and (20,12) complexes. Those with 20 H_2O reported by Yang and Castleman (1989) could correspond to H^+ guest species.

Figure 9.16. The channel in the hydrogen-bonded urea host lattice (from Smith, 1952, reprinted with permission).

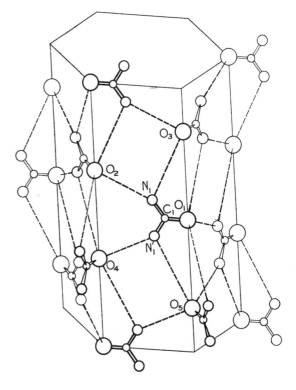

The Urea and Thiourea Inclusion Compounds

Urea, thiourea, and selenourea molecules hydrogen-bond to form channel-type host lattices. These inclusion compounds can be used for separating hydrocarbons and were the subject of patents in 1940. The structures were determined by Schlenk (1949, 1951) and Smith (1950) and were shown to have channels formed by the N—H···O═C hydrogen-bonding of the urea molecules, as shown in Figure 9.16. Three interpenetrating spirals of hydrogen-bonded urea molecules form the walls of channels with a diameter of about 5 Å. These channels can include hydrocarbons from n-hexane to pentamethylheptadecane, which has a length of 22 Å. The compounds are nonstoichiometric and the mole ratio of urea/hydrocarbon deduced by Smith (1952) is 0.684 $(N - 1) + 2.175$, where N is the number of carbon atoms in an extended zigzag chain. They can also include alcohols, ketones, carboxylic acids, alkylamines, and polyesters. The guests are generally disordered, as shown by Forst, Jagodzinski, and Frey (1990).

The thiourea channel structure has a larger diameter of about 6 Å. It can include branched alkanes, haloalkanes, and some ring systems, such as 1,5-cyclo-octadiene, but not the simple ethanes. The host lattices of these compounds occur in two enantiomorphic forms, depending on the chirality of the spirals, and can be used for optical resolution. The urea and thiourea inclusion compounds with polymers as guests have been studied more recently by Chenite and Brisse (1991, 1992, 1993). An interesting thiourea 1-5 cyclooctadiene clathrate structure is reported by Garneau, Raymond, and Brisse (1995).

9.4 | HYDRATE LAYER COMPOUNDS

In these compounds, hydrogen-bonded water molecules or water molecules plus anions or cations, form hydrogen-bonded layers which enclose other molecules, ions, or parts of molecules in a *sandwich* type of inclusion structure. In some structures, such as pinacol hexahydrate or piperazine hexahydrate, hydrogen bonds to or from the functional groups of the molecules link the water layers to form a semi-clathrate-like structure which encloses the hydrophobic moiety of the molecule, as shown in Figure 9.17.

In the hydrates of the alkane diols, 2,3-dimethyl-2,3-butanediol·4H₂O and 2,7-dimethyl-2,7-octanediol·4H₂O, the water layers are too far separated to be linked by hydrogen bonds, forming a true sandwich hydrate with hydrogen-bonded layers of $(OH·H_2O)_n$, as shown in Figure 9.18. A similar structural type was observed by Mootz and Wussow (1981) in pyridine trihydrate with $(N·H_2O)_n$ layers.

The water-OH or water-anion layer may consist of buckled pentagons, since flat layers of pentagons do not form a repeating pattern, or combinations of quadrilaterals, pentagons and hexagons, and very rarely heptagons, as shown in Table 9.6. Some examples are illustrated in Figure 9.19.

The crystal structure analysis of the hydrate and deuterate of trifluoroacetic acid tetrahydrate by Mootz and Schilling (1992) provides an interesting and unusual

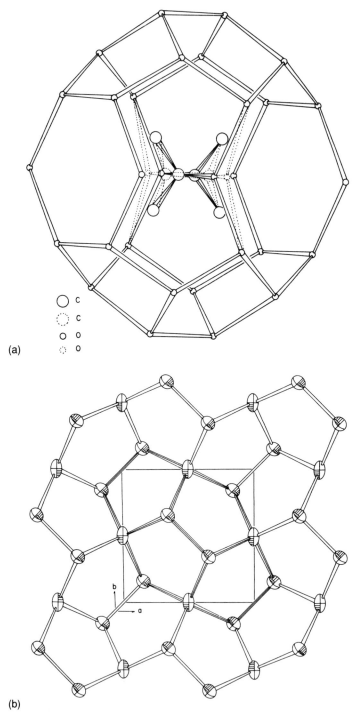

(a)

○ C
⊙ C
○ O
⊙ O

(b)

Figure 9.17. (a) pinacol molecule in void formed by O—H····O$_W$ and O$_W$—H····O hydrogen bonds linking layers of hydrogen-bonded water molecules (from Jeffrey, 1984, reprinted with permission); and (b) pattern of buckled pentagons of hydrogen-bonded water molecules observed in the crystal structures of pinacol and piperazine hexahydrate.

176

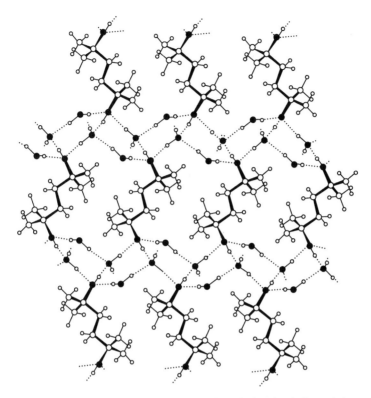

Figure 9.18. Sandwich inclusion structure with (OH·4H₂O) as the bread and alkane chains as the meat in the crystal structure of 2,5-dimethyl-2,5-hexanediol tetrahydrate (from Jeffrey and Shen, 1972).

Table 9.6.
Configuration of Hydrogen-Bonded Layers in Hydrate Crystal Structures.

Edge-sharing pentagons only
Piperazine 6H₂O
Pinacol 6H₂O
2,5-Dimethyl-2,5-hexanediol 4H₂O
2,7-Dimethyl-2,7-octanediol 4H₂O
CH₃COH·2H₂O
(CH₃)₄N⁺[F⁻5H₂O]

Edge-sharing quadrilaterals (Q), pentagons (P), hexagons (H), heptagons (Hp) or octagons (O)

	Q	P	H	Hp	O
CH₃COH·7H₂O	Q	P	H		
Pyridine·3H₂O	Q	P	H		
CF₃COOH·4H₂O	Q		H		
CsOH.2H₂O		P	H		
(C₂H₅)₄N⁺CH₂COO⁻·4H₂O		P		Hp	
(C₃H₇)₄N⁺F⁻·xH₂O	Q	P	H		
NN′-Dicyclohexyl piperazine NN′-dioxane 8H₂O	Q		H		O
(C₂H₅)₄N⁺X⁻(NH₂)Co·2H₂O	Q	P	H		
(X = Cℓ, Br, CN)					

177

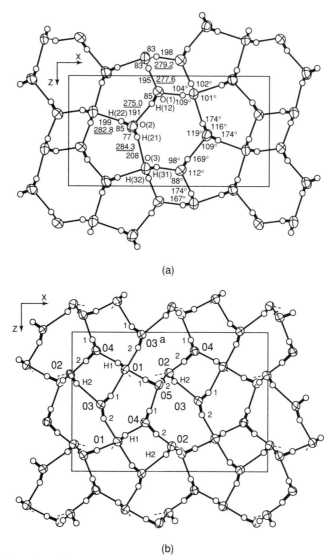

(a)

(b)

Figure 9.19. Examples of hydrogen-bonded layer structures containing 4-, 5-, and 6-membered rings. (a) $(H_2O \cdot OH)_n$ in $CsOH \cdot 2H_2O$ and $CsOH \cdot 3H_2O$; (b) $(H_2O)_n$ in pyridine trihydrate; and (c) $(H_2O \cdot COO^-)_n$ in $(C_2H_5)_4N^+CH_2COO^- \cdot 4H_2O$.

isotope effect, from an ionic to a molecular structure, from $(CF_3COO)^- (H_3O \cdot 3H_2O)^+$ to $CF_3COOD \cdot 4D_2O$ shown in Figure 9.20. In (4-dimethylaminopyridine)$_5$ (benzoic acid)$_3$ $(H_2O)_{10}$ from Biradha et al. (1995), the layers of hydrogen-bonded water-benzoate ions shown in Figure 9.21 includes layers of dimethylaminopyridine molecules packed in herringbone fashion and linked by $O_WH \cdots H$ hydrogen bonds.

As remarked previously by Jeffrey (1984), this work, despite its substantiality, is still only the tip of the iceberg in hydrate inclusion chemistry.

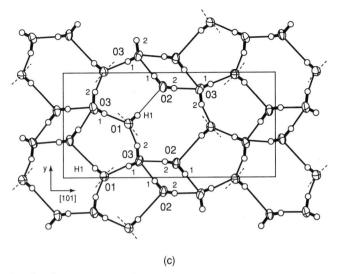

(c)

Figure 9.19. (*continued*).

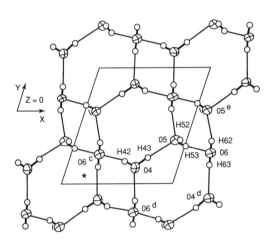

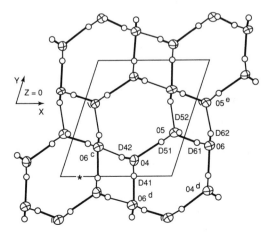

Figure 9.20. Change in hydrogen bonding on deuteration in $CF_3COOH \cdot 4H_2O$. Upper: $(H_2O \cdot COO)_n^-$; Lower; $(D_2O \cdot COO)_n^-$ (from Mootz and Schilling, 1992).

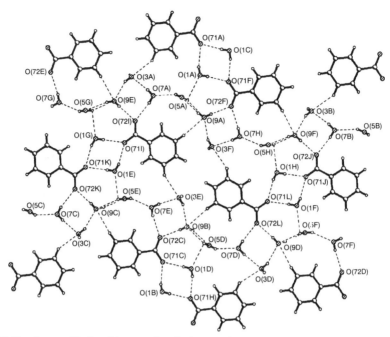

Figure 9.21. Layers of hydrated benzoate ions in the crystal structure of (4-dimethylaminopyridine)$_5$ (benzoic acid)$_3$ (H$_2$O)$_{10}$ (from Biradha et al. 1995, reprinted with permission).

Figure 9.22. Modes of packing in cyclodextrin crystal structures. (a) herringbone; (b) brickwall; and (c) channel (from Jeffrey and Saenger, 1991, reprinted with permission).

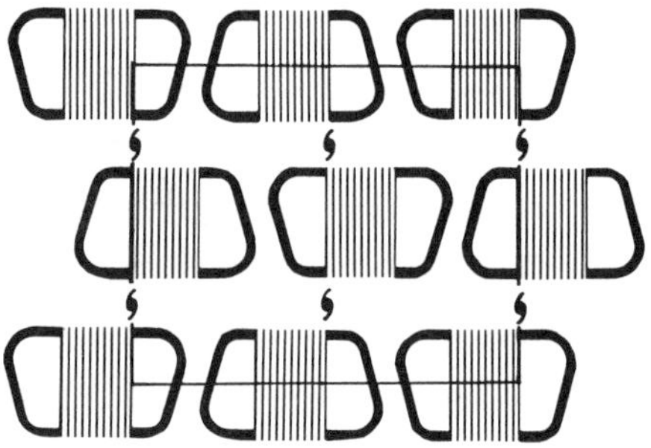

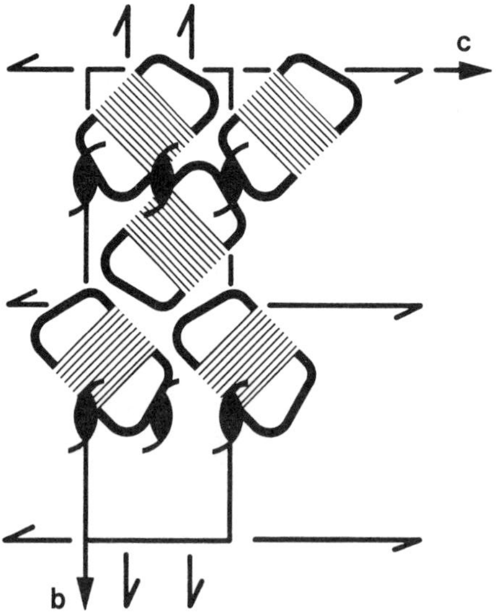

(a)

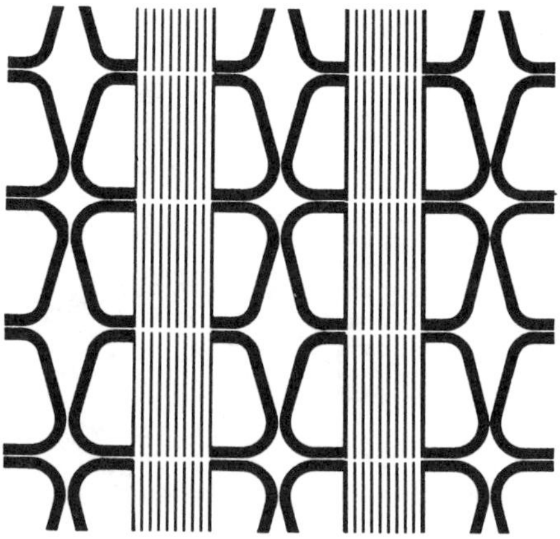

(b)

(c)

180

9.5 | THE CYCLODEXTRIN INCLUSION COMPOUNDS

These compounds were produced from the enzymatic partial hydrolysis of starch and were known as Schardinger dextrins. Their chemical and physical properties were extensively studied by Cramer (1954). They are cyclic polymers of α-glucopyranose $(C_6H_{12}O_6)_n$, where n = 6(α), 7(β), and 8(γ). The host lattices are the doughnut-shaped molecules, which pack in three different modes; the cage type *herringbone* and *brick-wall*, and a *channel* type, illustrated in Figure 9.22. Each macromolecule has 3n donor hydroxyls and 5n acceptor oxygens, that is, 3n hydroxyls and 2n ring or glycosidic oxygens. For this reason, the molecules form many intermolecular hydrogen bonds and one intramolecular hydrogen bond per glucose unit, as shown in Figure 9.23. Guest species occupy the hole in the doughnut. When the guests have hydrogen-bond functionality, they are hydrogen-bonded to the cyclodextrin host, an example of which is shown in Figure 9.24. The structural aspects of the inclusion are described by Saenger (1984a). When crystallized from water, hydrogen-bonded water is present both in the

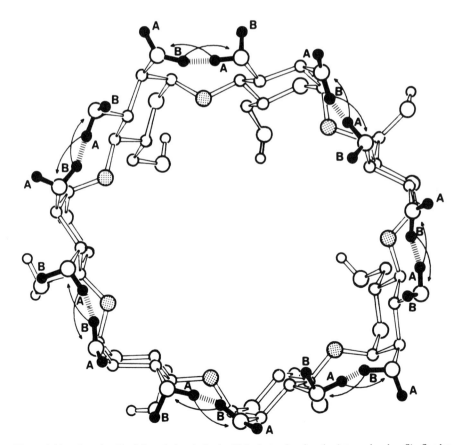

Figure 9.23. A molecule of β-cyclodextrin in the 11-hydrate, showing the *intramolecular, flip-flop* hydrogen bonds between the primary CH_2OH groups (from Jeffrey and Saenger, 1991, reprinted with permission).

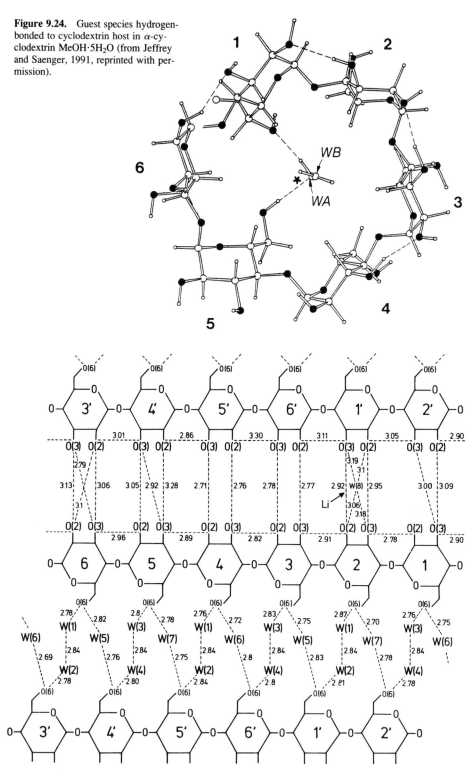

Figure 9.24. Guest species hydrogen-bonded to cyclodextrin host in α-cyclodextrin MeOH·5H₂O (from Jeffrey and Saenger, 1991, reprinted with permission).

Figure 9.25. Schematic diagram of the hydrogen bonding in (α-cyclodextrin)₂ Li I₃ I₂ 8H₂O (from Jeffrey and Saenger, 1991, reprinted with permission).

holes and between the cyclodextrins, forming very complex cooperative hydrogen-bonding patterns, as illustrated schematically in Figure 9.25. They are reviewed from the hydrogen-bonding point of view by Jeffrey and Saenger (1991).

A number of these hydrates have been crystallized and their structures analyzed by both X-ray and neutron diffraction. The neutron diffraction analyses were particularly important from the point of view of hydrogen bonding, since they permitted reliable location of the water hydrogen atoms. It was from these analyses that came the concepts of *homodromic, antidromic,* and *heterodromic* patterns of cyclic hydrogen bonding by Saenger (1979) and of *flip-flop* disorder (Saenger et al., 1982), discussed in Chapter 7. Those neutron diffraction analyses were also used as a basis for the studies of the geometrical characteristics of O—H----O and C—H----O hydrogen bonds, described in Chapters 4 and 5.

Hydrogen Bonding in Biological Molecules

10

10.1 ## THE IMPORTANCE OF HYDROGEN BONDS

The potential importance of hydrogen bonding in the structure and function of bio-logical macromolecules was predicted by the earliest investigators. However, the con-vincing evidence of this came with the hydrogen-bonded helical and sheet structures proposed for proteins by Pauling, Corey, and Branson (1951) and Pauling and Corey (1951) and verified by the long succession of protein crystal structure analyses, and by the hydrogen-bonded base-pair in the structure of deoxynucleic acid by Watson and Crick (1953).

At that time, neither proteins nor nucleic acids could be studied at a level that re-vealed the details of the hydrogen bonding. This prompted the many studies of hy-drogen bonding in the small molecule components of these macromolecules that are described in Chapter 4. Hydrogen bonding is now regarded as an essential component of the structure and function of biological molecules. It is now rare that a paper or pre-sentation on these subjects does not contain a reference to hydrogen bonding, however ill-defined the interaction may be. The latest text devoted to hydrogen bonding in bi-ological molecules is by Jeffrey and Saenger (1991).

Strong hydrogen bonds are rare in biological structures since they are too rigid and not easily broken. The strongest hydrogen bonds are the salt bridges, $\bar{N}$—H----O$=$$\overset{+}{C}$ in proteins and P—OH----O$=$P bonds in nucleic acids. These bonds are generally in-terrupted by water molecules which do not form very strong hydrogen bonds, either as donors or acceptors. Biological interest in strong bonds has, however, developed re-cently as a consequence of hypotheses concerning the role of low-barrier hydrogen bonds in enzyme catalysis, discussed later in this chapter. The role of very weak hy-drogen bonds such as the minor components of three-center bonds and C—H----O hy-drogen bonds in biological structures is only recently being considered.

The rapid development of methods for growing protein crystals which are good

Table 10.1
Hydrogen-Bonding Functional Groups in the Protein Side Chains.[a]

Donors		Acceptors	
$\overset{+}{N}(H_2)H$	lysine	$\diagdown N$	histidine, tryptophane
$\diagdown \overset{+}{N}(H)H$	arginine	$O{=}\overset{+}{C}{=}O$	glutamic acid, aspartic acid
$\diagdown \overset{+}{N}{-}H$	lysine, arginine	$C{=}O$	glutamine, asparagine
$\diagdown N{-}H$	tryptophane, proline, histidine	$O{-}(H)$	tyrosine, threonine, serine
$N(H)H$	glutamine, asparagine	$\diagdown S$	methionine
$C{-}O{-}H$	tyrosine, threonine, serine	⬡	tyrosine, tryptophane
$S{-}H$	cysteine		
$C{-}H$	all side chains		

[a]Hydrogen bonds between charged groups, *i.e.*, lysine or arginine to glutamine or aspartic acid, are referred to as *salt bridges*.

enough for single-crystal X-ray diffraction and more powerful X-ray sources has made protein crystallography the principal source of information concerning the hydrogen bonding in proteins.[1] For the nucleic acids and polysaccharides which form polycrystalline fibers, the information concerning hydrogen bonding comes primarily from the crystal structure analysis of smaller oligomers.

10.2 | IN PROTEIN STRUCTURES

The main polypeptide chains of proteins consist of peptide units, (**I**), interspersed by the Ca carbon atoms, (**II**), to which are attached the side groups, the variety of which distinguish the 20 naturally occurring amino acids.

(**I**)   (**II**)

The peptide N—H and C=O groups form intramolecular N—H---O=C hydrogen bonds which determine the conformation of the peptide main chain, being responsible for the formation of helical or sheet structures. The formation of these hydrogen bonds results in additional π character to the peptide C—N bond, which increases its resistance to twisting from the planar conformation. The side-groups contain the variety of hydrogen bond donor and acceptor groups shown in Table 10.1, which form the hydrogen bonds between the polypeptide chains.

[1]It is interesting to note that in the section on Applications to Various Sciences in *Crystallography in North America* by McLachlan and Glusker (1983), there is only one mention of protein crystallography.

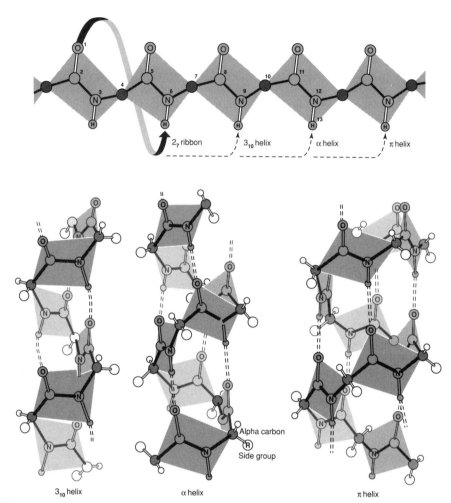

Figure 10.1.   The polypeptide helices of Pauling (from Dickerson and Geis, 1969).

There is much water in these crystals, sometimes as much as 70% by weight or volume. These water molecules act as both donors and acceptors to the functional groups on the main chain and on the side chains.

The Helical, Pleated Sheet, and Barrel Structures

The helical structures of a polypeptide chain proposed by Pauling, Corey, and Branson (1951) are formed by twisting the chains so that the N—H and C=O groups on successive amino acid residues form intramolecular hydrogen bonds, while retaining a *trans* planar $\underset{C_\alpha}{\overset{C_\alpha}{\diagdown}}$ C—N $\diagdown$ peptide bond conformation. The rotation is only about the C_α—C and C_α—N bonds. The peptide C—N bond is kept planar and rigid. The

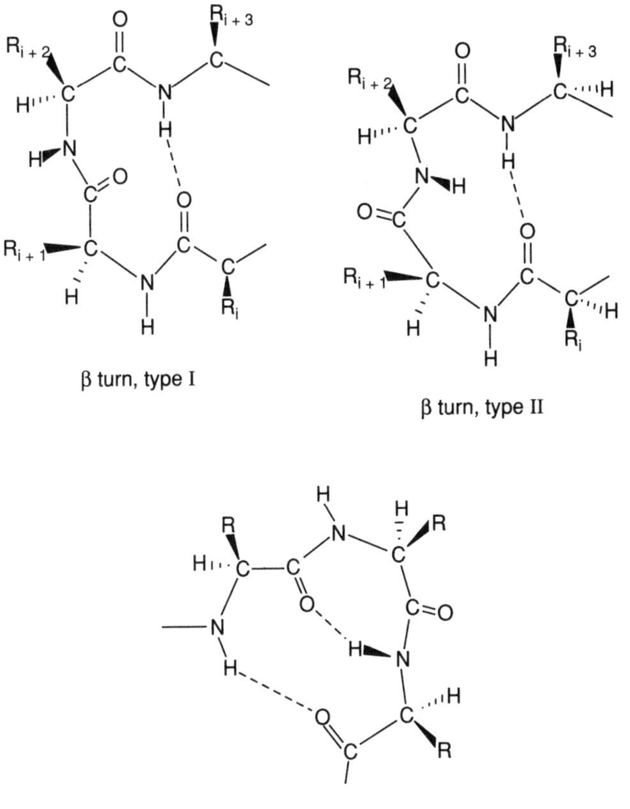

β turn, type I

β turn, type II

γ turn

Figure 10.2.   Some turns observed in protein crystal structures at the termini of helices.

helical chains are defined by numbers, which are the number of peptide units in the loop; the subscript is the number of bonds in the loop counting the hydrogen bond, as shown in Figure 10.1.[2]

The 3.6_{13}, α-helix, is observed in the majority of crystalline protein structures and can extend over long stretches of polypeptide chain. The 3_{10} helix is only observed in short segments in proteins, frequently at a carboxy terminus of an α-helix or at a turn which reverses the direction of the chain. These are known as β-turns or β-bends. A variety of β-turn conformations are observed which differ in their N—H---O=C torsion angles. The H---O distances and N—H---O valence angles are similar, but H---O=C acceptor angles varying from 115 to 140° have been observed in 3_{10} helices in some alanine-based peptides by Baldwin and Magusee (1987).

Single segments of the 2_7 helix, known as a γ-turn are rare examples. Examples of β turns and a γ turn are shown in Figure 10.2. In these the N—H---O=C bonds are

---

[2]If the polypeptide main chain is made into a strip and folded into a cylinder, the helices can be formed by taping together the NH and CO groups to form hydrogen bonds. It is said that Pauling developed the concept of the helical structures by this method in 1948 while recuperating in Oxford from the effects of the English weather and absence of heating, as reported by Goertzel and Goertzel (1995).

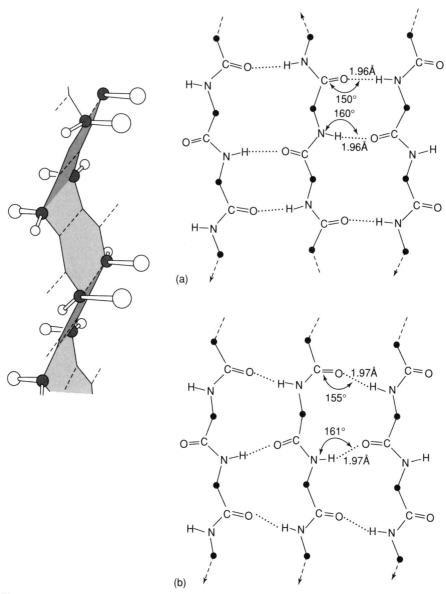

Figure 10.3.   The folding of polypeptide chains to form the pleated sheet structure: (a) anti-parallel; (b) parallel.

less linear with N—H---O ~140° and H---O=C ~100°. The 4.4_{16} (π) helix is rarely observed.

The parallel and antiparallel β-pleated sheet structures proposed by Pauling and Corey (1951), shown in Figure 10.3, are formed by NH---O=C hydrogen bonds which are lateral to the axis of the polypeptide chains. In both configurations, the side chains point alternately above and below the hydrogen-bonded sheets. The antiparallel sheets

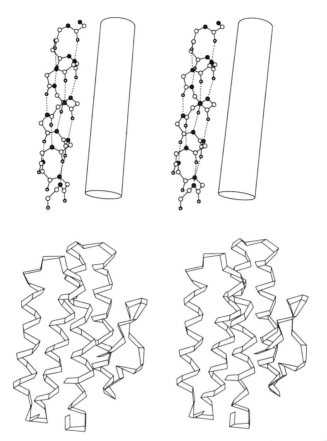

Figure 10.4. Symbolic cylinder and ribbon diagrams used in describing protein structure (from Lesk, 1991, reprinted with permission).

are more common, although the hydrogen-bond geometry is very similar. Although the peptide unit is planar, or approximately so, the tetrahedral C_α atoms result in a rippled or pleated conformation. In popular representations of protein architecture, α-helices and pleated sheets are often represented symbolically, as shown in Figure 10.4.

Related to the β-sheet is the β-barrel, so named because it is formed by joining two sides of the sheet to make a barrel-shaped conformation, shown in Figure 10.5. This is the equivalent of the linear and cyclic chains of hydrogen bonds observed in the carbohydrate crystal structures. It is interesting to note that the β-pleated sheets and β-barrel have continuous chains of ----O=C---N—H---O=C---N—H---- bonds, which are stabilized by the resonance-assisted hydrogen bonding described in Chapter 6. This is not true for the helices, where the chains are interrupted by the single N—C_α and C_α—C bonds.

Another variation of the β-sheet is a β-bulge, shown in Figure 10.6, in which the hydrogen bonding in a section between the polypeptide chains does not occur. Hydrogen bonds are sometimes absent in a-helices, particularly at a proline residue, allowing the helices to bend or unravel at their termini. A general discussion of hydrogen bonding in amino acids, peptides, and proteins is in Vinogradov (1980).

Figure 10.5. A β-barrel, surrounded by α-helices (from Lesk, 1991, reprinted with permission).

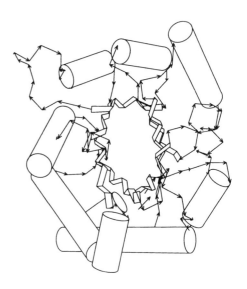

A monumental survey by Baker and Hubbard (1984) of hydrogen-bonding in the 15 *best* globular protein crystal structure analyses was the first attempt to analyze protein crystal structure data in order to investigate the hydrogen bonding. The analysis examined all aspects of hydrogen bond geometry. Hydrogen atoms were included in calculated po-

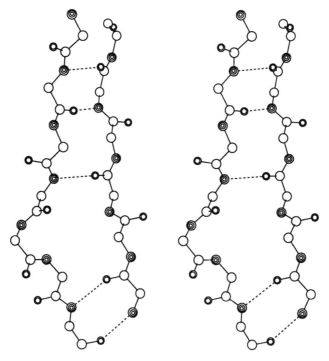

Figure 10.6. The β-bulge; a variation of the β-sheet (from Lesk, 1991, reprinted with permission).

sitions with N—H = 1.0 Å when this was possible. No hydrogens could be added to hydroxyl groups and amino groups where there could be free rotation. It was not possible to distinguish between NH_2 and C=O groups in some asparagine and glutamine residues.

Three-center bonds were recognized, but the analysis was somewhat compromised by the use of a 2.5 Å cut-off for H---O distances. This means that a number of three-center bonds would be identified as two-centered.

The following general results were obtained.

1. Ninety percent of the hydrogen bonds observed were main-chain to main-chain N—H---O=C.

2. Eighty-five percent of these bonds are in α-helices or β-pleated sheets.

3. A significant number—12% of N—H groups and 11% of C=O groups—are not involved in hydrogen bonding. More than half the C=O groups, 53%, only accept one hydrogen bond, rather than the expected two.

These statistics, when compared with the results from small molecule studies, suggest that not all the hydrogen bonds are identified. The mean bond lengths and angles for this study are given in Table 10.2. It is interesting to note that the hydrogen bonds in the β-sheets are shorter than in the α-helices, which is consistent with the Resonance Assisted Hydrogen Bonding hypothesis discussed in Chapter 6.

The question of three-center bonding is addressed more recently by Preissner, Enger, and Saenger (1991). They used 13 well-refined protein crystal structures, which provided data on 4974 NH---O=C and NH---O_W hydrogen bonds. With the same criterion for three-center bonds as used with the small molecules described in Chapter 4, they concluded that 24% of the hydrogen bonds are three-centered, which is approximately the same percentage as found in the small-molecule crystal structures. These three-center bonds occur systematically and most frequently (90%) in the α-helices. They add a

Table 10.2.
Mean Hydrogen Bond Geometries From the Baker-Hubbard (1984) Analysis of Hydrogen-Bonding in Proteins.[a]

N—H---O=C	No. of bonds	H---O (Å)	N---O (Å)	N—H---O (°)	C=O---H (°)
α-helices					
main chains	577	2.06(16)[b]	2.99(14)	155(11)	147(9)
N-terminus	17	2.25(19)		140(16)	
C-terminus	17	2.26(13)		152(13)	
β-sheets					
parallel	96	1.97(15)	2.29(14)	161(9)	150(11)
antiparallel	420	1.96(16)	2.91(14)	160(12)	150(8)
3_{10} helices	102	2.17(16)	3.09(14)	153(10)	114(10)
β-turns	118	2.13(15)	3.06(12)	154(9)	122(13)
N—H---O_W	389	2.06(20)	2.97(21)	156(15)	
O_WH---O=C[c]	1138		2.94(29)		

[a]Numbers in parentheses are the spread of values ($\pm x$)

[b]Compare with the mean value of 1.934 Å for 597 $>$N—H---O=C—NH_2 bonds in small molecules by Taylor, Kennard, and Versichel (1984a).

[c]Hydrogen positions not determined.

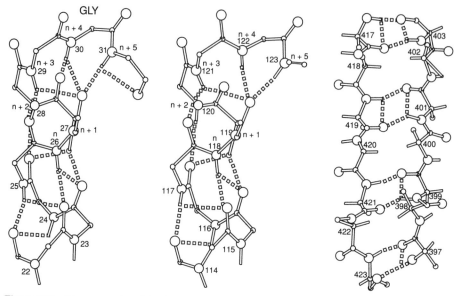

Figure 10.7. Three-center bonds in β-pleated sheets (from Preissner, Enger, and Saenger, 1991).

minor component to the N—H···O=C bonds of the α-helix. Three-center bonds are also observed to a lesser degree (40%) in the β-pleated sheets where the minor component is to the adjacent peptide unit in the polypeptide chain, as shown in Figure 10.7.

The relatively large number of nonbonded N—H and C=O groups could be explained by hydrogen-bonding to disordered water molecules which were not identified in the crystal structure analyses used. An ingenious approach to this problem was made by Savage et al. (1993). They compared the number of hydrogen bonds observed in 236 crystallographically independent protein molecules with the maximum number of hydrogen bonds possible, based on counting the number of X—H donor bonds and lone-pairs on acceptor atoms for the hydrogen bonding functional groups in the protein structures selected. Although the definition of a hydrogen bond interaction was more liberal, there were similar conclusions. The majority of *lost* hydrogen bonds were to the C=O acceptors, with 24% not forming the expected two bonds. The lost N—H donor bonds were much less, between 1 and 6%, and almost negligible for the OH bonds. The number of lost hydrogen bonds was then correlated with the sum of *stability factors* for the particular proteins. The stability factors are the hydrophobic contribution, HP; the ion-pair contribution, IP; and the disulfide contribution, SS. The HP contribution is based on estimates of the hydrophobic Gibbs free energy as described by Chothia (1974) and Richards (1977). Values of 16 and 21 cal per Å *buried* of Eisenberg and McLachlan (1986) were used. Oppositely charged side-chains were considered to form salt bridges if their distance apart was less than 4 Å. The stabilization energies of 1 kcal/mol^{-1} for surface accessible bridges and 3 kcal/mol^{-1} for internal bridges proposed by Fersht (1972) were used for the ion-pair (IP) contribution. For the covalent disulphide bonds, the formula of Rashin (1984) was used for the SS contribution. A high correlation between HP + IP + SS and the lost hydrogen bonds (LHB) was obtained, as shown in Figure 10.8.

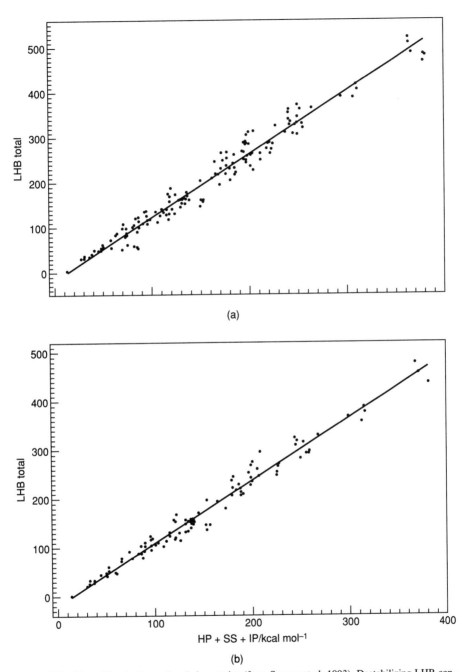

Figure 10.8. Plot of *lost* hydrogen bonds in proteins (from Savage et al. 1993). Destabilizing LHB contribution vs the sum of the stabilizing contributions of HP (hydrophobic energy), IP (ion pairs), and SS (disulfide bonds): (a) LHB (total) vs HP + IP + SS for proteins better than 1.9 Å resolution; and (b) for proteins better than 1.7 Å resolution. A least-squares regression line is included on both these plots. The correlation coefficients are 0.983 and 0.990 for (a) and (b), respectively.

Increasing attention is being focussed on the hydrogen bonding in protein crystal structures in recent years. As instrumental and computing methods improve, it is reasonable to believe that much more detailed information will be forthcoming.

10.3 LOW BARRIER HYDROGEN BONDS AND ENZYME CATALYSIS

There are two types of hydrogen bonds which can have low barriers. One is the strong hydrogen bonds discussed in Chapter 3, where the variable position of the proton implies a flat potential surface. The other is moderate hydrogen bonds, where there is a large difference in energy between the two unsymmetrical potential minima, shown in Figure 2.13. The former clearly facilitates proton transfer while the latter does not.

Cleland (1992), Cleland and Kreevoy (1994), and Frey, Whitt, and Tobin (1994) argue that although the initial hydrogen bond between enzyme and substrate may be normal, equalization of the pK values in a transition state results in a strong bond formation, making available some 10–20 kcal/mol^{-1} to catalyze the reaction.

Gerlt and Gassman (1993) also invoke low-barrier hydrogen bonds involving the thermodynamics of proton transfer involving the acidic C_α protons coupled with keto-enol tautomerism, as in (**II**).

(**II**)

Both these hypotheses have been criticized by Warshel, Papazyan, and Kollman (1995), Scheiner and Kar (1995), and Guthrie and Kluger (1993). Absent from these discussions is the question whether resonance-enhanced cooperativity plays a role in strengthening otherwise normal hydrogen bonds in the catalysis transition states.

10.4 HYDROGEN BONDING IN THE NUCLEIC ACID STRUCTURES

The discovery of the hydrogen bonding between the bases adenine and uracil and between guanine and cytosine, shown in Figure 10.9, by Watson and Crick (1953), together with the crystal structure analyses of the proteins hemoglobin and myoglobin by Perutz and Kendrew and their co-workers, initiated a new science, known as *mol-*

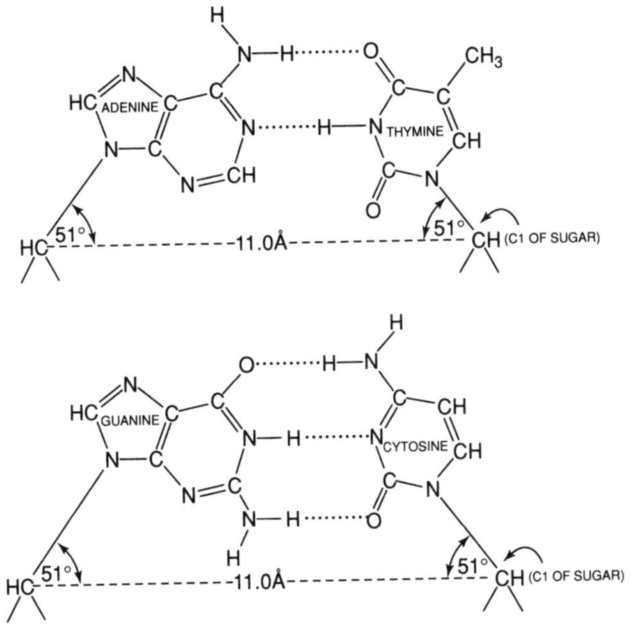

Figure 10.9. The famous Watson-Crick base pairing in DNA.

ecular biology[3], as described in Perutz (1963) and Kendrew (1963). Hydrogen bonding is the major factor in determining the structure of the nucleic acids as described by Saenger (1984b).

The hydrogen bond functionality of the major bases and some minor bases is shown in Table 10.3. For the pyrimidines, the acceptor functionality (counting both C=O lone pairs) exceeds the donor functionality; in purines they are equal. With so many hydrogen bonding functional groups, there are many possible modes of homo-base pairing, as shown in Figure 10.10. Examples of these are found in the crystal structures of the purines, pyrimidines, nucleosides, and nucleotides. A review of the crystal structures of purines, pyrimidines, and their complexes is provided by Voet and Rich (1970) and by Jeffrey and Saenger (1991). Despite the large number of different ways in which purines and pyrimidines might hydrogen-bond, co-crystallization to form hydrogen-bonded complexes is not common. There are a large number of crystal structures of purines, and pyrimidines, but many less of their complexes. Some that are observed are shown in Figure 10.11. Nature selects only two for the purpose of genetic coding; guanine to cytosine (G-C) and adenine to uracil (A-U) or to thymine (A-T), shown in Figure 10.11.

DNA and RNA provide fiber diffraction patterns from which only the average structure of the base pairs can be determined. The confirmation of the hydrogen bonding and information concerning the geometry is inferred from crystal structure

---

[3]W. T. Astbury has been described as the father of molecular biology, but I could not discover who gave him that accolade.

Table 10.3.
Hydrogen-Bond Functionality of Some Pyrimidine and Purine Bases.

	Donors	Acceptors
Major bases		
cytosine	3: $\rangle$N—H, —N$\langle^H_H$	3: C=O, N$\langle$
5-hydroxymethyl cytosine	4: $\rangle$N—H, $\rangle$N—H, OH	4: C=O, N$\langle$, OH
uracil	2: $\rangle$N—H, $\rangle$N—H	4: C=O, C=O
thymine	2: $\rangle$N—H, —N$\langle$H	4: C=O, C=O
adenine	3: $\rangle$N—H, —N$\langle^H_H$	3: N$\langle$, N$\langle$, N$\langle$
guanine	4: $\rangle$N—H, $\rangle$N—H, —N$\langle^H_H$	4: C=O, N$\langle$, N$\langle$
Minor bases		
5-methylcytosine	3: $\rangle$N—H, —N$\langle^H_H$	3: C=O, N$\langle$
1-methylguanine	3: $\rangle$N—H, $\rangle$N—H, —N$\langle^H_H$	4: C=O, N$\langle$,
dimethylguanine	2: $\rangle$N—H, $\rangle$N—H	4: C=O, N$\langle$, N$\langle$
6-methylaminopurine	2: $\rangle$N—H, $\rangle$N—H	3: N$\langle$, N$\langle$, N$\langle$
hypoxanthine	2: $\rangle$N—H, $\rangle$N—H	4: C=O, N$\langle$, N$\langle$

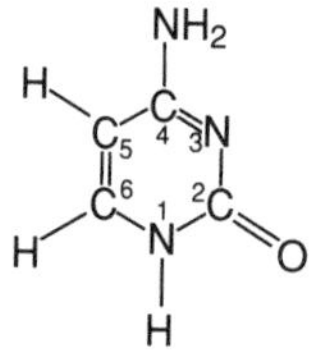

cytosine

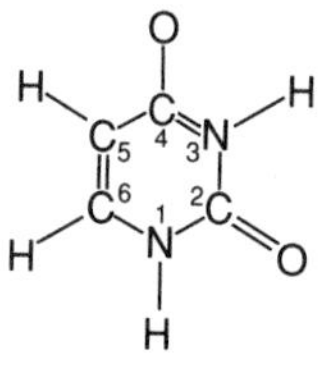

uracil

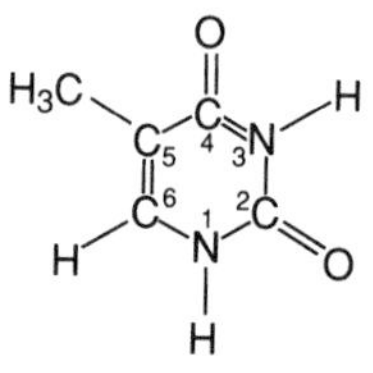

thymine

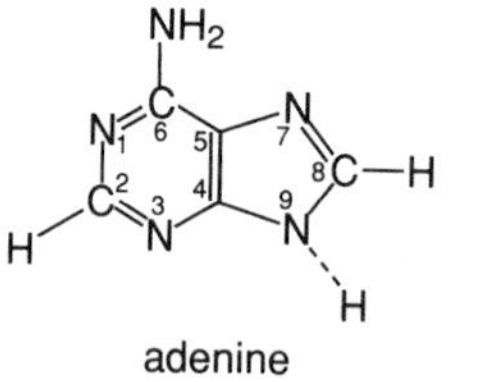

adenine

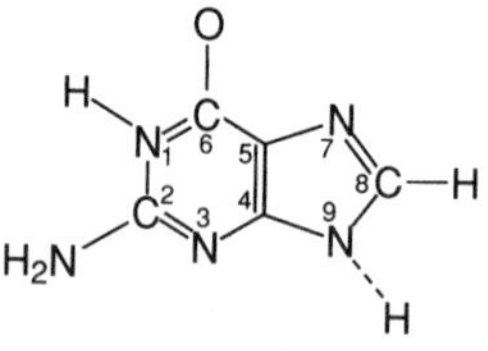

guanine

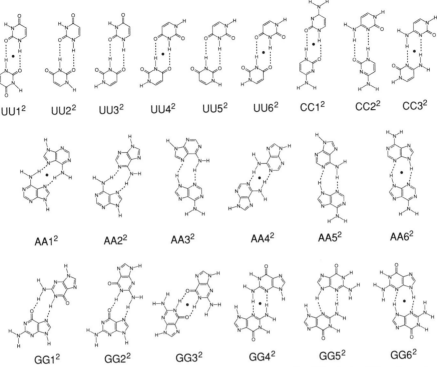

Figure 10.10. Modes of homo-base pairing between uracil (U), cytosine (C), adenine (A), and guanine (G) (from Jeffrey and Saenger, 1991).

analyses of purine-pyrimidine adducts and of oligonucleotides. Fortunately, in the late 1970s it became possible to synthesize and crystallize oligonucleotides with long sequences of base pairs which provided good crystal structure analyses. Some examples are shown in Table 10.4. These oligonucleotides have the important features of the natural polymers, including the biologically important major and minor grooves. These grooves arise from the difference in hydrogen bonding on the two sides of the base-pairs, as illustrated in Figures 10.12 and 10.13.

The Watson-Crick base-pairing, which occurs in both DNA and RNA, as observed in the crystal structures of sodium guanylyl-3′,5′-cytidine monohydrate by Rosenberg et al. (1976), and sodium-adenylyl-3′,5′-uridine by Seeman et al. (1976), is shown in Figure 10.14.

Another important base-pairing is the *Hoogsteen* type, shown in Figure 10.15, from the neutron diffraction refinement of 9-methyladenine 1-methylthymine by Frey et al. (1973) of a crystal structure determined by Hoogsteen (1963). The possibility of additional stabilization from a weak A----U, C(2)—H---O==C(2) hydrogen bond is discussed in Chapter 5.

In the nucleic acids, the hydrogen bonds of the bases are isolated by being spaced along the polynucleotide chain. In DNA, it is solely the hydrogen bonding of the base-pairs that keeps the two polynucleotide chains of the double helix aligned, since there are no other hydrogen-bonding donor groups. However, in RNA the C(2′)—OH groups

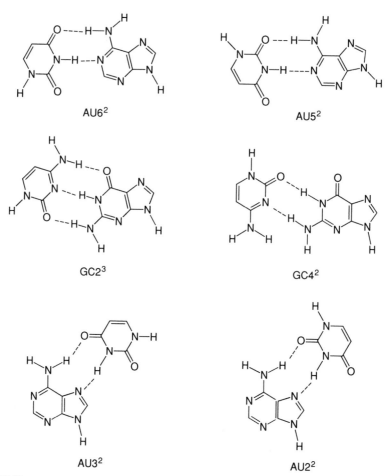

Figure 10.11. Examples of some hetero-base pairings. AU6[2] and GC2[3] is Watson-Crick. AU5[2] and GC4[2] is reversed Watson-Crick. AU3[2] is Hoogsteen and AU2[2] is reversed Hoogsteen (from Jeffrey and Saenger, 1991).

Table 10.4.
Some Oligonucleotide Crystal Structure Analyses.

Base-pair sequence	Helix type[a]	No. of water molecules located per nucleotide
d(GG^BrUA^BrUACC)	A	8.5
d(GGGGCCCC)Mg	A	13.3
d(CGCGAATTCGCG)	B	9.5
d(CGCGCG)	Z	16.7

[a]The A and B double helices have a right-handed twist, the Z is left-handed.

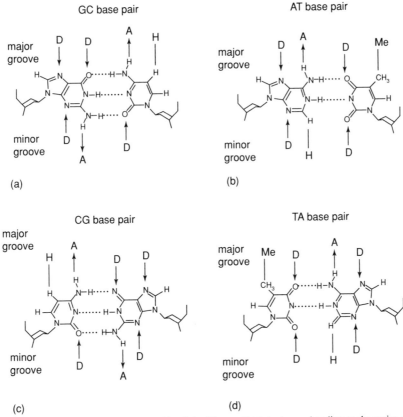

Figure 10.12. Diagram showing relationship of the Watson-Crick hydrogen bonding to the major and minor grooves in β-DNA (from Glusker, Lewis, and Rossi, 1994). The arrows point to acceptors (A), donors (D). (H) refers to possible C—H hydrogen bonding.

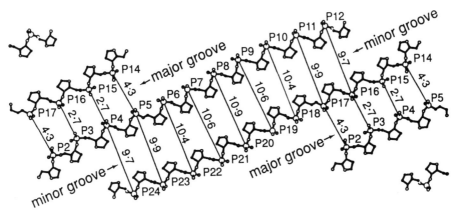

Figure 10.13. Cylindrical projection of sugar-phosphate backbone atoms in the A-DNA dodecamer d(CCGTACGTACGG). The widths are obtained by subtracting 5.8 Å from the P·····P separations to account for the van der Waals radius of the phosphate groups. The width of the minor groove is a consequence of the Watson-Crick base-pairing and is wider (from Bingham, Zon, and Sundaralingam, 1992, reprinted with permission).

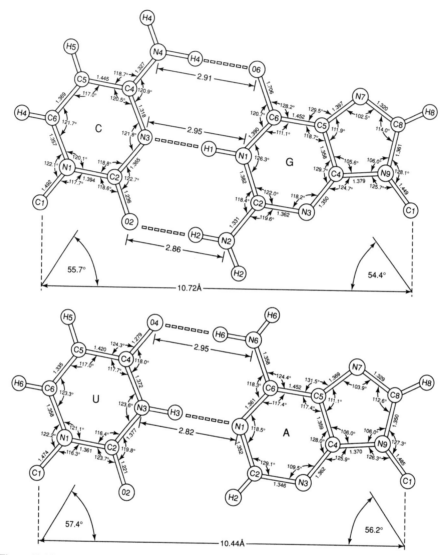

Figure 10.14.   The Watson-Crick base pairing as observed in the crystal structure of quanylyl-3′, 5′-cytidine by Rosenberg et al. (1976), and in the crystal structure of adenylyl-3′,5′-uridine by Seeman et al. (1976).

of the ribofuranose moieties are available to hydrogen bond. They form rather weak hydrogen bonds in the axial direction of the polynucleotide chain to the ring oxygen O(4′) of the adjacent nucleotide, with O----O distances of ~3.3 Å. These bonds have been observed in RNA-type oligonucleotides [U(U-A)$_6$A] by Doch-Bregon et al. (1989).

Mismatches in base-pairing are sometimes observed and were named *wobble* base pairs by Crick (1966). They occur when one of the bases is twisted about the N—C(1′) bond, so that the hydrogen bonding occurs through a different combination of donor-acceptor groups, as illustrated in Figure 10.16. Recent crystal structure analyses of nucleic acids and their complexes have been reviewed by Wahl and Sundaralingam (1995).

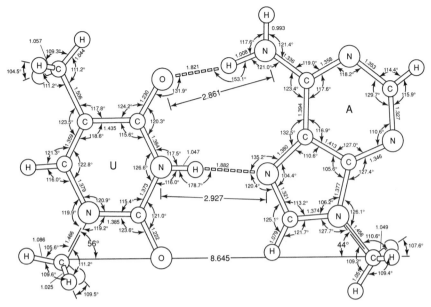

Figure 10.15. The Hoogsteen-type base pairing from the neutron diffraction crystal structure analysis of 9-methyladenine-1-methylthymine by Frey et al. (1973) (from Saenger, 1984b).

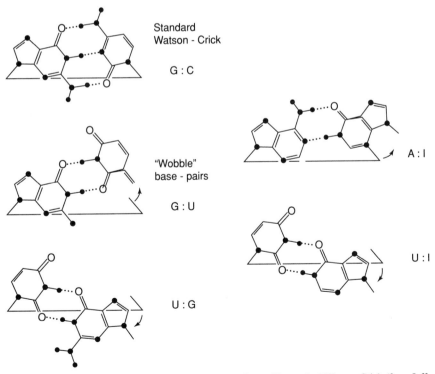

Standard
Watson - Crick

G : C

"Wobble"
base - pairs

G : U

U : G

A : I

U : I

Figure 10.16. Examples of *wobble-base pairing*. Comparison with standard Watson-Crick (from Jeffrey and Saenger, 1991).

10.5 | IN POLYSACCHARIDES

Here the problem is the same as for the DNA and RNA nucleic acids since the source of structural information is from fiber diagrams. The molecular packing to fit the fiber diffraction data is inferred from information derived from smaller oligosaccharide structures. The fit to the fiber diffraction data is necessarily an underdetermined problem, since the number of variables, i.e., atomic parameters, exceeds the number of observations. In addition, the diffraction spectra are diffuse, leading to ambiguities in assigning intensities due to overlapping reflections and correcting for background scattering.[4] The empirical force field methods described in Chapter 4 are generally applied as a refinement procedure to get the stereochemically best fit to the diffraction data.[5]

Hydrogen bonding schemes for cellulose II have been proposed from fiber diffraction data and packing considerations by Kolpak and Blackwell (1976), Stipanovic and Sarko (1976), Blackwell et al. (1980), and Pertsin and Kitaigorodskij (1987), but identifying the hydrogen bonding is made difficult by the possibility of rotation about C—O—H bonds, so that the position of the hydrogen atoms cannot be inferred, as is the case with $\rangle$N—H groups.

Recent X-ray crystal structure analysis of methyl β-cellotrioside monohydrate 0.25 ethanolate by Raymond et al. (1995a) and of cellotetraose hemihydrate by Raymond et al. (1995b) and Gessler et al. (1994, 1995) show a molecular packing similar to that proposed for cellulose II by Stipanovic and Sarka (1976) and Kolpak and Blackwell (1976). In both crystal structures, adjacent residues are linked by three-centered intramolecular inter-residue hydrogen bonds, as also observed in the crystal structures of the disaccharide cellobiose by Chu and Jeffrey (1968) and methyl β-cellobioside methanolate by Ham and Williams (1970). A proposed hydrogen bonding scheme for the cellulose II structure based on that of cellotetraose is shown in Figure 10.17. An analysis of the inter- and intramolecular hydrogen bonding in the methyl cellotrioside showed all the features observed in other oligosaccharide crystal structures, i.e., infinite chains, finite chains, three-centered bonds, and flip-flop disorder.

10.6 | WATER IN BIOLOGICAL MACROMOLECULES

Water is present in all macromolecular biological systems and is regarded as an essential component of both their structures and functions. Both proteins and nucleic acids are greatly hydrated in the crystalline state, where between 20 and 50% of the

---

[4]Disagreement factors between intensities measured from apparently similar diffraction patters have values between 20 and 50%, as pointed out by Jeffrey and French (1978).

[5]A similar procedure is used in protein crystal structure analysis when bond lengths and valence angles are normalized after each refinement cycle.

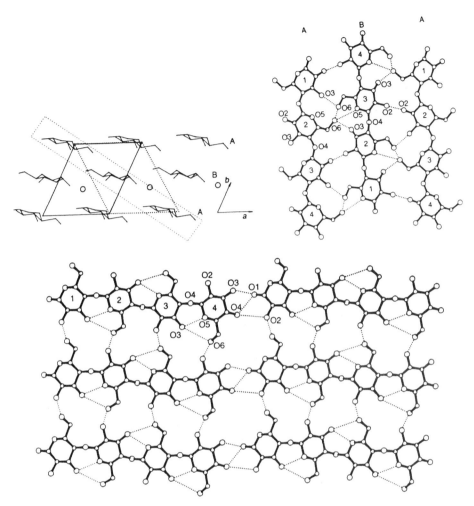

Figure 10.17.   Proposed chain packing and hydrogen bonding in cellulose (II) (from Gessler et al. 1994, reprinted with permission). Views are down chain axis and perpendicular to chain axis.

crystal volume can be occupied by water molecules. The water is more disordered than ordered. With the poor resolution of most macromolecular crystal structures, it is generally impossible to locate the hydrogen atoms and unequivocally identify systems of hydrogen bonds. Since water is a small molecule with considerable flexibility in its mode of hydrogen bonding, as described in Chapter 8, it can cover irregular surfaces and occupy channels or pockets in irregularly shaped macromolecules,[6] as illustrated in Figure 10.18.

The high dielectric constant of water, which arises from its mobility, serves to attenuate the attractive links between charged groups which form the stronger hydrogen bonds known as *salt bridges*. Such strong intramolecular bonding could otherwise impose an

[6]It is noticeable that small regularly-shaped molecules such as monosaccharides, which can pack efficiently, form much fewer hydrates than the more awkwardly shaped oligosaccharides or nucleosides and nucleotides.

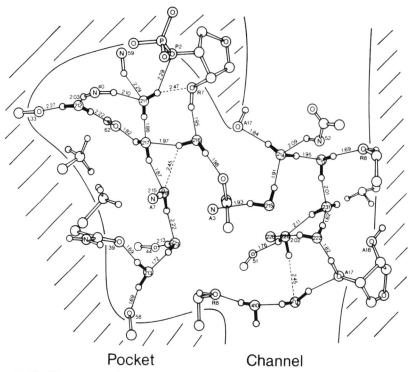

Pocket   Channel

Figure 10.18. Water structure in pockets and channels in the crystal structure of vitamin B_{12} coenzyme (from Savage, 1986a, reprinted with permission).

unacceptable rigidity on the molecules, as pointed out by Watson (1965) in the chapter entitled "The Importance of Weak Interactions," in The *Molecular Biology of the Gene.*

In proteins, there is a layer of water molecules close to the surface of the protein molecule where there is some evidence of order derived from the X-ray diffraction data. This is known as *bound water.* In the Baker and Hubbard (1984) analysis, it was observed that about 75% of the water molecules were directly hydrogen-bonded to the protein molecules, and of these more were bonded to C=O groups than to N—H groups. The proportions were 42% to main chain C=O, 14% to main-chain N—H, and 44% to side-chains.

Water molecules are frequently observed inserted between the N—H and O=C groups in α-helices and have been implicated in the bending and unfolding of helices by Blundell et al. (1983), and in the occurrence of a variety of turns by Sundaralingam and Sikhardu (1989). In their study using 35 well-refined protein crystal structures, they reported 312 examples of *hydrated segments* of helix structure of the type shown in Figure 10.19. Of these, 262 were external, 33 were internal, and 17 were three-centered. These are examples of indirect hydrogen bonding through water molecules analogous to those observed in some hydrated oligosacchardies and illustrated in Figure 4.5 and Table 4.15.

Beyond the layer close to the surface of the protein molecule, the water structure shows no order and is equivalent to liquid water or amorphous ice. Even in the bound water

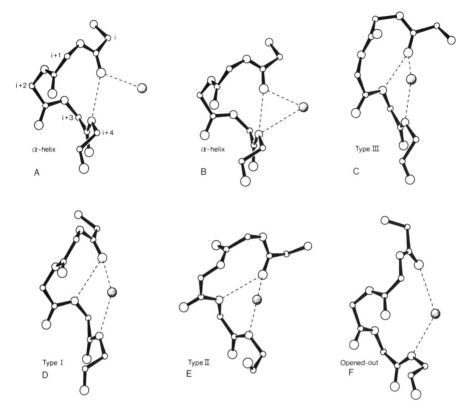

Figure 10.19.   Hydrated segments of α-helices. (a) external water; (b) three-center bonded water; (c) inserted water forming type II turn; (d) inserted water forming type I turn; (e) inserted water forming type II turn; and (f) description of helix forming open turn (from Sundaralingam and Sikhardu, 1989, reprinted with permission).

region, interpretation of the hydrogen bonding is hampered by the inability to locate the hydrogen atoms of the water molecules in most protein crystal structure analyses.[7,8] Despite the large number of protein crystal structure analyses, examples of well-organized hydrogen-bonded water structures, such as observed in the clathrate and cyclodextrin hydrate structures described in Chapter 9, are rare. One exception is the crystal structure analysis of the small plant protein, crambin, with molecular weight 4720. In this structure, Teeter (1984, 1991) observed fused pentagons of water structure with a marked resemblance to the structures observed in the clathrate hydrates, as shown in Figure 10.20. A *cap* of three fused pentagons of water molecules covers a valine side chain in the crystal structure of insulin by Baker et al. (1985), and fused pentagons and quadrilaterals of hydrogen-bonded

[7]An ingenious method of locating hydrogen atoms in proteins structures is to obtain difference Fourier synthesis maps with fully deuterated proteins, as described by Harrison, Wlodawer, and Sjölin (1988).

[8]A recent crystal structure analysis of $\gamma\beta$-crystallin, by Kumaraswamy et al. (1996), a structural protein of eye lens which preserves lens transparency, has revealed an unusually ordered water structure which covers 97% of the protein structure. The hydrogen bonding forms networks which contain 3-, 4-, 5-, and 6-membered rings.

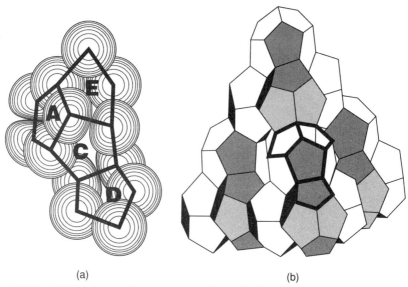

(a)   (b)

Figure 10.20. (a) fused pentagonal water structure observed in the small plant protein crambin; and (b) analogous structure in the type II cubic clathrate hydrate (from Teeter, 1991).

water molecules, shown in Figure 10.21, are observed in the cavity between two subunits in the crystal structure of glutathione reductase by Karplus and Schulz (1987).

Water structure involving fused pentagons was observed in the octanucleotide d(GGTATACC) by Kennard et al. (1986), and in a deoxynucleotide phosphate d(CpG)·proflavine complex with 27 H_2O by Neidle, Berman, and Shieh (1980), as shown in Figure 10.22. In a preliminary report of the crystal structure of an Arg-Gly-Asp peptide hydrate by Harlow (1993), the water structure contained a four-membered ring and several five-, six-, and seven-membered fused rings, with a six-ring backbone resembling the Ice I structures. It remains to be seen whether more organized water structures will be found as the resolution improves with the advances in X-ray diffraction technology, as suggested by Teeter (1991).

The Structure of the Nucleic Acids is Very Sensitive to Hydration

From the earliest X-ray studies by Franklin and Gosling (1953), it was realized that the conformation of DNA was particularly sensitive to humidity and the presence of salts and small organic molecules, as demonstrated, for example, by the circular dichroism studies of Hanlon et al. (1975) and Wolf and Hanlon (1975). The hydration of the nucleic acids has therefore attracted considerable attention and some excellent reviews are published by Texter (1978), Westhof (1987, 1988), Westhof and Beveridge (1989), and Schneider, Cohen, and Berman (1992), with two short updates by Berman (1991, 1994). A method for analyzing the distribution of water molecules around nucleic acid bases from the Nucleic Acid Data Base has been proposed by Schneider et al. (1993).

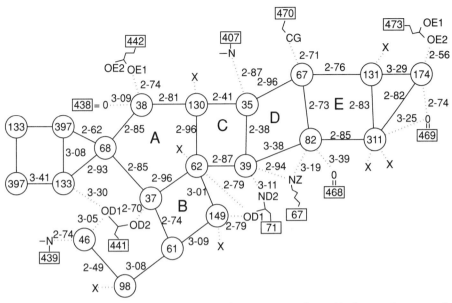

Figure 10.21. Fused pentagons and quadrilaterals of water structure observed in the crystal structure of glutathione reductase (from Karplus and Schulz, 1987).

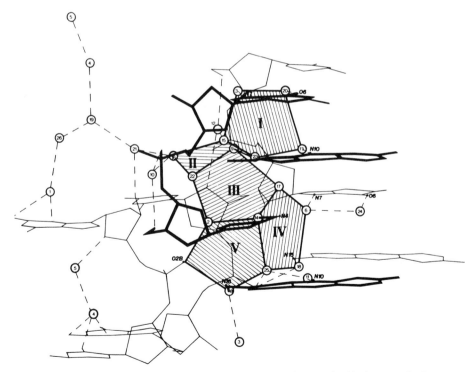

Figure 10.22. Water structure involving fused pentagons in the deoxynucleotide-drug complex by Neidle, Berman, and Shieh (1980, reprinted with permission).

The Nucleic Acids Also Have Three Levels of Water Structure

The primary level consists of about 12 water molecules per nucleotide, which are generally observed in oligonucleotide crystal structure analyses. This level is impermeable to cations and does not form ice on freezing. The secondary level is permeable to cations and forms ice on freezing, has fast exchange of water molecules and ions, and is partially ordered. The tertiary layer is completely disordered, so-called *bulk water*.

Since fiber X-ray diffraction patterns, however well-resolved, depend on modeling for interpretation, they cannot provide the hydrogen-bonding geometries associated with the hydration. Much of the information relating to water hydration comes from the crystal structures of synthetic oligonucleotides. These represent segments of the natural polynucleotides, and are extensively hydrated, as illustrated in Figure 10.23.

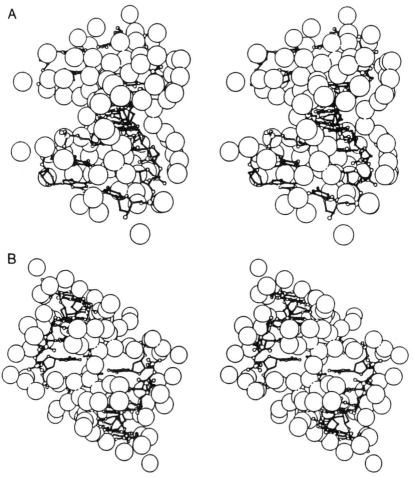

Figure 10.23. Ordered water (open circles) in A-type octamer duplex d(GTGCGCAC). (a) view into minor groove; and (b) view with molecular two-fold axis in plane of figure (from Binghan et al. 1992, reprinted with permission).

An extensive review of nucleic acid hydration provided by Westhof and Beveridge (1989) includes comparison with small molecule nucleoside and nucleotide hydrates and a discussion of computational approaches.

The availability of crystal structure analyses of oligonucleotides permitted Schneider, Cohen, and Berman (1992) to make an analysis of the information available from 28 crystal structures. They examined the patterns of water molecules both

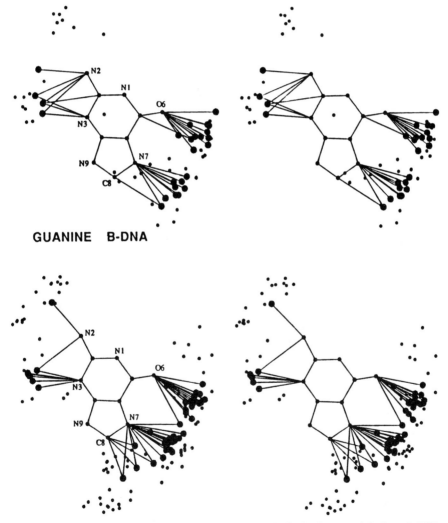

Figure 10.24.   Stereoviews of scattergrams of the water molecule distribution around the bases in DNA configurations. Large dots are closer than 3.2 Å, small dots are between 3.2 and 4.0 Å (from Schneider, Cohen, and Berman, 1992, reprinted with permission).

GUANINE   Z-DNA

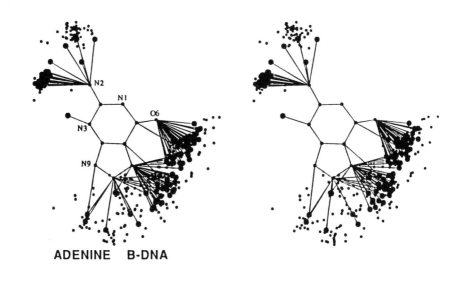

ADENINE   B-DNA

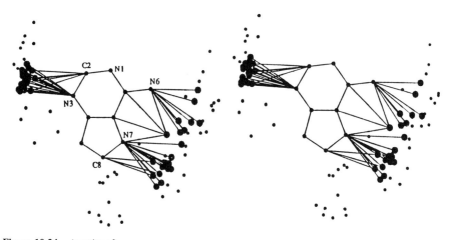

Figure 10.24.   (*continued*).

with respect to the four bases—guanine, cytosine, adenine, and thymine—and with respect to the type of DNA—A-, B-, Z-, and U-—which refers to single-stranded or non-
standard conformations. The number of contacts between the base atoms (including
carbon but excluding hydrogen) and the water oxygens were counted, out to 4.0 Å and
to 3.2 Å, respectively. The average number of contacts per base was also recorded.
The contacts below 3.2 Å correspond to potential hydrogen bonding, those between
3.2 and 4.0 Å corresponding to hydrogen bonds and hydrophobic interactions. This
corresponds well with the conclusions of Steiner and Saenger (1994) regarding

O—H····O hydrogen bonds, referred to in Chapter 4. The scattergrams obtained in this analysis are shown in Figure 10.24.

The Spine of Hydration, a Hydration Network

Water molecules are frequently observed to form *intranucleotide hydrogen-bonded bridges* in nucleic acid structures. The most famous of these occur in the minor groove of DNA, known as a *spine of hydration*, as illustrated in Figure 10.25. This sequence of hydrogen-bonded water molecules was first observed in the central AATT section in the crystal structure of the oligonucleotide dodecamer d(CGCGAATTCGCG) by Drew and Dickerson (1981) and Kopka et al. (1983). In that structure, a water molecule is hydrogen-bonded to O(2) of thymine and the N(3) of adenine with an additional close contact to the ring oxygen of the ribofuranoside residue.

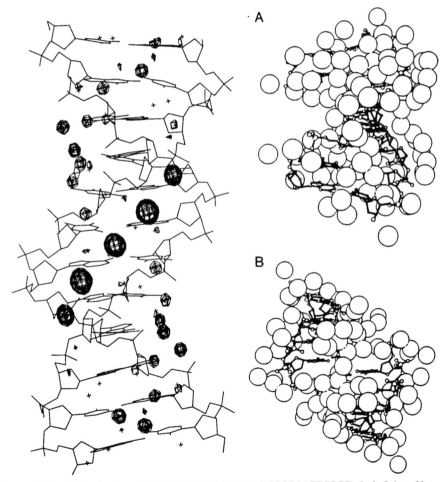

Figure 10.25.   Ordered water in oligonucleotide dodecamer d(CGCGAATTCGCG). Left: Spine of hydration (from Schneider et al., 1993). Right: (a) view into minor groove; and (b) view with molecules two-fold across in plane of figure (from Bingham et al., 1992, reprinted with permission).

Figure 10.26. Diagrammatic representation of movement of water molecules over N—H surface using two-center (····) and three-center(----) bonds.

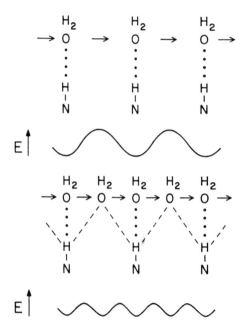

The Role for Three-center Hydrogen Bonds in Biological Water[9]

When substrates attach to surfaces or enter pockets or channels at the active sites of enzymes, the water molecules already present have to be displaced. As shown in Figure 10.26, three-center hydrogen bonding provides a mechanism for flattening the potential energy surface over which the water molecules must move to make room for the reacting molecule.

---

[9]Biological water is not pure water and thinking solely in terms of the hydrogen bonding of pure water may be misleading. From humans, the concentration of Na^+, K^+, Ca^+, Mg^+, $C\ell^+$ and $CO_3^=$ ranges from ~250 mM/liter of Na^+ in urine to 0.2 mM/liter of Mg^+ in sweat. The pH ranges from 7.8 in urine to 4.0 in sweat (from *Water and Aqueous Solution*, R. A. Horne, ed. Wiley-Interscience, 1972).

Methods

11

11.1 | INTRODUCTION

Hydrogen-bonding can affect the physical properties of gases, liquids, and solids. In the absence of hydrogen bonding, the physical properties of gases, liquids, and solids are said to be *normal*. That is, they are predictable from laws that relate physical properties solely to the chemical composition of the molecules involved. For example, melting and boiling points normally increase with molecular weight. H_2O as compared with other hydrides shown in Figure 11.1 is the classical example. Were it not for hydrogen bonding, water would freeze at about $-90°C$ and boil about $10°$ higher.

In gases, deviations from Raoult's law for ideal gas properties,

$$P_V = (W/M)(RT),$$

is evidence of, but not proof of, hydrogen bonding. In liquids, deviations from Trouton's Rule,

$$\Delta H_{vap}/T_{bp} \approx 22 \text{ cal deg}^{-1} \text{ mol}^{-1},$$

is also evidence. In crystals, hydrogen bonding in hydrates increases molar volume, due to the low—four-fold—coordination of the water molecules. In organic compounds, it tends to decrease molar volume, because intermolecular hydrogen bonding generally brings the molecules into closer contact than van der Waals interactions.

Hydrogen bonding generally becomes weaker with an increase in the thermal motion of the atoms involved. Therefore the measurement of any physical properties that are sensitive both to hydrogen bonding and to temperature, in principle, provides access to the thermodynamics of hydrogen bonding by measurement of the equilibrium constant for the reaction

$$A—H + B \rightleftharpoons A—H\cdots B.$$

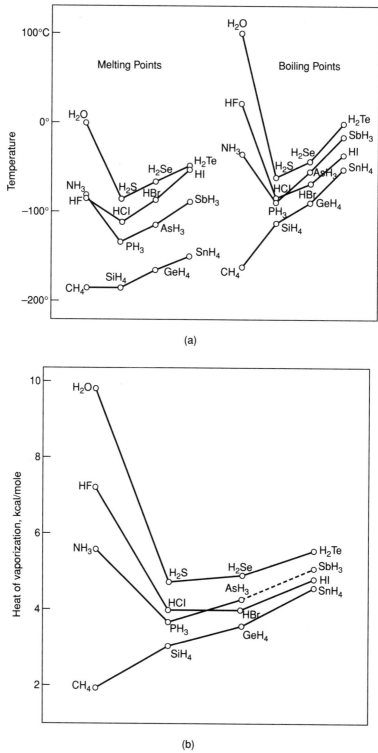

Figure 11.1. Effect of hydrogen bonding on (a) the melting and boiling points, and (b) the enthalpies of vaporization of some hybrides (from Pauling, 1939).

Table 11.1.
Properties Sensitive to Temperature Which Can Be Used As a
Basis for Thermodynamic Measurements of Hydrogen Bond
Enthalpies (From Pimental and McClellan, 1960).

Acoustic absorption	Freezing point
Adsorption	Heat conductivity
Absorption of light	Heat of mixing
Boiling point elevation	Heat of solution
Band spectra	Heat of vaporization
Cryoscopy	Infrared spectra
Calorimetry	Nuclear magnetic resonance
Heat capacity	Pressure-volume-temperature
Clausis-Clapeyron equation	Raman spectra
Conduction of electricity	Specific heat
Distribution or partition	Ultrasonic absorption
Dielectric absorption	Ultraviolet spectra
Dielectric constant	Second and higher virial coefficient
Density	Vapor density
Dipole moment	Vapor pressure
Electromotive force	
Fluorescence	

Pimental and McClellan (1960) list the 32 properties shown in Table 11.1 which can form the basis for the measurement of equilibrium constants. In an Appendix of that book, more than 150 equilibrium constants are reported for one-, two-, and three-component systems, with references to the methods used, with infrared spectroscopy the most common. By measurement of change of K with temperature, the enthalpy $-\Delta H^\circ$ and entropy ΔS° can be obtained from the Van't Hoff equation,

$$2.3 \log K = \frac{-\Delta H^\circ}{RT} + \frac{\Delta S^\circ}{RT}.$$

There were many studies in the 1960s and 1970s based on acid-base equilibria. Joesten and Schaad (1974) give an Appendix containing K_{25C}, $-\Delta H^\circ$ and $-\Delta S_{25^\circ C}$ and infrared $\Delta \nu_{X-H}$ frequency shifts, when appropriate, for nearly 2000 binary acid-base combinations.

With the discovery by Liddell and Wulf (1933), Hilbert et al. (1936) and Hendricks et al. (1936) that infrared spectroscopy provides a sensitive and, at that time, the most convenient tool for identifying hydrogen bonding, many examples of chelation in pure liquids and association in solutions were studied using differences in the A—H----B spectra associated with A—H stretching and bending vibrational modes. This large volume of research led to the identification of the hydrogen bond donor and acceptor groups given in Table 2.2, and the recognition of the strong, moderate and weak bonding shown in Table 2.1. Infrared spectroscopy has continued to be a major tool for studying hydrogen bonding up to the present time, with nearly 100 papers published each year.[1] An early review of the spectroscopy of the hydrogen bond is by Hadzi and Bratos (1976). The replacement of the grating spectrometer by the interferometer and the development

---

[1]A convenient bibliography of infrared and Raman spectroscopy under the category of topics investigated, including hydrogen bonding, is provided in the *Journal of Molecular Structure* from 1985 to 1994 in volumes 132, 155, 171, 191, 215, 241, 261, 290, 302, and 329.

of Fourier Transform Infra-Red (FTIR) spectroscopy, as described for example by Koenig (1981), has greatly improved the speed and resolution of the method.

NMR spectroscopy is another extremely sensitive method for identifying hydrogen bonding. This was first recognized by Liddell and Ramsey (1951) and is well described in the authoritative text *NMR Spectroscopy* by Pople, Schneider, and Bernstein (1959). It is less widely applied than infrared spectroscopy, because of the complexity of hydrogen bonding in solution. A review by Tucker and Lippert (1976) describes some results and discusses some of the problems involved in their interpretation. The development of solid-state NMR, through ^{13}C the cross-polarization magic angle spinning methods (CP-MAS) and the more difficult ^{1}H combined rotational and multiple pulse spectroscopy (CRAMPS) provides very effective tools for studying hydrogen bonding in solids, the full potential of which has not yet been realized. A review of the application of both ^{13}C and ^{1}H solid-state NMR to crystal structural problems including hydrogen bonding is provided by Etter, Reutzel, and Vojta (1990).

Crystal structure analysis by means of single-crystal neutron diffraction is the most definitive method for locating hydrogen atoms in hydrogen bonds. This is particularly important for strong hydrogen bonds due to the elusive hydrogen described in Chapter 3. The neutron-scattering power of the hydrogen or deuterium atoms is comparable to that of other atoms and as a result their atomic positions can be determined with comparable accuracy, together with the tensor parameters that define their thermal motion. This is important since hydrogen atoms experience greater thermal motion than the atoms to which they are covalently bonded and this can significantly affect the observed bond lengths. Since the scattering of hydrogen has the reverse sign to that of other elements, particularly deuterium, fractional occupancy and disorder can be defined more quantitatively.

The greatest advances in the neutron diffraction studies have come not only from the development of automatic diffractometers and better parameters refinement methods, as with X-ray analyses, but also from the availability of cryostats that allow controlled temperatures down to 5 K.[2]

When the automatic diffractometer displaced film methods for recording X-ray diffraction spectra and liquid nitrogen cooling became routine, the increase in precision of intensity measurements was such that it became regularly possible to locate the hydrogen atoms. The ability to do so has steadily increased with the increased sophistication of the recording equipment and the computer software for analyzing and refining the atomic parameters. Even so, because of the low X-ray scattering of hydrogen or deuterium, the accuracy of the positional parameters is an order of magnitude less than that for nonhydrogen atoms. The majority of the 100,000 crystal structure analyses in the organic, metal-organic, and inorganic databases that do contain hydrogen-bonding functional groups were carried out for purposes unconnected with the study of hydrogen bonds. For this reason, and perhaps from the old tradition of *not seeing hydrogen atoms*, the reporting of hydrogen positions in many recent X-ray crystal structure analyses leaves much to be desired. The atomic coordinates of the hydrogen atoms, even when precise enough to be useful to someone interested in hydrogen bonding, are not always published.

---

[2]Since most metals are transparent to neutrons, neutron diffraction cryostats are easier to construct than those for X-ray diffraction.

Both X-ray and neutron diffraction crystal structure analyses have one important advantage over every other method of structure analyses. Both methods are over-determined. Except for macromolecules such as proteins, the number of observations exceeds the number of variable parameters, generally by a factor between five and ten. It is rare that a crystal structure analysis is proved to be incorrect unless the investigator has made the initial mistake of selecting the wrong space group or wrong unit cell. In contrast, most spectroscopic methods are under-determined. There are 3N-6 fundamental frequencies for the translational, oscillation and vibrational motions for a molecule with N-atoms which must be distinguished from overtones; a number that generally exceeds the number of observations.

The most difficult quantities to measure precisely are hydrogen bond energies. Microwave spectroscopy provides good experimental values for gas-phase, hydrogen-bonded binary complexes, and modern ab-initio molecular orbital calculations provide theoretical values that generally agree within the limits of experimental accuracy.

Excellent comprehensive texts exist for all the principal methods of investigation. The methods of infrared and NMR spectroscopy, X-ray crystallography, and theoretical calculations have advanced so rapidly in recent years with the development of computer-controlled equipment and computer software that many otherwise excellent texts become out of date as practical guides within a few years of publication.

Hydrogen bonding changes the potential energy surface for an A—H group from a single minimum to a symmetrical or unsymmetrical double minimum, as shown in Figures 2.10 and 2.11. The hydrogen bond vibration modes used for normal and strong hydrogen bonds are shown in Table 11.2. The vibrational frequency criteria for diag-

Table 11.2.
Hydrogen Bond Vibration Modes For Normal Hydrogen Bonds.[a]

Vibration mode			Frequencies (cm^{-1}) for O—H····O
A—H̄····B (R)	A—H stretch	ν_s, ν_{OH}	3700–1700
A—H····B (R)	A—H in-plane bend	ν_b, ν_δ	1700–1800
A—H̄····B (R)	A—H out-of-plane bend	ν_σ, ν_r, ν_t	900–400
A—H····←B→ (R)	Hydrogen-bond stretch	ν_g, ν_σ	600–50
A—H····B (R)	Hydrogen-bond bend	ν_b, ν_β	<50

[a]Symbols differ in different texts and there is some variation in ranges quoted.
A better model for interpreting the vibration modes or strong, nearly centric hydrogen bonds is shown below:

Vibrational modes for strong, nearly symmetrical hydrogen bonds. (There are $3 \times 3 - 5 = 4$ normal modes, of which ν_{2a} and ν_{2b} are double degenerate.)

←F—H—F→	symmetric stretch	ν_1	
F—H—F	in-plane bend	ν_{2a}	
F̄—H—F̄	out-of-plane bend	ν_{2b}	ν_r
←F—H—F→	anti-symmetric stretch	ν_3	ν_a

Table 11.3.
Infrared Spectral Criteria For Hydrogen Bonding.

1. A—H stretching frequency, ν_s, moves to lower frequencies.
2. This is accompanied by an increase in intensity and band width.
3. A—H bending frequencies, ν_b, move to higher frequencies.
4. Upon cooling, ν_s shifts to high frequencies with increase in intensity and decrease in band width; ν_b moves to lower frequencies with decrease in band width.
5. Substitution of H by D lowers ν_s frequencies by a factor of ~0.75.

nosing the involvement of A—H in hydrogen bond formation is given in Table 11.3. In addition to the frequency shifts, there is generally a dramatic increase in intensity and band width with the development of substructure.

The source of the band broadening which accompanies the shift in the ν_s O—H frequencies on the hydrogen bond formation has been a subject of inquiry and disagreement among spectroscopists for many years. To quote from Bratos, Ratajczak, and Viot (1991), "Leaving a number of rapidly abandoned proposals aside, three mechanisms merit examination." The authors delve more deeply in the mystery of spectroscopic theory than is appropriate for an introduction to hydrogen bonding.

The methods for studying hydrogen bonding can be categorized as follows: (1) spectroscopy, (2) diffraction, (3) thermochemical, and (4) theoretical. The spectroscopic methods include infrared and Raman microwave, NMR, and neutron inelastic scattering, using the wave-lengths and frequencies shown in Table 11.4. Diffraction includes X-ray and neutron diffraction. Thermochemical includes calorimetry of heats of mixing or dilution and the determination of enthalpies directly or through the measurement of equilibrium constants. Theoretical includes ab-initio, semi-empirical, and empirical methods. In all these methods, the development of computer technology and the associated electronics has revolutionized the instruments available and the software associated with the use of them.

Table 11.4.
Wavelength and Frequency Ranges of UV, Visible, Infrared, Raman, Microwave, NMR, and Neutron Spectroscopy.

Spectral range		Wavelength λ	Frequencies ν (cm^{-1})
UV	Electronic vibration		
Far		10–200 nm	50,000–1,000,000
Near		200–400	25,000–50,000
Visible	Electronic vibration	400–800 nm	13,000–25,000
Infrared	Molecular vibration		
Near		0.8–2.5 μm	4000–13000
Middle		2.5–50	200–4000
Far		50–1000	10–200
Raman	Molecular vibration	2.5–1000 μm	10–4000
Microwave	Molecular rotation	0.01–10 cm	0.1–100
NMR	Nuclear precession	10–100 cm	0.01–0.1
Neutron	Nuclear vibration	1–5 Å	1–100 meV

Two of the most dramatic advances are in X-ray crystal structure analysis and in ab-initio quantum mechanics. A crystal structure analysis that might have taken three years 50 years ago can now be completed in three hours with a 100 times increase in precision. With theoretical calculations, there was a time a few years ago when ab-initio quantum mechanical calculation of molecular structures became obsolete before they could be published.

Spectroscopy methods depend on exciting the vibrational or rotational energy levels of bonds, groups, or small molecules, resulting in the absorption, or emission of the incident radiation at specific frequencies. This radiation can be electromagnetic or neutrons. Diffraction methods depend on the three-dimensional periodicity of the atoms in crystals to provide a diffraction grating for X-rays or neutrons of wave lengths comparable to the interatomic distances. Diffraction by liquids gives much less information, even when X-ray and neutron diffraction results are combined for simple liquids such as water.

Thermodynamic methods involve either direct calorimetry or using the effect of hydrogen bonding on physical properties at different concentrations or temperatures to determine the equilibrium constants for the formation of the hydrogen bond. Pulsed-ion cyclotron resonance spectroscopy provides one of the most accurate means of determining strong hydrogen bonding energies involving anions as described by Lehman and Bursey (1976) and Kebarle (1977). The substitution of hydrogen by deuterium can result in changes in the physical properties of hydrogen-bonded solids. This is known as the *isotope effect*. The most dramatic example is the change in the Curie temperature of hydrogen-bonded ferroelectrics, discussed in Chapter 7.

Different Methods have Different Time-Scales and Sensitivities

It is important to realize that different methods observe atomic structure and processes on different time scales. Spectroscopic methods provide information relating to structure and processes on a picosecond time scale (10^{-10}–10^{-15} sec). Diffraction and thermodynamic methods are at the other end of the time-scale (10–10^3 sec). Crystal structure analyses give an average structure both with respect to both time and space. NMR spectroscopy is midway at 10–10^{-4} sec. It is not surprising therefore that spectroscopists and diffractionists sometimes view hydrogen-bonded structures from different perspectives and their conceptual models do not always correspond. When discussing the theories of water structure, Eisenberg and Kauzman (1969) distinguish very clearly between these two views, referring to one as a V-structure and the other as a D-structure.[3]

Methods also differ in their sensitivities. The spectroscopic methods are more general and more sensitive than diffraction methods. They are commonly used to identify the presence of hydrogen bonding in all states of matter. As discussed in Chapter 5, weak C—H⋯O hydrogen bonds were identified by spectroscopists long before they were recognized by the crystallographers.

---

[3]V is vibrationally averaged, D is diffusionally averaged. There is also the concept of the I, instantaneous structure which is calculated theoretically.

11.2 | INFRARED AND RAMAN SPECTROSCOPY

Infrared spectroscopy uses the electromagnetic frequency ranges of 10,000–4000 cm^{-1} (wavelengths = 1–2.5 μm) near; 4000–200 cm^{-1}, middle; and 200–10 cm^{-1} far. Raman spectra are in the range 4000–10 cm^{-1}. Studies of hydrogen bonding are in the middle range. These methods obtain information about hydrogen bonding from the spectra arising from transitions between the vibration energy levels of the bonds involved in hydrogen bonding.

An important advance in infrared spectroscopy was the introduction of interferometers in place of spectrometers, which permits the acquisition of data from all frequencies at the same time. This method, known as Fourier transform infrared spectroscopy (FTIR), also delivers more radiation with greater stability (Koenig, 1981). A good review of high-resolution instrumental techniques, is provided by Robiette and Duncan (1983) and includes the use of lasers with applications to some individual molecules.

Polarized Fourier transform infrared spectroscopy is very effective when used with crystals of known crystal structure. The crystals have to be cut into plates several millimeters thick parallel to known crystallographic planes; a requirement which places some limitations on the use of the method. An excellent example of the use of polarized FTIR spectroscopy with single crystals of known structure is the study of the hydrogen bonding in potassium hydrogen maleate by Bezyszyn et al. (1992) discussed in Chapter 3.

The importance of correlating infrared frequency shifts with other measurements relating to hydrogen bonds was recognized very early. A relationship between A—H infrared stretching frequencies $\nu_{A—H}$ and hydrogen bond energies was proposed by Badger and Bauer (1937) and Badger (1940). The absence of reliable values for hydrogen bond energies hampered the development of this relationship. Bellamy and Pace (1969) observed linear relationships between the *relative* changes in A—H stretching frequencies for a variety of donor-acceptor combinations. Each line has a different slope. Measurements by Purcell and Drago (1967) and Drago, Vogel, and Needham (1971) also indicated that approximately linear relationships existed but were different for different acid-base combinations. Interest in this important correlation seems to have had an early and possibly premature death.

Similar relationships between $\nu_{A—H}$ and the A····B bond distances, shown in Figure 11.2, were observed by Rundle and Parasol (1952), Lord and Merrifield (1953), Nakamoto, Margolis, and Rundle (1955), Ratajczak and Orville-Thomas (1967–68), and Novak (1974). These relationships have been pursued with much enthusiasm up to the present time, as shown in Figures 11.3 and 11.4. The earlier data gave promise of a definite ν_s vs A····B or A—H algebraic relationship. A number of such relationships were suggested by Pimental and Sederholm (1956) and Bellamy and Owens (1969).

As more crystal structural data became available, it was apparent that for the O—H····O hydrogen bonds a linear relationship only held for the strong bonds with ν_s

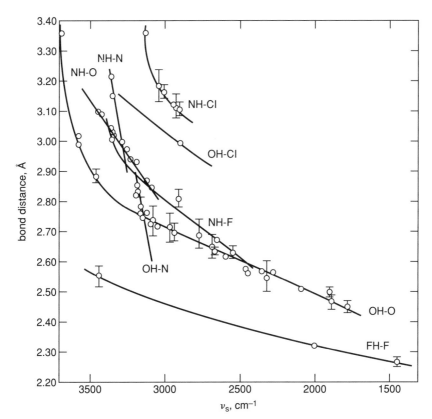

Figure 11.2. Correlations between stretching frequencies and hydrogen-bond geometries. (a) ν_s vs A—(H)····B distances (from Nakamoto, Margolis, and Rundle, 1955).

2700 cm^{-1} to 750 cm^{-1} and O····O from 2.60 to 2.45 Å. For weaker bonds with O····O > 2.6 Å, the relationship curved, the agreement deteriorated, and the frequency shifts became increasingly insensitive to changes in O····O distances. A more precise relationship with less scatter is reported for the N—H ν_s vs N····N distance for normal hydrogen bonds, but there are many less data points (Figure 11.5).

More recently, polarized FTIR spectroscopy using interferometry has been applied to slices of single crystals cut parallel to specific crystal faces. This method provides a more exact identification of frequencies with specific hydrogen bonds. It has been applied to two carbohydrates, sucrose by Giermanska and Szostak (1991), Lutz et al. (1992), and β-fructopyranose by Baran et al. (1994) where single crystal neutron diffraction data are available. Even with the more precise crystal structural data, where the O—H and H····O distances are corrected for thermal motion effects, the deviations from a smooth relationship between $\nu_{O—H}$ and O····O, H····O, and O—H distances far exceed the experimental uncertainties. The ab-initio calculations of vibrational O—H frequencies vs H····O bond lengths by Ojamäe and Hermansson (1992) show a similar scatter, as shown in Figure 11.6.

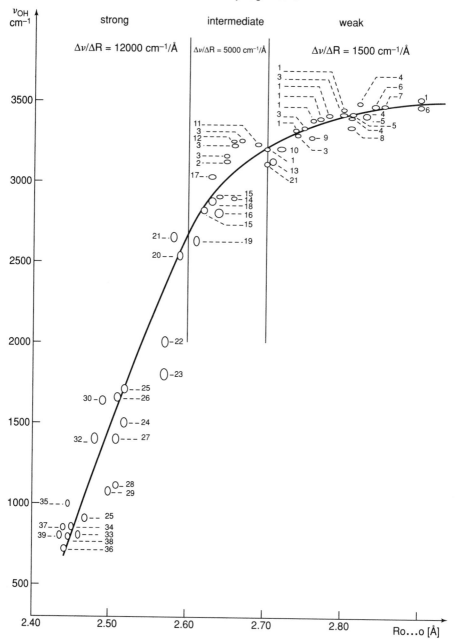

Figure 11.3. Correlations between O—H ν_s and O····O hydrogen bond distances (from Novak, 1974). Points 1–6 are in the salt hydrates; 7 is oxalic acid dihydrate; 8 and 9 are ices VI and I; 33–38 are strong bonds in organic hydrogen anions.

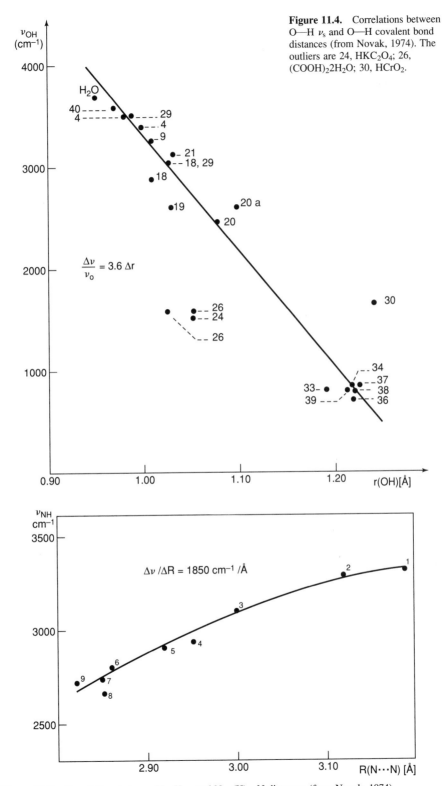

Figure 11.4. Correlations between O—H ν_s and O—H covalent bond distances (from Novak, 1974). The outliers are 24, HKC_2O_4; 26, $(COOH)_2 2H_2O$; 30, $HCrO_2$.

Figure 11.5. Correlations between N—H ν_s and N—(H)····N distances (from Novak, 1974).

223

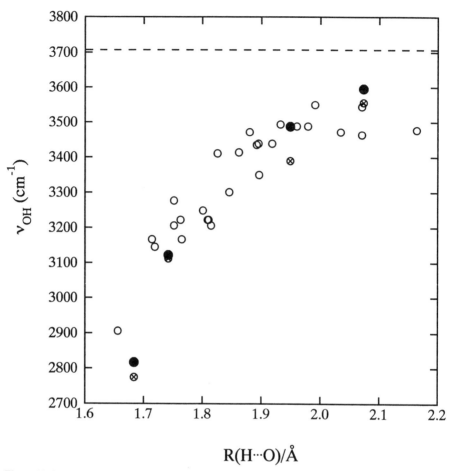

Figure 11.6.   Correlation between experimental H----O hydrogen bond distances and calculated ν_{O-H} stretching frequencies. Open circles are experimental values. Solid circles calculated for $LiClO_4 \cdot 3H_2O$, $LiHCOO \cdot H_2O$, $LiOH \cdot H_2O$ by Ojamäe and Hermansson (1992, reprinted with permission).

A study of the O—D(H)----F stretching frequencies from uncoupled ν(OH) values in isotopically dilute HDO in 16 crystal structures of metal fluoride hydrates by Mikenda and Steinbock (1994) showed a similar trend and scatter, as shown in Figure 11.7.

Raman spectroscopy has been overshadowed by infrared methods for studying hydrogen bonding. Of the nearly 500 spectroscopy papers relating to hydrogen bonding reported in the annual bibliographies in the *Journal of Molecular Structure* for the period 1990–1994, fewer than ten used Raman spectroscopy, presumably because it is more difficult.

Infrared and Raman spectroscopy depend on different electrical properties: changes in dipole moment for infrared and changes in polarization for Raman. Raman spectra are only readily accessible for liquids. With gases, the spectra are very weak and with solids, the Rayleigh scattering, i.e., with no change in frequency, is too strong. Thus

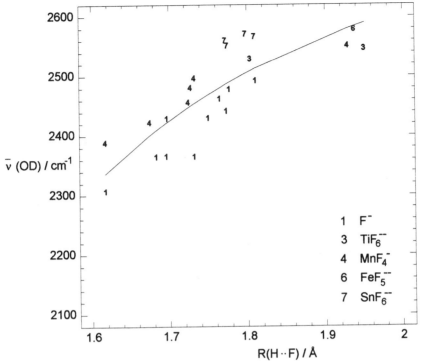

Figure 11.7. Uncoupled $\bar{\nu}$(OD) stretching frequencies *vs* R(H---F) hydrogen bond distances in fluoride hydrates (from Mikenda and Steinbock, 1994, reprinted with permission).

the useful correlation with microwave spectroscopy, ab-initio theoretical calculations, and crystal structure analyses is not possible.

The selection rules are different for the two methods and for this reason Raman spectroscopy is often used to supplement infrared spectral assignments.

11.3 | GAS-PHASE MICROWAVE ROTATIONAL SPECTROSCOPY

This method uses electromagnetic radiation in the frequency region 10^9–11^{11} Hz to record the vibrational and rotational spectra of hydrogen-bonded dimers and 1:1 adducts in the gas phase. The method is very specialized and is described in excellent reviews by Legon (1983), Millen (1983), Legon and Millen (1986), and Legon et al. (1987). The 1986 paper contains a useful bibliography of hydrogen-bonded adducts studied prior to 1986, with the hydrogen halides, HCN, HCF_3 and HC≡CH as donors and the $(H_2O)_2$ dimer.

Analysis of such spectra provides a rich source of information by measuring rotational constants, centrifugal distortion constants, nuclear quadrupole and nuclear spin-nuclear spin coupling constants, and the Stark and Zeeman effects.[4] Molecular geometries, bond energies, force constants, electric dipole moments, electric charge distributions, and electric quadrupole moments are derived from these measurements. The method provides information concerning hydrogen bonds which is not compromised by solvent effects or crystal field effects. It is sufficiently sensitive to give the information for the very weak hydrogen bonds discussed in Chapter 4.

There are two experimental methods. One uses Stark modulated microwave spectroscopy with binary gas mixtures at temperatures $\geq 175°C$ and pressures -50 m Torr, as described by Legon, Millen, and Rogers (1980). The second uses Fourier transform microwave spectroscopy of a pulse of gas mixture diluted in Argon and expanded supersonically into an evacuated Fabry-Perot cavity. The two methods are complementary, depending upon which properties of the hydrogen-bonded dimers are of primary interest.

Since the method measures the distance between centers of mass of the donor and acceptor molecules, and hydrogen atoms make only a very small contribution to the molecular mass, there may be some ambiguity in measurements of geometry for weakly bonded dimers where the A—H---B interaction is not linear. For example, in HCN----HF, a knowledge of the angles α_1 and α_2 is necessary to obtain N----F and H----N distances from the distances between the centers of mass r.

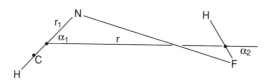

Characteristics of the vibrational satellite spectra can be used to establish that the dimer is linear for the stronger hydrogen bonds, but for weak hydrogen bonds it is necessary to distinguish carefully between hydrogen-bonded and van der Waals adducts.

11.4 | NEUTRON INELASTIC SCATTERING

Neutron diffraction depends on the elastic scattering of slow neutrons, in which the scattering neutrons have the same energy as the incident neutrons. In inelastic neutron scattering (INS), energy is transferred to or from the scattering object to the incident neutrons. This energy transfer depends upon the vibrational energy levels of the scattering system. The measurement of this energy gain, or loss, therefore provides information concerning transitions between the vibrational energy states. In molecular or crystal systems, these are in the range of milli-electron volts (1 mV$=$8 cm^{-1}). This measurement resembles Raman spectroscopy and is on the same short time scale

---

[4]The *Stark effect* is the change in frequencies produced by application of a strong electric field. This provides values for the dipole moments. The *Zeeman effect* is the change in frequencies produced by the application of a magnetic field.

$(10^{-11}–10^{-14}$ sec) and the methods refer to the same vibrational process. Therefore it provides information about hydrogen bonding that elastic neutron scattering cannot because of the averaging over a long time scale and space.

Hydrogen has a very large incoherent scattering cross-section of $a_{inc} = 2.52 \times 10^{-12}$, forty times larger than the coherent scattering and more than ten times larger than that of any other element. This is a source of inconveniently large background scattering in the coherent diffraction experiments. In inelastic scattering experiments, it provides an excellent tool for studying the potential energy surfaces of protons in hydrogen bonds, particularly for strong hydrogen bonds.

An excellent account of the mathematics of inelastic neutron scattering is to be found in Hamilton and Ibers (1968). The method can be applied both to powders and to oriented single crystals. A detailed account of the application of INS to the cyclic dicarboxylic acid dimer, the [H—F—H]⁻ ion, and the hydrated proton species is provided by Tomkinson (1992). When the incoherent elastic scattering is studied in the region of a Bragg elastic intensity peak, it is known as *quasi-elastic scattering*. A quasi-elastic scattering study of the orientational and positional disorder of the water molecules in β-cyclodextrin $11H_2O$ was studied by Steiner, Saenger, and Lechner (1991). They found that they could interpret the spectra shown in Figure 11.8 in terms of a simple two-site jump model with orientational jumps of hydroxyl groups and water

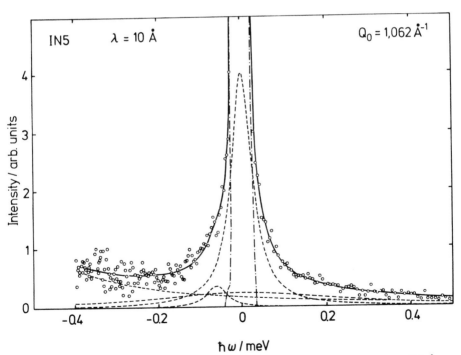

Figure 11.8.   Example of a quasi-elastsic spectrum of β-cyclodextrin $11H_2O$ taken with $\lambda = 10.0$ Å. (—·—·—) the elastic (Bragg) component of the spectrum; and (———) two Lorentzians and flat corrections used to interpret the data (from Steiner, Saenger, and Lechner, 1991, reprinted with permission).

molecules involving H----H distances of about 1.5 Å together with diffusive motions of the water molecules with H----H distances of about 3 Å. The jump rates were around 10^{-10}–10^{-11} sec^{-1} and persisted to about $-30°C$.

11.5 | NMR SPECTROSCOPY

NMR spectroscopy measures the degree to which the proton is shielded by its electronic environment in terms of proton chemical shifts. These shifts provide evidence of hydrogen bonding in liquids and solution and their magnitude is quantitatively proportional to the strength of the hydrogen bond. One of the same authors who recognized the significance of the effect of hydrogen bonding on A—H infrared frequency shifts was also the first to recognize that proton chemical shifts could provide a tool for recognizing and studying hydrogen bonding. Liddell and Ramsey (1951) observed that the ^{1}H NMR resonance of the hydroxyl proton in phenol was temperature sensitive. As with infrared spectroscopy, the change in chemical shift with concentration or temperature can give equilibrium constants and hence thermodynamic data. There are only two references to the use of NMR spectroscopy in the table of thermodynamic measurements in Pimental and McClellan (1960). These are temperature variation studies of CHCℓ_3-acetone and CHCℓ_3 by Huggins, Pimental, and Schoolery (1956) and concentration variation studies of trimethylamine and acetone in CHCℓ_3 by Huggins and Schoolery (1956). Other examples are to be found in reviews by Kollman and Allen (1972) and Tucker and Lippert (1976). The application of NMR spectroscopy to the different views of the structure of liquid water are discussed by Deverell (1969).

The sensitivity of ^{1}H NMR to changes in electronic environment make it a useful probe for detecting hydrogen bonding from weak donors, such as S—H and C—H and weak acceptors, such as multiple bonds and aromatic rings, as described in a review by Foster and Fyfe (1969).

Hydrogen bonding is complex in the liquid state because of the uncertainty in identifying the particular bonds and the number of molecules involved. Possibly for this reason, solution NMR spectroscopy, unlike solution infrared spectroscopy, has never become a major investigative tool in this field. The development of multi-dimensional methods has made NMR spectroscopy as powerful a tool for elucidating molecular structure in solution as X-ray crystallography is in crystals. Nevertheless it has had relatively little impact on the study of hydrogen bonds. It is notable that a recent text on multidimensional NMR by Croasman and Carson (1994) does not mention hydrogen bonding as one of the applications. As with Raman spectroscopy, there are relatively few papers on hydrogen bonding using NMR spectroscopy reported from 1990 to 1994, and those that do combine it with infrared spectroscopy. An interesting description of the use of ^{1}H NMR spectroscopy to examine the persistence of the intramolecular inter-residue hydrogen bond in methyl β-cellobioside dissolved in H$_2$O—CD$_3$OD and Me$_2$SO—D$_6$ is described by Leeflang et al. (1992). They found that the hydrogen bond persists in the nonaqueous solvent, but not in the aqueous solvent.

Solid State NMR Provides a New Tool for
Studying Hydrogen Bonding

The more recent development of solid-state NMR spectroscopy has prompted a greater interest in the application to hydrogen bonding. This is certainly because the results can be correlated with those of crystal structure analyses. As with polarized FTIR, solid state NMR has great potential in the future when combined with neutron diffraction crystal structure analysis.

In the absence of the tumbling motion of molecules in the liquid state and in solution, the dipolar interactions between like nuclei in solids make the NMR spectra very broad, so much so that they cannot be resolved. This problem is overcome by mechanical macroscopic spinning, known as *magic angle spinning*. A powdered, crystalline, or amorphous specimen is aligned at an angle θ to the magnetic field and spun at speeds in excess of 5 k Hz. The angle is $54°44'$, such that $(1-3\cos^2\theta) = 0$. It is applied to ^{13}C resonances in a method known as $^{13}C.CP.MAS$ (cross-polarization magic angle spinning). The cross-polarization refers to the enhancement of the ^{13}C by the flow of magnetization from the protons in the molecules. A general review of this field is provided and its applications are given by Fyfe (1983). A review of the application of this method to studies of hydrogen bonding in solids is given by Etter, Hoye, and Vojta (1988).

Solid-state ^1H NMR is more difficult to study than ^{13}C due to the abundance of the ^1H isotope resulting in much more proton homonuclear dipolar interactions. The problem has been resolved by a method known as CRAMPS (Combined Rotational and Multiple Pulse Microscopy) developed by Gerstein et al. (1977) and Ryan et al. (1980). A similar result is obtained by reducing the number of protons by deuterium substitution known as *deuterium spin dilution*, which is also combined with magic angle spinning.

As with infrared A—H stretching frequencies, a correlation has been observed between proton ^1H chemical shifts and distances in crystals, by Berglund and Vaughan (1980), Rohlfing, Allen, and Ditchfield (1983), Jeffrey and Yeon (1986), and Harris et al. (1988). The first two publications sought a correlation between O----(H)----O distances and the chemical shifts and the third used neutron diffraction data to extend this to H----O distances, giving the plot shown in Figure 11.9. The anisotropy $\Delta\sigma = \sigma_{33} - \sigma_{11}$ of chemical shift tensor[5] also shows a trend toward higher values for shorter hydrogen bond distances. The last paper gives values of the carboxylic acid isotopic proton shifts for 83 acids, for which 48 have O----O distances and 22 have H----O distances. These show a definite trend from $\sigma_{iso} = 21.0$ ppm in KH maleate with O----O = 2.437 Å to 4.8 ppm in tartaric acid with O----O = 2.909 Å. As with infrared stretching frequencies and bond lengths, the correlation is not exactly linear. Both solid-state ^{13}C and ^{15}N chemical shifts have been used to study hydrogen bonding, although the shifts are small. Imashiro et al. (1983) compared ^{13}C chemical shifts of hydroxybenzaldehydes in DMSO solution and in the crystalline state and observed small downfield shifts in the crystals inversely proportional to the O—(H)----O distances. An approxi-

---

[5]The chemical shift tensor is σ_{11}, σ_{22}, σ_{33}, σ_{12}, σ_{13}, σ_{23}. The familiar isotropic chemical shift is the trace $1/3(\sigma_{11} + \sigma_{22} + \sigma_{33})$.

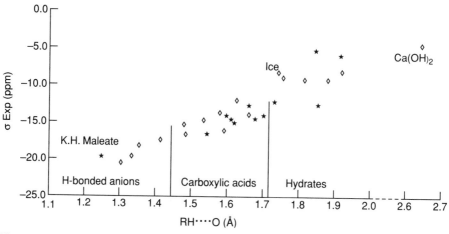

Figure 11.9. $\bar{\sigma}$ versus RH----O for all data ($\diamond$: neutron data, *: X-ray data).

mately linear correlation between ^{13}C MAS chemical shifts and the differences in the carboxylic acid C—OH and C=O bond lengths was observed by Kalsbeck, Schaumberg, and Larsen (1993) for a series of acid salts with asymmetric and crystallographically symmetric hydrogen bonds. Asakawa et al. (1992) used ^{13}C MAS to study the relationship between N—(H)----O=C hydrogen bond distances and chemical shifts of the carbonyl carbons in the L-alanine residues in a number of peptides. They also obtained an approximately linear relationship.

Since the time scale of NMR spectroscopy is much shorter than that of diffraction, variable temperature ^{13}C.CP.MAS can be used to distinguish between static and dynamic disorder in the crystal structure, as in 3,5-dimethylpyrazole by Baldy et al. (1985). Kuroki et al. (1991) measured ^{15}N chemical shifts for the glycine residue in some solid oligopeptides and found regular downfield shifts of both σ_{iso} and σ_{33} with a decrease of N----O distances.

For crystals of relatively simple molecules, the complete NMR anisotropic chemical shift tensor can be measured as described by Veeman (1981). Since this tensor has the same form as the anisotropic thermal motion parameter used in crystal structure analysis, it can be displayed using the ORTEP program of Johnson (1970a, 1970b), as shown in Figures 11.10 and 11.11. This method has been applied to ^{13}C and ^{31}P resonances. These experiments require relatively large crystals ($\sim$20.0 mm^3) with good morphology and are tedious because of the absence of the NMR equivalent of a computer-controlled X-ray, single-crystal NMR diffractometer.

Full tensor analysis of ^1H solid-state resonances is more difficult but could become a very useful tool for studying hydrogen bonding in the future.

The ^{13}C tensor elements can also be measured from the shape of the spectral envelope of side-band intensities from magic angle slow spinning (MASS). Gu and McDermott (1993) give the tensor elements for 35 amino acids and find that $\sigma_{11} + \sigma_{33} - \sigma_{22}$ gives a measure of protonation changed by varying crystallization conditions. The experiments suggest that solid-state NMR spectroscopy with its sensitivity

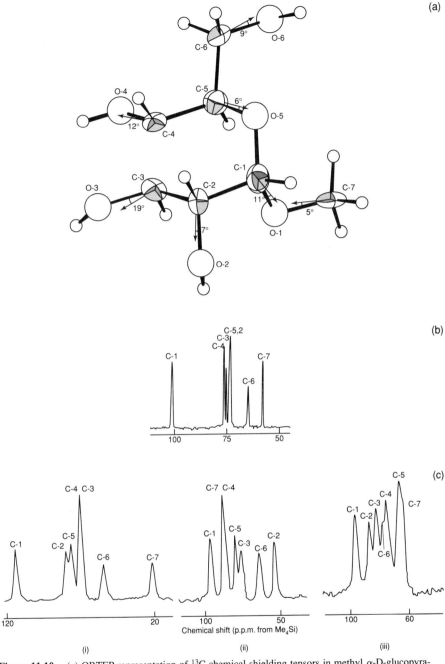

Figure 11.10. (a) ORTEP representation of ^{13}C chemical shielding tensors in methyl α-D-glucopyranoside; (b) Solid-state ^{13}C NMR spectra of ^{13}C in methyl α-D-glucopyranoside single crystal; and (c) Solid-state ^{13}C CP/MAS spectrum from powder specimen (from Sastry, Takegoshi, and McDowell, 1987). [Key: (i) $\mathbf{H_o}$//a-axis, (ii) $\mathbf{H_o}$//b-axis, (iii) $\mathbf{H_o}$//c-axis]

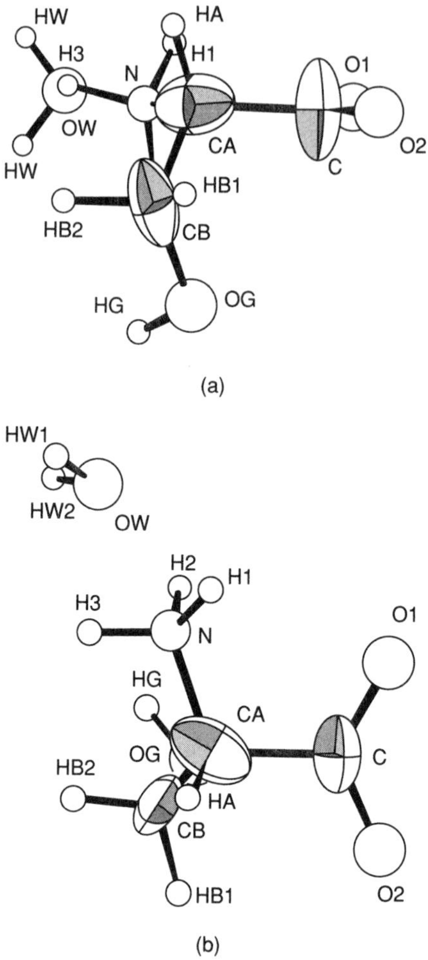

Figure 11.11. An ORTEP drawing showing the representation of the ^{13}C chemical shielding tensor deduced for *l*-serine monohydrate. The lengths of ellipsoid axes are proportional to the principal values shifted to make all three values positive. (a) viewing down from the direction perpendicular to the C_α—C bond and parallel to the O—C—O plane; and (b) viewing down from the direction perpendicular to the O—C—O plane (from Naito et al., 1983).

(a)

(b)

and time-scale of microseconds makes it an increasingly powerful tool for exploring proton mobility in biological systems.

11.6 | DEUTERON NUCLEAR QUADRUPOLE COUPLING

Nuclear quadrupole coupling constants are a measure of the electric field gradient close to the nucleus and therefore are sensitive to hydrogen bond formation. As with NMR spectroscopy, NQR is studied by radio-frequency nuclear resonance.

The quadrupole coupling constant is

$$e^2 q_{ij}.Q.h^{-1}$$

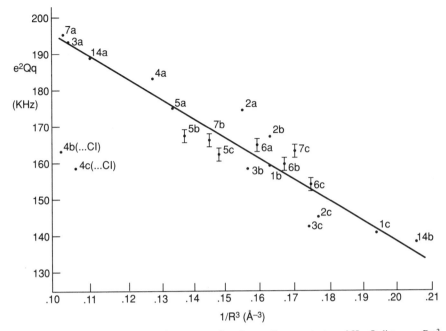

Figure 11.12. Correlations between deuteron quadrupole coupling constants and H⋯O distances, R^{-3}. For $\overset{+}{N}$—H⋯O=$\overline{C}$ hydrogen bonds in amino acids. (7) is serine hydrate, (14) is hydroxyproline (from Hunt and Mackay, 1974).

where e is the electronic charge; Q is the nuclear quadrupole moments; q_{ij} is the electric field gradient near the nucleus; and h is Plancks constant.

$$q_{ij} = -\delta^2 V/\delta x_i \delta x_j$$

where V is the electrostatic potential at the nucleus and x_i and x_j are cartesion coordinates. q_{ij} is a traceless symmetric tensor with $q_{xx} + q_{yy} + q_{zz} = 0$.

Nuclear quadrupole moments can be measured in the gas phase by microwave spectroscopy as described by Legon and Millen (1986). Nuclear quadrupole coupling constants can be measured for single crystals or powders and have been correlated with hydrogen bond A—H stretching frequencies and hydrogen bond lengths.

The correlation between deuteron e^2qQ/h and hydrogen bond distances and O—H stretching frequencies was reported by Chiba (1964), Blinc and Hadzi (1966), and Soda and Chiba (1969). Hunt and Mackay (1974) obtained the linear relationships shown in Figure 11.12 between the deuteron quadrupole coupling constants and $(H⋯O)^{-3}$ in a series of hydroxy compounds, including oxalic acid dihydrate, KD_2PO_4, $Li_2SO_4D_2O$ and D_2O ice, and in some zwitterion amino acids.

Berglund, Lindgren, and Tegenfeldt (1978) correlated the deuteron quadrupole coupling constants with ν_s for O—H and O—D and H⋯O bond lengths for a series of inorganic salt hydrates to obtain the linear relationships shown in Figure 11.13.

A qualitative relationship between the deuteron coupling constants and type of compound shown in Figure 11.14 was presented by Clymer and Ragle (1982). The C—D⋯O bonds in complexes of chloroform-d and diethyl ether and acetone were

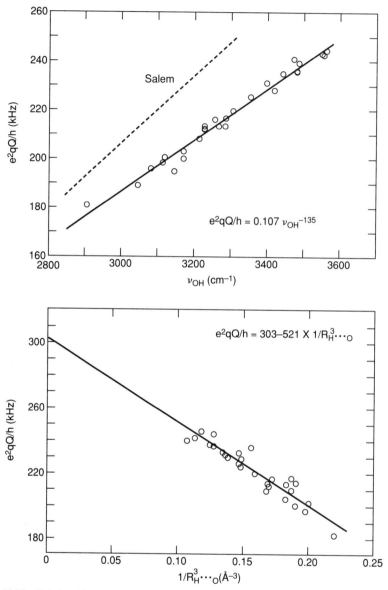

Figure 11.13. Relationship between deuteron quadrupole coupling constants and ν_{OH}, and H----O (from Berglund, Lindgren, and Tegenfieldt, 1978). Salem (1963) refers to a theoretical prediction.

studied by Ragle, Minott, and Morkarram (1974).The relationship between experiment and theory is discussed by Clymer et al. (1982).

A theoretical study by Huber (1985) uses a classical model to rationalize the r^{-3} dependence and the force constants. Since the barriers to rotation of water molecules in crystalline hydrates can be measured from the temperature dependence of the line width, this seems to be an underutilized method.

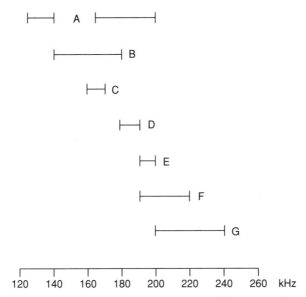

120   140   160   180   200   220   240   260   kHz

Figure 11.14. Qualitative representation of observed deuteron coupling constants in various simple organic compounds. A, amino acids; B, carboxylic acids; C, alkanes; D, aromatics; E, amides; F, alcohols; and G, hydrates (from Clymer and Ragle, 1982).

11.7 | DIFFRACTION METHODS: NEUTRON AND X-RAY

Single-crystal structure analyses have a special role in the study of hydrogen bonds because they provide direct information concerning the stereochemistry. Location of the hydrogen atoms is essential to understanding the nature of the hydrogen bond. Together with infrared spectroscopy, crystal structure analysis provides a basis for distinguishing between strong, moderate, and weak bonds. It also provides information necessary to distinguish between two-, three-, and four-center bonds. In strong hydrogen bonds where the A—H----B bonds are almost collinear, the covalent A—H bond length is extended to become nearly equal to that of the hydrogen bond. In moderate and weak hydrogen bonds the extension of the covalent A—H bond is small and is marginally observable, but the A—H----B angles may deviate significantly from 180°.

Electron diffraction and X-ray diffraction of liquids provide no information concerning the position of the hydrogen atoms. Even the neutron diffraction of water is difficult to interpret in terms of the hydrogen bond geometry, as discussed by Dore (1985).

In the early days of X-ray and neutron crystal structure analysis, atoms were located by means of Fourier syntheses in which the electron or neutron (Fermi) densities were calculated from the structure factors, $F_{hk\ell}$,

$$\rho_{xyz} = \frac{1}{v} \sum_{-\infty}^{h=\infty} \sum_{-\infty}^{k=\infty} \sum_{-\infty}^{\ell=\infty} F_{hk\ell} \exp -2\pi i (hx + ky + \ell z).$$

$F_{hk\ell}$ is a complex number with an amplitude $|F_{hk\ell}|$ and phase $\alpha_{hk\ell}$ of which only $|F_{hk\ell}|$ can be measured directly for each reflection since the intensity

$$I_{hk\ell} = kF_{hk\ell}^2.$$

Determining the other component is the *phase problem*, which was solved by the development of the *Direct Method*, for which H. Hauptman and J. Karle received the Nobel Prize in 1985.

In the mid-1950s, the development of automatic diffractometers replaced the measurement of X-ray diffraction intensities on film by direct photon counting and improved their accuracy by an order of magnitude. A discussion of the factors influencing the measurement of precise crystal structure intensities is given by Seiler (1992). The problems of converting measured intensities into structure amplitudes is discussed by Blessing (1987). The refinement of the atomic parameters by a least-squares procedure, first introduced by Hughes (1941), replaced Fourier synthesis and became increasingly sophisticated with the growth of computing power. Modern X-ray diffractometers with their associated commercial software packets have made X-ray crystal structure analysis almost a foolproof analytical procedure, providing not only the atomic coordinates, but also the equally important thermal motion parameters, conveniently illustrated by the ORTEP display of Johnson (1970a, 1970b).

The number of erroneous crystal structure analyses is small compared with the large number reported and even larger number not reported. A record of incorrect structures is published by Marsh (1986), Marsh and Schomaker (1979), Marsh and Herbstein (1988), and Marsh and Bernal (1995). The majority of these errors involve incorrect space group assignments such as described by Baur and Kassner (1992). Organo-metallic crystal structures with polar axes are well known to be particularly susceptible to error, as illustrated by the example provided by Murphy, Rabinovich, and Parkin (1995).[6] In most modern X-ray analyses, anisotropic temperature factors are determined, from which the ORTEP plot provides a valuable clue to an incorrect analysis, as illustrated in Figure 11.15.

With modern instrumentation and software, crystal structure analyses at liquid nitrogen temperatures can provide hydrogen atom positions with a precision an order of magnitude less than those of the nonhydrogen positions. If very high precision is required or disorder requires resolution, neutron diffraction is necessary and can be carried out conveniently at any temperature down to 10 K.

The advantages and disadvantages of single-crystal neutron diffraction vis-a-vis X-ray diffraction are shown in Table 11.5. A more detailed discussion is given by Finney (1995). The most serious disadvantage in this age of instant gratification is the relatively long time necessary to carry out the neutron diffraction measurements. As a result, there are less than 1000 neutron crystal structure analyses in the Cambridge Crystallographic Data Base, as compared with more than 100,000 X-ray analyses.

The use of pulsed neutrons, obtained by stopping an accelerated beam of electrons, i.e., a spallation source, provides an alternative in which the diffraction measurements use all available wavelengths, i.e., the Laue method. The radiation is pulsed,

---

[6]The classical case was that of bond-stretch isomerism reported in the *Journal of the American Chemical Society* in 1991; a consequence of inadequate chemistry plus inadequate crystallography.

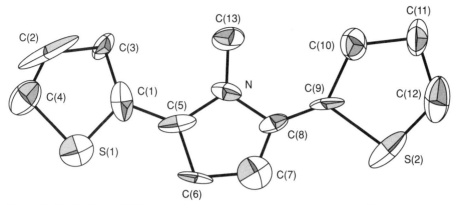

Figure 11.15. Published ORTEP indicating an incorrect crystal structure analysis of 2-(1-methyl-5-(2′-thienyl)-2-pyrrolyl thiophene. (Courtesy P. Harlow, who presents an annual prize for the worst published ORTEP of the year. This was from *J Chem Soc Chem Comm* 1989.)

Table 11.5.
Comparison of X-ray and Neutron Diffraction Single Crystal Analysis.

X-ray diffraction	Neutron diffraction
X-rays available on demand from laboratory instruments	Neutrons available only from national or international centers
Data collection time, a day or less (for routine work)	Data collection time a few weeks
Temperatures below 120 K not easily available	Temperatures down to 10 K conveniently available
Hydrogen atoms poorly located, especially O—H. Accuracy ~0.1 Å	Hydrogen positional parameters comparable in accuracy to C, N, and O, ~0.001 Å
Cannot analyze anisotropic thermal motion or disorder of H atoms	Analysis comparable to that of C, N, O
Small crystals can be used (~0.01 mm^3, ~0.01 mg)	Large crystals required (~1 mm^3, 1–2 mg)
Can be difficult to distinguish between thermal motion and disorder even for nonhydrogen atoms because of fall-off in intensity with scattering angle	Fall-off in intensity with scattering angle only due to its thermal motion. Therefore easier to distinguish from disorder
Number of variable parameters $9N_X + 4N_H + 1$, where N_X is the number of nonhydrogen atoms, N_H the number of hydrogen atoms	For comparable observation-to-parameter ratio, needs more observations. Number of variables $9N_T + 7$, where N_T is the total number of atoms and six anisotropic extinction parameters are used
	If the structure contains many hydrogens, deuterium substitution may be essential to reduce incoherent background
Careful absorption corrections necessary for other than first-row atom molecules	Absorption negligible, except for crystals containing B, Cd, Sm, Li (corrections for H advisable for molecules with large H content)
Extinction generally not serious for organic compounds	Extinction serious and pervasive. Careful corrections necessary
Radiation damage can occur and must be monitored by repeating selected measurements	No radiation damage

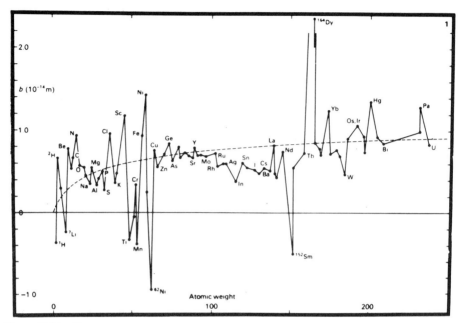

Figure 11.16. Neutron scattering lengths for elements as a function of atomic weight.

enabling a time-of-flight (TOF) measurement.[7] Hitherto there have been relatively few studies specifically aimed at hydrogen bonding using this method.

In neutron diffraction, unlike X-ray diffraction, the coherent scattering of the atoms is not a function of atomic number, but varies over a relatively narrow range, as shown in Figure 11.16. Hydrogen has a scattering power for neutrons about half that of carbon and oxygen as compared with 1/50 for X-rays. With X-rays, there is a fall-off with scattering angle due both to thermal motion and to the size of the electron cloud. Since nuclei are very small compared with the wavelength of the neutrons, the only fall-off with scattering analysis is that due to thermal motion. Hydrogen is therefore not at a disadvantage with regard to scattering power as in X-ray diffraction, but does have a very large incoherent neutron scattering factor. This is an inconvenience in diffraction analysis, since it contributes to a high background noise, but very useful in neutron spectroscopy, as discussed earlier. Substitution by deuterium, which has a small incoherent scattering factor, is often used to avoid this problem.

For the study of hydrogen bonding, an important feature is that hydrogen, with a cross-section of -3.74×10^{-13} cm^{-1} is opposite in sign to deuterium, with a cross-section of 6.67×10^{-13} cm^{-1}. This feature is very useful since the scattering power of H/D mixtures can be adjusted to any value between the two extremes by crystallizing from an appropriate H_2O/D_2O solution. By comparing X-ray diffraction data of a hydrogenated

---

[7]In place of Bragg's Law, $d_{hk\ell} = \lambda/2\sin\theta_{hk\ell}$, the TOF relationship is

$$d_{hk\ell} = t_{hk\ell} \cdot h/2m_n\ell \cdot \sin\theta_D,$$

where t is the time-of-flight in $\mu s^{-1} \cdot m^{-1}$, h is Plancks constant; ℓ is the flight path from source to detector; m_n is an instrumental constant; and $2\theta_D$ is the angle between the incident beam and the detector at the sample.

and deuterated protein crystal, information can be derived concerning the positions of the hydrogen atoms, as described by Harrison, Wlodawer, and Sjölin (1988).

Very few elements have significant absorption coefficients for neutrons. Fortunately, boron and samarium do have high absorption coefficients, otherwise neutron beams could not be stopped; an interesting thought. An advantage of this is that low-temperature diffraction experiments are much simpler, since normal metal cryostats can be used, permitting routine measurements down to 10 K or lower.

Neutron diffraction crystal structure analyses follow the same procedures as X-ray analyses. However, since a neutron analysis is a much greater commitment in time and money, an X-ray analysis usually precedes it to ensure that there are no complications, that might make the results of the neutron study ambiguous such as crystal twinning or disorder.

Thermal motion corrections are more important for bonds involving hydrogen atoms. Even in crystals at low temperatures, hydrogen atoms have greater thermal motion than the non-hydrogen atoms, as illustrated in Figure 11.17. This affects the observed bond lengths as pointed out by Coulson and Thomas (1971). One effect of the greater thermal motion is that of *riding motion*, which arises because the A—H bend-

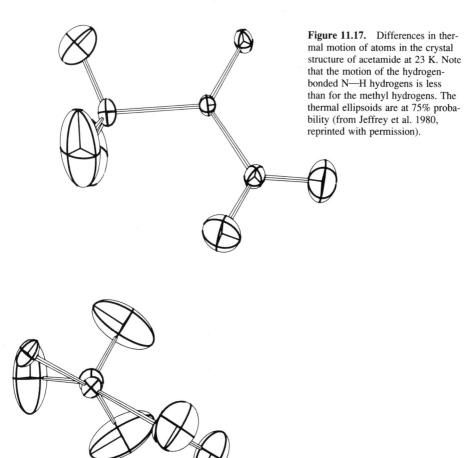

Figure 11.17. Differences in thermal motion of atoms in the crystal structure of acetamide at 23 K. Note that the motion of the hydrogen-bonded N—H hydrogens is less than for the methyl hydrogens. The thermal ellipsoids are at 75% probability (from Jeffrey et al. 1980, reprinted with permission).

Table 11.6.
Effect of Thermal Motion Corrections on Bond Lengths From Neutron Diffraction Data in Å.

	Uncorrected		Corrections		Corrected	
Bonds	X—H	H⋯A	Riding	Anharmonic	X—H	H⋯A
1,2,4-Triazole at 15 K (from Jeffrey, Ruble, and Yates, 1983)						
N(1)—H(1)	1.048(1)		+0.003	−0.017	1.033	
C(3)—H(3)	1.086(2)		+0.002	−0.016	1.072	
C(5)—H(5)	1.086(2)		+0.002	−0.015	1.073	
Sucrose at 290 K (from Brown, and Levy, 1973)						
O(4)H⋯	0.912		+0.070	−0.014	0.968	
O(4′)H⋯O(1)	0.976	1.760	+0.026	−0.014	0.960	1.771
O(1′)H⋯O(2)	0.974	1.851	+0.013	−0.022	0.944	1.845
O(6′)H⋯O(5)	0.972	1.895	+0.021	−0.028	0.973	1.890
O(2)H⋯O(6′)	0.972	1.892	+0.026	−0.017	0.967	1.892
O(3)H⋯O(3′)	0.959	1.907	+0.015	−0.021	0.966	1.907
O(3′)H⋯O(4)	0.969	1.908	+0.023	−0.018	0.979	1.908
O(6)H⋯O(3)	0.976	1.921	+0.011	−0.022	0.965	1.921
e.s.d. = 0.003						

ing force constants are weaker than those of covalent bonds. This effect was first discussed by Busing and Levy (1964). The second effect is *anharmonic stretching* motion. Correcting for these thermal motion effects is important if precise comparisons are made between hydrogen-bond geometries at different temperatures, or with theoretical calculations, which predict the geometries of molecules *at rest*.[8] These corrections can be made from the anisotropic thermal motion parameters obtained from neutron diffraction analysis.

These riding and anharmonicity effects can be corrected from the thermal motion parameters approximately by

$$\Delta r \text{ (riding)} = (U^2_H)_\perp - (U^2_A)_\perp / 2r_{A—H}$$

and

$$\Delta r \text{ (anharm)} = -[3a/2(U^2_H)_\| - (U^2_A)_\|]$$

where $(U^2)_\perp$ and $(U^2)_\|$ and are the components of the anisotropic tensor perpendicular and parallel to the A—H of length $r_{A—H}$. a is a constant taken from the work of Kuchitsu and Morino (1965). At low temperatures, the anharmonic correction is the larger, but at room temperature the reverse is true. Some examples of these corrections are shown in Table 11.6. For very precise neutron diffraction studies such as those conducted on the structure of Ice I by Kuhs and Lehman (1983), more sophisticated methods for correcting for thermal motion are necessary. An account of these methods is given by Johnson (1970b).

[8]Corrections for thermal motion is much more important in gas-diffraction structure analysis. An excellent mathematical treatment of the effects of vibrational motion is by Cyvin (1968). This led to different definitions of bond lengths, given in Appendix II.

Hydrogen atoms are generally located unambiguously in modern X-ray analyses, particularly if liquid nitrogen cooling is used. The accuracy is an order of magnitude less than for nonhydrogen atoms and only isotropic equivalent temperature parameters can be determined.

The hydrogen atoms are located at an intermediate stage in the structure refinement; at the end of the refinement of the nonhydrogen atoms with isotropic temperature factors and before going to the anisotropic temperature factors. If the hydrogen atoms are omitted in anisotropic refinements, the nonhydrogen thermal factors will compensate for this omission. The hydrogen atom positions are determined by searching for residual electron density in different Fourier syntheses, for which $|F_{obs}| - |F_{calc}|$ are the structure amplitudes with F_{calc} from nonhydrogen atoms only. The hydrogen positions from these difference maps are then included in the refinement with isotropic temperature factors. These factors are usually those of the atoms to which they are covalently bonded, or twice those values, or may be included as variable parameters in the refinement. Since part of the hydrogen electron density is displaced into the A—H bond, as described in Chapter 2, the X-ray A—H distances are usually ~0.2 Å shorter than the internuclear distances obtained from neutron diffraction analysis, and consequently the H⋯B hydrogen bond lengths are longer. This displacement of the electron density can be corrected by normalizing the A—H bond lengths to mean internuclear values, i.e., O—H = 0.97, N—H = 0.99, C—H = 1.01 Å. Where X-ray and neutron analyses are available for the same crystal structure, the improvement of the agreement in hydrogen bond lengths is significant, as shown in Table 11.7. A discussion of accuracy in neutron crystal structure analysis is given by Jeffrey (1992a).

Table 11.7.
Comparison of O—H and H⋯O Distances From Room Temperature X-Ray and Neutron Diffraction Crystal Structure Analyses For Two-Center Bonds (From Jeffrey and Saenger, 1991).

Crystal structure	O—H (Å)		H⋯O (Å)		
	X-ray	Neutron	X-ray	X_{norm}[a]	Neutron
Methyl α-D-altropyranoside	0.81	0.971	1.90	1.74	1.736
	0.88	0.961	2.00	1.91	1.922
Methyl α-D-glucopyranoside	0.87	0.985	1.84	1.74	1.738
	0.97	0.969	1.76	1.76	1.770
Methyl α-D-mannopyranoside	0.80	0.976	1.98	1.81	1.810
	0.67	0.957	2.22	1.96	1.998
	0.96	0.959	2.07	2.05	2.052
Methyl β-D-galactopyranoside	0.77	0.957	1.98	1.79	1.773
	0.86	0.958	1.87	1.78	1.739
	0.84	0.976	1.98	1.86	1.860
Methyl β-D-galactopyranoside	1.06	0.946	1.76	1.85	1.817
monohydrate	1.06	0.971	1.66	1.75	1.747
	0.95	0.976	1.75	1.73	1.706
	0.89	0.967	1.97	1.90	1.851
	0.92	0.958	2.03	2.08	1.983
Means	0.89	0.966	X-ray-Neutron, −0.11		
			X_{norm}-Neutron, −0.03		

[a] The X_{norm} values of H⋯O are obtained by normalizing the X-ray O—H distances to 0.97 Å.

For $\diagdown$N—H and $\diagdown$N—H bonds, the hydrogen atomic positions can be calculated with reasonable accuracy from the positions of the adjacent nonhydrogen atoms. These positions may be used with or without further refinement.

Some modern refinement programs have the computer automatically search for hydrogen positions, by selecting the region of greatest residual density. Substituting computer intelligence for human intelligence is less tiring but does not always give the best results.

Deformation Electron Density Distributions and Molecular Electrostatic Potentials

The abundance of very precise structure factor measurements that can be measured with modern single-crystal diffraction equipment makes it possible to extract much more information than the positional and thermal motion parameters of the atoms. Of particular interest with respect to hydrogen bonding are the deformation electron density and the electrostatic potential.

The deformation electron density measures the difference between the electron density in a molecule and that of unbonded atoms placed at the nuclear positions. It reveals the effects of chemical bonding on the electron distribution. This deformation density is obtained by subtracting from the experimentally observed electron density that calculated by placing nonbonded Hartree-Fock theoretical atoms at their nuclear or core electron positions. Neutron diffraction can give the nuclear positions, as in the X—N method,[9] while high 2θ X-ray data can give the core-electron positions, as in the X—X method. A more detailed discussion of these methods is given by Hirshfeld (1992). An X—N map is shown in Chapter 2, Figure 2.2, to demonstrate the electron density distribution in an O—H⋯O hydrogen bond. This method clearly revealed the shared electrons in single, double, and triple covalent bonds at the midpoints of the bonds. Crystallographers were excited by this, but there is an impression that chemists were not. After all, since Lewis and Sidgwick they had been taught that there were shared electrons in the covalent bonds. This was merely an experimental verification. Nevertheless, a large number of deformation density measurements have been made, as reviewed recently by Spackman and Brown (1995). A number of these compare the experimental studies with high-level ab-initio calculations, as illustrated by the study of a small peptide by Souhassau et al. (1992).

More interesting is the derivation of experimental electrostatic properties from X-ray diffraction data. The electrostatic properties that can be derived from the Fourier synthesis of X-ray crystal structure factors are given in Table 11.8. Of particular relevance to hydrogen bonding is the electrostatic potential since this reveals the electrophilicity of the hydrogen bonding functional groups. As shown in Figures 11.18 and 11.19, the donor hydrogen bonding groups have a region of positive electrostatic potential; for the acceptor groups, the electrostatic potential is negative. As pointed out in Chapter 2, the formation of hydrogen bonds between molecules is a consequence of the complementarity of their electrostatic potentials.

---

[9]Nonhydrogen thermal vibration parameters measured by X-ray and neutron diffraction on the same crystal structure, at the same temperature, but not the same crystal, sometimes show significant differences. Possible sources for these differences is discussed by Blessing (1995).

Table 11.8.
Electrostatic Properties That Can Be Derived From the Fourier Synthesis of X-ray Crystal Structure Factors With Different Dependences on $\sin\theta/\lambda$.

Property	type	Dependence on $\sin\theta/\lambda$
Electrostatic potential	scalar	-2
Diamagnetic shielding tensor	2nd rank	-2
Electrostatic energies	scalar	-2
Electric field	vector	-1
Electric field gradient	traceless 2nd rank tensor	0
Charge density	scalar	0
Diamagnetic current density	vector	0
Gradient of field gradient	3rd rank	1
Gradient of charge density	vector	1
Grad-grad of field gradient	4th rank	2
Hessian of charge density	2nd rank	2
Laplacian of charge density	scalar	2

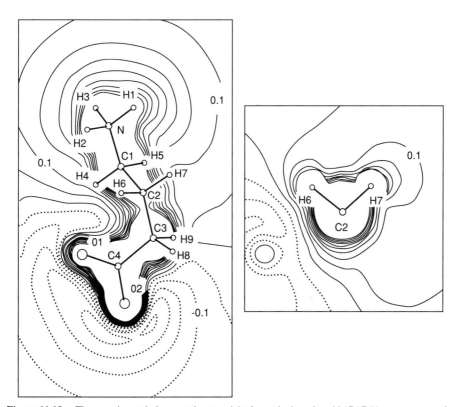

Figure 11.18. The experimental electrostatic potential of γ-aminobutyric acid (GABA), a neurotransmitter. The region around the hydrogen bond donor $\overset{+}{N}H_3$ group is positive, while that around the carboxylate acceptor group is negative. The potential around the CH_2 group is also positive (shown in insert), corresponding to intramolecular C—H····N and C—H····O hydrogen bonds at 2.64 and 2.54 Å. Contours are 0.05–0.50 eÅ^{-1}, with full lines positive, dotted lines negative (from Stewart and Craven, 1993).

243

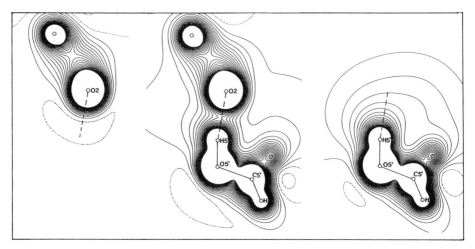

Figure 11.19. Electrostatic potentials around an O(5′)—H⋯O(2)=C hydrogen bond in 3′-O-aceteyl-2′-deoxy-5-methoxymethyluridine (from Wei, Barton, and Robertson, 1994, reprinted with permission).

In order to extract electrostatic potentials from the X-ray structure factors, it is necessary to have procedures to account for the deformation of the isolated atom electron density due to the chemical bonding in an analytical form. These procedures are therefore an extension of the deformation density analysis described earlier.

Methods commonly used are those of Hirschfeld (1971), Stewart (1982, 1991), and Hansen and Coppens (1978). In the *pseudoatom model* of Stewart (1976), the electron density distribution is described in terms of Hartree-Fock isolated atom cores to which are added deformation density terms. Each term is assigned a Slater-type *radial function* with a variable exponent, a, for each pseudoatom. *Angular functions* were obtained for each atom by a multipole expansion up to the octapole level for all atoms except hydrogen, for which a dipole only is used. This pseudoatom description of the electron density contains 16 parameters for each nonhydrogen atom, one population, three dipole, five quadrupole, and seven octapole parameters, and one population and three dipole for each hydrogen. These parameters, together with the six anisotropic thermal motion parameters, are refined by least squares to minimize the residual $\Sigma\omega(F_c{}^2 - F_o{}^2)^2$ or some equivalent as described by Craven (1988) and Craven and Stewart (1990). In the Hirshfeld (1971, 1992) approach, the deformation electron density is described in terms of a series of spherical symmetrical, and antisymmetrical exponential functions with adjustable parameters.

The electrostatic potential $\phi(r')$ is then obtained from the total charge density distribution $\rho(r)$ from the pseudoatom model by the equation

$$\phi(r') = \int \rho(r) \left|r' - r\right|^{-1} dr$$

in procedures described by Stewart (1982, 1991), Spackman and Stewart (1984), Su and Coppens (1992), Stewart and Craven (1993), and Ghermani, Lecomte, and Bouhmaida (1993). This work requires very precise experimental measurements at liquid nitrogen temperatures and preferably lower. The electrostatic potential for the molecule at rest consists of the nuclear point charges $Z_i|r_i|^{-1}$, and the sum of the special Hartree-Fock cores and the deformation. A general review of the experimental measurement of electron density distributions and electrostatic potentials is by Klein and Stevens (1988).

Electrostatic potentials can be calculated by ab-initio molecular orbital methods. A comparison of the electrostatic properties of L-alanine from X-ray diffraction data at 23 K with ab-initio calculations with different basis sets is given by Destro, Bianchi, and Morosi (1989). A study of the hydrogen bonding in L-arginine phosphate monohydrate is reported by Espinosa et al. (1996).

The Laplacian of the electron density can be derived from diffraction data. Another representation of the electron density distribution in a molecule is the Laplacian $-\nabla^2\rho(r)$ of the electron density $\rho(r)$, which contains much more detail than the electron density. The philosophy of describing atomic structure in molecules in terms of the Laplacian of the electron density is presented by Bader and Laidig (1991) and Bader (1992).

This distribution can be obtained from the pseudoatom description of the electron density by calculating at each point r, the sum of the three principal values of the curvatures on $\rho(r)$, i.e., $\delta^2\rho/dx$, $\delta^2\rho/dy$, $\delta^2\rho/dz$. The negative Laplacian identifies *critical points*, one with three negative curvatures $(3, -3)$ is an *atomic site*, and one with two negative and one positive curvature $(3, -1)$ is a *bond path*. A recent review of the calculation of electrostatic properties from X-ray diffraction data is by Lecomte (1995).

11.8 | COMPUTATIONAL CHEMISTRY

Calculating the structure, energetics, and other physical properties of the assembly of nuclei and electrons in molecules is an extraordinarily complex problem. It is more difficult than calculating the motion of planets, since quantum mechanics is more complex than classical mechanics, and came later in the history of science. Nevertheless, computational chemistry has become an important method for understanding hydrogen bonding as described in Chapter 2. With the super computers now available, ab-initio molecular orbital calculations are the methods of choice for hydrogen bonded complexes of simple molecules containing first and second row elements. The descriptor ab-initio implies seeking as accurate a solution to the Schrödinger wave equation as is possible with the computing facilities available with no input from chemical intuition. More complete descriptions are to be found in Dunning and Hay (1976, 1977), Hehre et al. (1986), and Clementi, Corongiu, and Stradella (1991).[10] These calculations refer to the isolated hydrogen bonded adducts *at rest*. To compare these values with experimentally measured enthalpies of gas-phase reactions, ΔH^{298}, quantities

---

[10]This 1200 page text weighing over five pounds provides a glimpse of the frontier of the field. Clearly, computational chemistry will soon be as obscure to ordinary chemists as is nuclear physics.

Table 11.9.

Enthalpies (kcal/mol) of Molecular Association Through Hydrogen Bonding (From Delbene, 1986).

Dimer	ΔEe^a		ΔEv^b						
	HF	MP	ΔE_ν^0	$\Delta(\Delta E_\nu)^{298}$	ΔE_r^{298}	ΔE_t^{298}	ΔnRT	ΔH_{calc}^{298}	ΔH_{exptl}^{298}
$(HCN)_2$	−4.2	−4.7	0.9	2.3	−0.6	−0.9	−0.6	−3.6	−4.4 ± 0.2^e
$(H_2O)_2^c$	−4.3	−5.4	2.2	2.2	−1.1	−1.1	−0.7	−3.9	−3.6 ± 0.5^d
$(H_2O)_2$	−4.3	−5.4	2.2	1.9	−0.9	−0.9	−0.6	−3.7	
FH—HCN	−6.3	−7.2	1.6	1.7	−0.6	−0.9	−0.6	−6.0	
$(HF)2^f$	−4.1	−4.8	1.8	1.0	−0.3	−0.9	−0.6	−3.8	−5.1 ± 1.0^g

^{a}Hartree-Fock and MP3 6-311+G(2d,p) energies.
^{b}Determined from Hartree-Fock 6-31G(d) frequencies.
^{c}Del Bene et al. (1983).
^{d}Curtiss, Frurip, and Blander (1979).
^{e}Mettee (1973).
^{f}Calculated at the optimized Hartree-Fock 6-31+G(d) geometry (Frisch, Pople, and Del Bene, 1983).
^{g}Del Bene (1986).

relating to the translational, rotational and vibrational energies have to be calculated, as described by Del Bene et al. (1983). For example, in the reaction

$$H_2O + H_2O \rightarrow (H_2O)_2$$

$$\Delta H_{calc}^{298} = \Delta E_e^0 + \Delta(\Delta E_e)^{298} + \Delta E_v^0 + \Delta(\Delta E_v)^{298} + \Delta E_r^{298} + \Delta E_t^{298} + \Delta Pv$$

where ΔE_e^0 is the calculated difference in the electronic energies of reactants and products at absolute zero.[11] $\Delta(\Delta E_e)^{298}$ is the change in electron energy difference between 298 K and 0 K. Δ_v^0 is the difference between the zero-point vibrational energies of reactants and products at 0 K. $\Delta(\Delta_v)^{298}$ is the change in vibrational energy differences between 298 K and 0 K. ΔE_r^{298} is the difference in rotational energies of reactant and product. ΔE_t^{298} is the translational energy change due to change in the number of degrees of translational freedom. ΔPv is a work term that relates ΔH and ΔE, ideally —RT. An example of the values assigned to these terms is given in Table 11.9.

After 1980, it was possible for theoreticians to aim for computations for simple dimers which reproduced the binding energies within the uncertainties of the experimental measurements.[12] A detailed review of these efforts with regard to the water dimer is given by Scheiner (1994).

The ab-initio molecular orbital method seeks minimum energy intra- and intermolecular geometry by solving the wave equation using a linear combination of atomic orbitals known as the LCAO approximation. Coefficients are attached to each of the one-electron atomic orbitals and varied to obtain the minimum energy. Polarization

[11]In theoretical papers, distances and energies are reported in unfamiliar units; 1 bohr = 0.52172 Å; 1 Hartree = 627.51 kcal/mol^{-1}.

[12]A Quantum Chemistry Literature Data Base is published in the *Journal of Molecular Structure* from 1970 to 1992 in volumes 91, 106, 119, 134, 148, 154, 182, 203, 211, 225, 252, 278, and 289. It contains 23,240 entries, increasing at the rate of more than 2500 per year. Unfortunately, unlike the spectroscopy database, it is not separated into topics.

Table 11.10.
Ab-initio Studies of $(H_2O)^2$ and $(HF)_2$ Showing Basis Set Dependency (From Dill et al., 1975).

	E_D (kcal/mol)	
Basis	$(H_2O)_2$	$(HF)_2$
STO-3G counterpoise	4.9	4.4
STO-3G optimized	6.0	5.5
STO-4G	6.1	5.2
Minimal Guassian	12.6	—
Bond orbital approximation	8.1	—
Minimal Slater	6.6	—
Minimal Slater	6.1	—
Minimal Slater	6.6	—
4-31G (at STO-3G geometry)	9.6	7.6
4-31G (partially optimized)	8.1	7.7
4-31G (fully optimized)	8.2	8.0
HFAO Gaussian	5.3	4.6
6-31G	7.8	7.5
Double ζ "split-out"	7.9	6.5
6-31G*	5.6	5.9
6-31G*	5.6	6.0
6-31G**	5.5	6.0
Extended Gaussian + polarization	4.7	4.6
Extended Gaussian + polarization	5.0	4.5
Extended Gaussian + polarization	4.8	3.5
Extended Gaussian + polarization	5.1	—
Hartree-Fock limit	3.7	—
SCF + CI	6.0	—
PNO-CI (correlation)	—	3.5
CEPA (correlation)	—	3.4
IEPA (correlation)	—	—
Experiment	5.1	5.7

effects are introduced by adding p-orbitals to the s-orbitals of hydrogen, d-orbitals to the s- and p-orbitals of first row elements and so forth. To facilitate the calculations, these atomic orbitals have to be expressed in an analytical form, most often by using Gaussian functions.[13] This is known as the *basis-set*. The more sophisticated the basis-set, the greater the computer power and time needed. The computing time is roughly proportional to the fourth power of the number of basis functions. With the development of the super-computing facilities, this is not a serious problem for most gas-phase adducts. The greater the basis-set the closer the result comes to the Hartree-Fock limit. However, sometimes simpler basis-sets give better agreement with experimental data, due to cancellation of errors, as illustrated in Table 11.10.

Apart from the best possible choice of basis-set, there are three problems to overcome:

1.   Starting with the correct model. The methods will seek the minimum energy potential closest to the starting model. This may not be that of lowest energy, espe-

---

[13]A description of the meaning of the acronyms used in the Gaussian programs is given in an appendix by Del Bene (1986).

cially in weakly bonded adducts. Generally, it is not possible to explore all configurational and conformational space This is known as the problem of *global searching*. It applies to all methods of structure determination but is most serious for weakly hydrogen bonded adducts where there might be a number of van der Waals complexes of comparable energy.

2. The LCAO assumption ignores the correlation between electrons of opposite spin. The most common method for correcting for electron correlation is the first, second, third, or fourth order Møller-Plesset approximation (MP1-4). There are other methods. A comparison of these methods for some charged and uncharged binary adducts is given by Del Bene and Shavitt (1989). This correction is essential, particularly for weak interactions.

3. For calculations involving two or more molecules, or molecules and ions, there is the basis-set superposition error (BSSE). It is a consequence of the necessarily finite nature of basis sets. It is described as one of the true challenges of quantum chemistry by Clementi, Corongiu, and Stradella (1991). It arises from the fact that the basis-set used for a donor molecule lowers the calculated energy for the acceptor molecule and vice versa, thereby leading to binding energies that are too large. The larger the basis-set, the less serious is the correction. A common way of correcting is by means of the *counter-poise* method of Boys and Bernardi (1970).[14]

A good example of the effects of basis-set size, correlations and BSSE errors in calculating the hydrogen bonding energies of the Watson-Crick and Hoogsteen base pairs is given by Gould and Kollman (1994). The results are compared with experimental values and molecular mechanics calculations. A good general account of the use of ab-initio methods for calculating molecular structures is given by Boggs (1992) and for ionic hydrogen-bonded complexes, in particular, by Deakyne (1987). The latter paper contains tables of calculated hydrogen bond energies for a number of molecule-ion adducts and for proton-molecule and anion-molecule clusters including $H^+(H_2O)_n$ and $OH^-(H_2O)_n$. An excellent analysis of the many recent ab-initio attacks on the water dimer is provided by Scheiner (1994).

Prior to 1970, ab-initio methods were limited mainly to small single molecules. Hydrogen-bonded adducts were studied by means of semi-empirical quantum mechanics. Joesten and Schaad (1974) contains a table of nearly 500 results from and references to theoretical calculations on hydrogen-bonded adducts, the majority of which are semi-empirical. A review of this period is by Murthy and Rao (1970).

The semi-empirical methods such as CNDO, INDO, MNDO, and AM1 are considered to be inappropriate for simulating moderate or weak hydrogen bonding due to an overestimation of the exchange repulsion at hydrogen bond distances, although some recent attempts have been made to improve them by Rodriguez (1994).

---

[14]The basis set superposition error by the counterpoise method of Boys and Bernardi (1970) is computed as follows: Compute E(A<u>H</u>····B) using basis set for AH; Compute E(AH····<u>B</u>) using basis set for B; Compute E′ (AH) using basis set for AH; and Compute E′ (B) using basis set for B. Then BSSE = E(A<u>H</u>····B) + E(AH····<u>B</u>) − E′(AH) − E′(B). For 6.31G** computations on the water and methanol linear dimers, the corrections by this method reduced the binding energies by 17 and 16% respectively (Lii and Allinger, 1994).

The semi-empirical program PM3 of Stewart (1989) is said to give better results, at least for the H_2O dimer. A new version of AM1, SAM1, by Dewar, Jie, and Yu (1993), is said to correct the deficiency. A comparison of AM1, PM3, SAM1, and ab-initio methods for O—H····O, O—H····N, O—H····S, and S—H····O bonded complexes is reported by Zheng and Merz (1992). A similar comparison for weakly bonded C—H····OH complexes is reported by Turi and Dannenburg (1993a). An evaluation of SAMI for modeling hydrogen bonds is by Holder and Evieth (1994).

At the other end of the spectrum from ab-initio calculations are the empirical force-field methods, which are based on classical mechanical concepts. These are much less computer intensive. There are two forms: *molecular mechanics* and *molecular dynamics.*

Molecular mechanics is used primarily to simulate the structures of larger molecules and macromolecules. It can also be extended to arrays of hydrogen-bonded molecules. Molecular dynamics, which uses the same force fields, permits the exploration of transitions between different conformations separated by torsional energy barriers. It can also be extended to clusters of molecules and to simulate the effect of solvation.

A good account of the past, present, and future of these methods, particularly molecular dynamics, is given by Van Gunsteren and Berendsen (1990).

All empirical force field methods have parameters which are adjusted to give the best possible results when compared with experimental observations of energies or molecular dimensions. Some programs are designed for general use, such as the continuously upgraded MMX series and the popular programs AMBER, CHARMM, and GROMOS. Most of these were parameterized initially for particular classes of molecules and were later modified for other classes or for more general use. A list of some of the more commonly used and more recently parameterized programs is given in Table 11.11.

In the empirical force field methods, atoms in molecules are conceived as being linked by springs which obey Hooke's Law, with force constants for stretching and compressing and bending and twisting from equilibrium bond lengths, valence angles, and torsion angles derived from experimental data. That is, for the purpose of describing these structures, they obey the laws of classical mechanics. The energies of pair-wise covalently bonded atoms is modelled with stretching, binding, and torsional force constatns. The nonbonding intermolecular interactions are modeled using a Lennard-Jones potential with repulsion and attraction components of the form

$$\sum_{ij} Ar_{ij}^n - Br_{ij}^{-6}, \text{ where } n = 9, 10 \text{ or } 12$$

where A and B are adjustable parameters. When the intermolecular interactions include hydrogen bonding, most programs add a nondirectional monopole-monopole term of the form

$$Zq_iq_j \cdot (4\pi\epsilon r_{ij})^{-1}$$

where q_i and q_j are the charges on the donor and acceptor groups and ϵ is a dielectric constant which may be dependent on the distance between charges, r_{ij}. The balancing exchange repulsion and any nonelectrostatic attractive terms are modelled using a Lennard-Jones type potential such as $C_{ij}r^{-12} - d_{ij}r^{-10}$. The exception is the MM2,3

Table 11.11.
Some Commonly Used Molecular Mechanics and Molecular Dynamics Programs.[a]

CFF, CFF(93)	(Consistent force field) for carboxylic acids and amides, peptides and proteins. Hagler, Huler and Lifson (1974); Lifson, Hagler, and Dauber (1979); Hwang, Stockfish, and Hagler (1994); Maple et al. (1994).
PEF422	An ongoing series of PEF programs for carbohydrates. Rasmussen (1982).
MM1, MM2, MM3	A molecular mechanics program originally parameterized for hydrocarbons but later extended for general use. MM3 has a molecular dynamics extension. Allinger (1976, 1977, 1992); Burkert and Allinger (1982); Allinger, Yuh, and Lii (1989); Allinger, Rahman, and Lii; Lii and Allinger (1994).
CHARMM	A program designed primarily for macromolecules using both molecular mechanics and dynamics. Brooks et al. (1983).
CHARMM-H	A modification of CHARMM for use with carbohydrates. Ha et al. (1988).
CHARMM-CHEAT	A modification of CHARMM for carbohydrates using extended atoms for hydroxyl group. Grootenhaus and Haasnoot (1993).
AMBER	A molecular mechanics program designed primarily for proteins and nucleic acids. Weiner and Kollman (1981); Weiner et al. (1984, 1986).
AMBER-OPLS	Above program modified for molecular dynamics solution simulations. Jorgensen (1981); Jorgensen and Tirado-Rives (1988).
AMBER-H	AMBER modified to include polysaccharides and glycoproteins. Homans (1990).
AMBER-united atom	Parameterized for carbohydrates. Senderowitz, Parish, and Still (1996).
ECEPP	A molecular mechanics program for small peptides and proteins. Dunfield, Burgess, and Scheraga (1978), Nemethy, Pottle, and Scheraga (1983)
GROMOS	Primarily a molecular dynamics program for small and large biological molecules. Van Gunsteren (1987).
TRIPOS	A molecular mechanics program for small organic and bio-organic molecules. Clark, Cramer, and van Opdenbosch (1989).
DREIDING	A molecular mechanics program for organic, biological and inorganic molecules. Mayo, Olafson, and Goddard (1990)

[a]Many programs are available at nominal charges for academic users from the Quantum Chemistry Exchange Program, Indiana University, Bloomington, IN, U.S.A. References to published force field parameters are provided by Osawa and Lipkowitz (1995).

program of Allinger which includes a directional hydrogen bond potential in the form of a dipole-dipole interaction but no monopole-monopole term (Lii and Allinger, 1994).[15]

Expressions for hydrogen bonding potentials have been developed by Hagler, Huler, and Lifson (1974), Taylor (1981), and Pertsin and Kitaigorodskij (1987). These incorporate a van der Waals component, a dipole-dipole electrostatic component, and a Morse component to accommodate the charge transfer contribution. However, these multiparameter functions are considered to unnecessarily complicate the programming of most molecular mechanics and dynamics programs currently in common use.

Other than the MM3 series, most programs include hydrogen atoms by appropriately extending the atoms to which they are covalently bonded. This simplifies the calculations, especially for $C-OH$, NH_2 and $\overset{+}{N}H_3$ groups where the location of the hydrogen atoms is flexible. The advantages and disadvantages of this extended atom representation are discussed by Brooks et al. (1983) and more recently by Grootenhaus and Haasnott (1993).

[15]When positive and negative charges are placed on A and H, this operates as a dipole. The same is true if different charges are placed on the lone-pairs and nuclei of B.

These methods are based on atom-pair potentials, but because they are parameterized on experimental data, the nonadditive or cooperative effects are at least partially included.[16] Computations that specifically take into account the cooperative effects in the water-water potential and water-ion clusters are reported by Caldwell, Dang, and Kollman (1990).

While gas-phase calculations are invaluable for the insight they provide into intermolecular interactions, most chemistry, biochemistry, and molecular biology takes place in an aqueous medium. In water or a polar solvent, the molecules are subject to distortions or conformational changes due to hydrogen bonding to or from adjacent solvent molecules or ions. Gas-phase calculations, by whatever method, are particularly academic for biological molecules such as carbohydrates, nucleotides, and peptides and related macromolecules which do not exist in the gas phase.

Conformational transitions and the effects of hydrogen bonding by solvent water molecules are most commonly explored using *molecular dynamics*. In a molecular dynamics simulation, the same molecular mechanics formulations are used. The lowest energy conformations are explored by simulating thermal motion in short time intervals, usually femtoseconds. Random directions and velocities are applied to each variable parameter and Newton's equations of motion are solved numerically. The trajectories of these motions are followed over several picoseconds displaying the dynamics of molecular motion over energy barriers. The number of water molecules bonded to a particular functional group on the molecule can also be displayed by means of a theoretical radial distribution curve.

In typical simulations such as those of Brady (1989) for α-D-glucose in water, Bagley et al. (1994) for sucrose in water, and DMSO or Kulinska and Laaksonen (1994) for cytidine and $2'$-deoxycytidene, the molecule is placed in a box containing 200–300 solvent molecules, heated to 300 K and then allowed to equilibrate. The kinetic energy of the system will overcome the torsional barriers associated with single bonds and thereby achieve a measure of global searching. The conformational distribution can be obtained from a Bolzman distribution from the energies. The number of water molecules associated with a particular functional group can be derived from a radial distribution curve relating to that group. The method requires a water force field of which there are several choices. Popular force fields are the TIP3P of Jorgensen (1981) and Jorgensen et al. (1983) and the TIP4P of Berendsen, Grigera and Straatsma (1987).

Carbohydrates provide a particularly good test for the ability of theory to model hydrogen bonding since they have the highest ratio of hydrogen bonding functional groups to monomer units, as shown in Table 11.12. They present a problem with isolated molecules, i.e., gas-phase calculations, because of the proximity of the donor and acceptor groups, the lowest energy conformation of the hydroxyl groups will always be that of maximum intramolecular hydrogen bonding. In solution or in the solid state, intermolecular hydrogen bonding to solvent or adjacent molecules takes precedent. Therefore, restrictions on the hydroxyl torsion angles have to be applied to avoid complete intramolecular bonding as described by Jeffrey and Taylor (1980).[17]

---

[16]References to a wide variety of special functions to represent hydrogen bonding potentials are in a paper by Damewood et al. (1990).

[17]A program specially parameterized to carbohydrates by Glennon et al. (1994) adds an additional monopole-monopole term for the 1-4 interactions which are common in these structures.

Table 11.12.
Ratio of H-bonding Functional Groups Per Monomer Unit
in Biological Polymers.

	Donors	Acceptors
Polysaccharides	4	6
Polynucleotides	2	6
(excluding base-pairing)	(1 in DNA)	(5 in DNA)
Proteins (main chain)	1.0	1.0
Proteins (side chain)	0.7	0.5
(average)		

The flexible inter-residue bonds in oligosaccharides provide a good test of empirical force field methods and have been used by French and coworkers to study the conformation of a number of disaccharides, as described by French and Miller (1994). A comparison between ab-initio and molecular mechanics calculations on a trisaccharide, raffinose, is reported by Van Alsenoy et al. (1994).

The use of ab-initio methods to predict solvation and crystal field effects is still in its infancy but is expected to develop rapidly. An approach is to add to the molecular Hamiltonian an external coulombic field as described by Pisani, Dovesi, and Roetti (1988). A program, CRYSTAL, described by Dovesi, Saunders, and Roetti (1992) has been applied to calculate the vibration frequencies of the water molecules in $LiC\ell O_4 \cdot 3H_2O$, $LiHCOO \cdot H_2O$, and $LiOH \cdot H_2O$ giving good agreement with experiments by Ojamäe and Hermansson (1992). This program was also used by Dovesi et al. (1990) to predict the crystal structure of urea, by Ojamäe et al. (1994) to predict the structure and vibration frequencies of $LiOH \cdot H_2O$, a crystal structure with a short H----O hydrogen bond, and by Starikov (1995) to investigate the intermolecular interactions of a number of deprotonated mononucleotides. Another approach is to carry out ab-initio calculations on solute molecules in a solvent cavity modelled by a polarizable dielectric medium as discussed by Keith and Frisch (1994).

An extension of the semi-empirical method PM3 to include solvation effects is described by Cramer and Truhlar (1992). Since empirical force field methods are less computing intensive, molecular dynamics have been used since the 1970s to study the structure of water by Stillinger and Rahman (1974), nucleotides by Mezei et al. (1983), cyclodextrins by Koehler, Saenger, and Van Gunsteren (1987), and some methyl glycopyranosides by Cheetham and Lam (1996). An analysis of the problems involved in dealing with solvation particularly is given by Wilcox et al. (1987).

More recently, van Eijck, Mooij, and Kroon (1995) used the GROMOS program of Van Gunsteren (1987) in the molecular mechanics mode to test its predictive powers for the crystal structures of six hexapyranoses which had the same space group symmetry, $P2_1 2_1 2_1$, with one molecule in the unit cell. By allowing rotation about the primary alcohol C—C bonds and all the C—C—O—H bonds, the search generated about 1000 structures within a limit of 10 kcal/mol^{-1} above the minimum. As shown in Table 11.13, the observed crystal structures came first or second in lowest energy only using an all-atom force field with a cut-off at 10 Å, except for β-D-glucose and β-D-galactose, which were consistently poor. This calculation would suggest that polymorphism was common in the hexapyranoses, which generations of carbohydrate

Table 11.13.
Tests of Crystal Structure Predictions of Monosaccharides Using GROMOS.[a]

	U7		U10		A10		A30	
	R	E	R	E	R	E	R	E
α-D-galactose	32	3.7	8	2.5	2	0.2	1	
α-D-glucose	21	2.3	8	1.5	2	0.2	3	1.1
α-D-talose	58	3.6	20	2.4	1		1	
β-D-allose	3	0.5	6	0.7	2	0.1	2	0.9
β-D-galactose	29	4.8	105	4.9	231	6.0	88	4.9
β-D-glucose	356	5.6	213	5.7	374	5.9	384	5.1

[a]U is the standard GROMOS united-atom force field. A is an all atom force field. Numbers following give cut-off radii in Å. R is ranking of observed crystal structure in E, energy in kcal/mol^{-1} (from van Eijck, Mooij, and Kroon (1995).

chemists have failed to discover. These calculations appear to show that something more sophisticated than the extended atom approach with monopole changes is needed to deal with highly hydrogen-bonded assemblies of molecules in crystals.

Using a different approach, French, Miller, and Aabloo (1993) used the CRYS-TAL program in MM3(92) by Allinger, Yuh, and Lii (1989) to construct miniature crystals to study the polymorphs of cellulose. Since the hydrogen bonding in these structures is unknown, it is not possible to critically evaluate their results in the same way as for the monosaccharides. The increasing power of NMR spectroscopy to define structure in solution and the extension of ab-initio and empirical force field methods to crystal structures is likely to provide more stringent tests of the structure predictive powers of these methods in the near future. A recent molecular dynamics study of maltose in water by Ott and Meyer (1996) using a modified GROMOS program gave reasonable agreement with the time-averaged NMR data.

Undoubtedly, the emphasis in all methods will be to better predict the structure and stability of biological macromolecules and to simulate the interactions between drug molecules and enzymes. A very lucid overview of the problems involved is presented by Kollman (1989).

11.9 | THERMOCHEMICAL MEASUREMENTS

The formation of hydrogen bonds liberates heat hence the calorimetric measurement of heats of mixing or dilution of donor-acceptor liquid mixtures can provide access to hydrogen bond energies. Increase in temperature breaks hydrogen bonds, absorbing heat and increasing the kinetic energy of the molecules involved. It is therefore possible to study the thermodynamics of hydrogen bond formation either directly by calorimetry or by measuring the equilibrium constants of any property that changes linearly with temperature or with concentration in an inert solvent.

In the system

$$A\text{—}H + B\big\langle \rightleftharpoons A\text{—}H\cdots B\big\langle$$

the equilibrium constant

$$K = \frac{conc[A\text{—}H\cdots B]}{conc[A - H] = conc[B]}.$$

Then the change in free energy

$$\Delta G^\circ = RT \log_n K$$

and the enthalpy of hydrogen bond formation

$$\Delta H^\circ = -RT \log_n + \Delta S^\circ/R \qquad -RT \log_n + \Delta S^\circ/T$$

and $\log_n K$ is a linear function of T^{-1}, if ΔH° and ΔS° are independent of temperature.

Joesten and Schaad (1974) report over 1500 equilibrium constants for acid-base combinations with the $\Delta \nu_{A\text{—}H}$ frequency shifts and in many cases the values of the enthalpy and entropy. The $-\Delta H$ values generally lie in the range of 2–10 kcal/mol^{-1}. When the same system is measured by different investigators, the values may differ by as much as 100%. It was this type of result that no doubt prompted the statement by Pimental and McClellan (1960) that additional accurate and systematic studies are needed.

Considerable effort was, in fact, directed in the 1960s to the measurement of hydrogen bond enthalpies by direct calorimetry and from infrared shifts and intensity changes with temperature or concentration. Generally, a standard acid such as phenol in carbon tetrachloride was combined with a variety of bases. Excellent accounts of these methods and their results are given in Vinogradov and Linnell (1971) and Joesten and Schaad (1974). Some typical results are shown in Table 11.14.

A notable series of calorimetry studies of hydrogen-bonded complex formation was made by Arnett et al. (1967, 1970). Attempts were made to correlate the results with those from infrared spectroscopy. The later paper provides an extensive table of frequency shifts, $\Delta \nu_s$, equilibrium constants, and enthalpy values from spectroscopic and calorimetric measurements of the hydrogen-bonded adducts between p-fluorophenol, phenol, and methanol with a variety of bases in an inert solvent. Good correlations were obtained but in one sense the results were disappointing. To quote, "although the trend is clear, no good general correlation is found" (p. 2365). As pointed out with hydrogen bond enthalpies falling between 3 and 20 kcal/mol^{-1}, discrepancies amounting to 2 or 3 kcal/mol^{-1} from different measurements of the same system reduce the utility of the measurements for obtaining structural correlationship. The methods used as a source of absolute heats of formation of hydrogen bonds were, in fact, criticized by Duer and Bertrand (1970). Nevertheless, empirical correlations for particular acid-base adducts were proposed by Drago and co-workers, e.g., Drago and Wayland (1965) and Drago, Vobel, and Needham (1971).

For certain hydrates, thermodynamic data can be obtained from equilibrium vapor pressure measurements, as described, for example, by Harmon, Toccalino, and Janos (1989).

The original incentive for measuring hydrogen bond energy was probably the concept that for a particular donor-acceptor combination, hydrogen bond energies are, at least approximately, additive and transferable, as are covalent bond energies. For reasons discussed in Chapter 6, this is very far from true; hydrogen bond energies are nonadditive and nontransferable. The $O\text{—}H\cdots O$ bond energy in $(H_2O)_2$, for example,

Table 11.14.
Some typical values of —ΔH° From Infrared Spectroscopy Measurements (From
Vinogradov and Linnell, 1971).

Hydrogen bond	System	Phase or solvent	—ΔH°
O—H⋯O=C	formic acid	gas	7.4 ± 0.5
	acetic acid	gas	7.3 ± 0.5
O—H⋯O—H	methanol	gas	7.6 ± 1.4
O—H⋯O<	phenol dioxane	$CC\ell_4$	5.0 ± 0.2
O—H⋯S<	phenol n-butyl sulfide	$CC\ell_4$	4.2 ± 0.3
O—H⋯Se<	phenol n-butyl selenide	$CC\ell_4$	3.7 ± 0.3
O—H⋯N⟨	phenol pyridine	$CC\ell_4$	6.5 ± 0.4
O—H⋯N←	phenol triethylamine	$CC\ell_4$	8.4 ± 0.5
N—H⋯S<	thiocyanic acid n-butyl sulfide	$CC\ell_4$	3.6 ± 0.2
N—H⋯Se<	thiocyanic acid n-butyl selenide	$CC\ell_4$	3.7 ± 0.2
N—H⋯π	aniline-benzene	C_6H_{12}	1.6

is significantly less than that in $(H_2O)_3$, which depends on the configuration of the bonding in the trimer. O—H⋯O_W hydrogen bonds are stronger than O_W—H⋯O bonds. The bond energy of N—H⋯O=C will be different in an α-helix to that in a β-sheet because of the RHHB effect discussed in Chapter 6.

Only in special groups of hydrogen-bonded systems, such as the gas-phase molecule-ion or molecule-molecule adducts is it possible to obtain a meaningful sequence of bond energies, either experimentally or by theoretical computations.

The nonadditivity and nontransferability of hydrogen bond energies, together with the experimental difficulties of the thermochemistry, probably account for the lack of enthusiasm in the field in recent years. The only sources of reliable bond energies that have been developed are those for molecule-ion binary gas-phase adducts by ion-cyclotron resonance as described by Baldeshwieler and Woodgate (1971), McMahon and Beauchamp (1972), Cunningham, Payzant, and Kebarle (1972) for $H^+(H_2O)_n$, Lehman and Bursey (1976), Larson and McMahon (1983, 1984, 1987) for hydrogen bonding to F^-, $C\ell^-$ and CN^- ions, and by mass spectrometry by Meot Ner (1987, 1988). Some values from cyclotron-ion resonance are given in Table 3.2.

The thermochemistry of $\overset{+}{N}$—H⋯O, $\overset{+}{N}$—H⋯N and $\overset{+}{O}$—H⋯O in a variety of gas phase adducts is reported by Meot Ner (1984a, 1988) using pulsed high pressure mass spectrometry to obtain the standard enthalpies of the reaction

$$HA + B^+ \rightarrow AH^+ + B.$$

Hydrogen Bond Energies from Proton Affinities

Proton affinity is defined as the decrease of molar enthalpy accompanying protonation in the gas phase, i.e.,

$$B + H^+ \rightarrow BH^+.$$

Table 11.15.
Morokuma Decomposition of Computer Proton Affinity Energies
For Water, Methanol, Ammonia, and Methylamine (From
Umeyana and Morokuma, 1976).

	H_2O	CH_2OH	NH_3	CH_2NH_2
ΔE	-182	-194	-222	-230
ES	-79	-76	-100	-97
PL	-29	-44	-27	-40
CT	-71	-75	-88	-92
MIX	-3	—	-7	-2
Exper.	-164	-180	-253	-233

As discussed by Umeyana and Morokuma (1976) there is no exchange repulsion component to the energy, since the proton has no electrons. For H_2O, the three major components are electrostatic (43%), charge transfer (39%), and polarization (16%), as shown in Table 11.15.

The energy of ionic hydrogen bonds in the gas phase is related to the difference in proton affinities for the donor and acceptor; one of which will be a molecule, the other an ion. A small proton affinity for a molecule has a correspondingly large value for the corresponding ion, e.g., HF = 112 kcal/mol^{-1}, F$^-$ = 372 kcal/mol^{-1}.

Correlations between proton affinity differences and hydrogen bond energies have been reported for several different classes of ion-molecule adducts by Meot-Ner (1984a, 1984b, 1987), Meot-Ner, and Sieck (1984, 1986), and by Davidson, Sunner, and Kebarle (1979). There is a large amount of experimental data on proton affinities from the *Journal of Physical Chemistry Reference Data Supplement* 17 (1988).

Proton affinities can be computed by ab-initio methods as described, by De Frees and McLean (1986) and Del Bene (1993). As shown in Table 11.16, the agreement is good because the numbers are larger. A discussion of the "plausibility" of numerical values of proton affinities is given by Liebman (1987).

Table 11.16.
Computed Proton Affinities (kcal/mol^{-1}). B + H$^+$ → BH$^+$, ΔH_{298}.

	DeFrees and McLean (1986)	Del Bene (1993) using different basis sets	Experimental[a]
NH_3	204	210–220	204
H_2O	165	169–181	167
HF	117	118–128	117
NH_2^-	404	407–451	404
OH^-	391	391–442	391
F^-	373	370–420	371

[a]From J. Phys. Chem. Ref. Data Supplement, 17 (1988).

Structural Databases

A gas-phase structural database, MOGADOC (Molecular Gas Phase Documentation), has 22,000 bibliographic entries, 6000 compound entries, and 3200 structural data entries. It dates back to 1930 for electron diffraction and to 1945 for microwave spectroscopy. It is available from Sektion für Spektren- und Strukturdokumentation, University of Ulm, D-89069 Ulm, Germany.

There are five computer accessible databases for crystal structure analyses:

metals and alloys (MCDF)
inorganic and minerals (ICSD)
organics and metalloorganics (CSD)
proteins (PDB)
nucleic acids (NADB)

MCDF contains ~12,000 entries of which about half contain atomic coordinates. It is available from the Canadian Institute of Scientific and Technical Information, National Research Council of Canada, Ottawa, Canada K1A 0S2.

ICSD contains ~35,000 entries with software for search, retrieval, and display. It is available from Fachinformationszentrum Energie, Physik, Mathematik, D-76344 Eggenstein, Leopoldshafen 2, Germany.

CSD contains ~100,000 entries and software for search, retrieval, display and analysis. Hydrogen atomic coordinates from X-ray analyses are not generally reported. There are about 1000 neutron diffraction analyses. It is available through 31 National Affiliated Centers to academic users.

PDB contains ~4600 entries with atomic coordinates. It is available from the Biology Department, Brookhaven National Laboratory, Upton, NY 11973–5000.

NADB contains ~320 entries with atomic coordinates. It is available from the Chemistry Department, Rutgers University, PO Box 939, Piscataway, NJ 08855-0939.

Effect of Thermal Motion on Observed Bond Lengths

Thermal motion affects the experimentally observed bond lengths. This is most serious in gas-diffraction studies where the oscillation motion is very large. For this reason, there are a number of definitions of internuclear bond lengths that differ with different ways of averaging over the effects of thermal motion. These are:

r_e distance between equilibrium nuclear positions at rest. This is the minimum in the potential energy surface from ab-initio molecular orbital calculations.

r_a distance between average nuclear positions at thermal equilibrium.

r_z distance between average nuclear positions in the ground vibrational state.

r_g thermal average of internuclear distances.

In neutron crystal structure analysis, only the correction from r_a to r_e is generally applied for comparison with theoretical calculations. In some empirical force field calculations where equilibrium bond lengths are taken from gas-phase measurements, these distinctions are important.

For a more detailed discussion see Cyvin (1968) and Kuchitsu (1992).

Distance Dependence of Energy Contributions

Electrostatic

$$\text{ion-ion} \quad q_i \, q_j \, r_{ij}^{-1}$$
$$\text{ion-dipole} \quad q_i \, \mu_j \, \cos\theta_{ij} \, r_{ij}^{-1}$$
$$\text{dipole-dipole} \quad \mu_i \, \mu_j \, f(\theta_{ij} \cdot \theta_{ji}) \, r^{-3}$$

Exchange repulsion

$$A \exp \alpha \, r_{ij} \text{ or } A \, r_{ij}^{-12}$$

Polarization (of i by j)

$$-\mu_i^2 \, \alpha_j \cdot r_{ij}^{-4}$$

Charge-transfer

$$C \exp(-Dr_{ij})$$

Dispersion

$$-B \, r_{ij}^{-6}$$

APPENDIX **IV**

Some Useful Conversions

Energy

$$1 \text{ kcal/mol}^- = 4.184 \text{ kj/mol}^{-1} = 349.755 \text{ cm}^{-1} = 1.048539 \times 10^7 \text{ MHz}$$
$$= 4.336411 \times 10^{-2} \text{ eV}$$
$$1 \text{ Hartree} = 627.75 \text{ kcal/mol}^{-1}$$

Length

$$1 \text{ Å} = 10^{-10} \text{ m}$$
$$1 \text{ } \mu = 10^{-6} \text{ m}$$
$$1 \text{ fermi} = 10^{-15} \text{ m}$$
$$1 \text{ Bohr} = 0.529177 \text{ Å}$$

Wave Number

$$1 \text{ cm}^{-1} = (2.998 \times 10^{10})^{-1} \text{ sec}^{-1}$$
$$= 2.997925 \times 10^4 \text{ MHz}$$
$$= 1.239842 \times 10^{-4} \text{ eV}$$
$$= 2.85914 \times 10^{-3} \text{ kcal/mol}^{-1}$$

For further information see *Quantities, Units and Symbols in Physical Chemistry*, 2nd Edition, I. Mills, T. Cvitas, K. Homann, N. Kallay, and K. Kuchitsu, eds. Oxford, UK: Blackwell Sci. Publ. Ltd. 1988.

References

Abrahams, S. C. Systematic prediction of new ferroelectrics on the basis of structure. *Ferroelectrics* 104: 37–50, 1990.

Abrahams, S. C. Structure relationship to dielectric, elastic and chiral properties. *Acta Cryst* A50: 658–85, 1994.

Abrahams, S. C. and E. T. Keve. Structural basis of ferroelectricity and ferroelasticity. *Ferroelectrics* 2: 129–54, 1971.

Adams, B. and L. Lerner. Observation of hydroxyl protons of sucrose in aqueous solution: No evidence for persistent intramolecular hydrogen bonds. *J Am Chem Soc* 114: 4827–29, 1992.

Akin, A. C. and K. M. Harmon. Hydrogen bonding Part 54. NMR study of the effects of anesthetics on hydration of choline, acetylcholine and tetraethylammonium halides in aqueous solution. *J Mol Struct* 319: 47–53, 1994.

Albrecht, G. and R. B. Corey. The crystal structure analysis of glycine. *J Am Chem Soc* 61: 1087–103, 1939.

Alder, R. W., P. S. Bowman, W. R. S. Steele, and D. R.Winterman. The remarkable basicity of 1,8-bis(dimethylamine) napthalene. *Chem Commun* 723–24, 1968.

Allen, F. H., O. Kennard, and D. G. Watson. *Crystallographic Data Bases: Search and Retrieval of Information from the Cambridge Structural Data Base*. In: *Structure Correlation*, vol. 1. Edited by H.-B. Burgi and J. D. Dunitz. Weinheim, Germany: VCH Publishers, 1994.

Allen, L. C. Simple model of hydrogen bonding. *J Am Chem Soc* 97: 6921–40, 1975.

Allerhand, A. and P. v. R. Schleyer. A survey of C—H groups as proton donors in hydrogen bonding. *J Am Chem Soc* 85: 1715–23, 1963a.

Allerhand, A. and P. v. R. Schleyer. Halide anions as proton acceptors in hydrogen bonding. *J Am Chem Soc* 85: 1233–37, 1963b.

Allinger, N. L. Calculation of molecular structure and energy for force field methods. *Adv Phys Org Chem* 13: 2–75, 1976.

Allinger, N. L. Conformational analysis 130. MM2. A hydrocarbon force field utilizing V1 and V2 torsional terms. *J Am Chem Soc* 99: 8217–34, 1977.

Allinger, N. L. "Molecular Mechanics," in *Accurate Molecular Structures*. Edited by A. Domenicano and I. Hargittai. UK: Oxford University Press, 1992.

Allinger, N. L., M. Rahman, and J. H. Lii. A molecular mechanics force field (MM3) for alcohols and ethers. *J Am Chem Soc* 112: 8293–8307, 1990.

Allinger, N. L., Y. H. Yuh, and J.-H. Lii. Molecular mechanics. The MM3 force field for hydrocarbons 1. *J Am Chem Soc* 111: 8551–66, 1989.

Almenninger, A., O. Bastiansen, and T. Motzfeld. Reinvestigation of the structure of the monomer and dimer formic acid by gas diffraction techniques. *Acta Chem Scand* 23: 2848–64, 1969.

Altman, L. J., D. Laugani, G. Gunnarsson, H. Wennerström, and F. Forsén. Proton, deuterium and tritium nuclear magnetic resonance of intramolecular hydrogen bonds. Isotope effects and the shape of the potential energy function. *J Am Chem Soc* 100: 8264–66, 1978.

Angell, C. A. Approaching the limits. *Nature* 331: 206–7, 1988.

Arnett, E. M., I. Joris, E. Mitchell, T. S. S. R. Murty, T. Gorrie, and P. v. R. Schleyer. Studies of hydrogen-bonded complex formation III. Thermodynamics of complexing by infrared spectroscopy and calorimetry. *J Am Chem Soc* 92: 2365–77, 1970.

Arnett, E. M. and E. J. Mitchell. Hydrogen bonding VI. A dramatic difference between proton transfer and hydrogen bonding. *J Am Chem Soc* 93: 4052–53, 1971.

Arnett, E. M., T. S. S. R. Murty, P. v. R. Schleyer, and I. Joris. Hydrogen bonding I. Two approaches to accurate heats of formation. *J Am Chem Soc* 89: 5955–57, 1967.

Asakawa, N., S. Kuroki, H. Kurosu, I. Ando, A. Shoji, and T. Ozaki. Hydrogen-bonding effect on ^{13}C NMR chemical shifts of L-alanine residue carbonyl carbons of peptides in the solid state. *J Am Chem Soc* 114: 3261–65, 1992.

Ashton, P. R., N. S. Isaccs, F. H. Kohnke, A. M. Z. Slawin, C. M. Spencer, J. Fraser Stoddart, and D. J. Williams. Towards the making of a [12] collarene. *Angew Chem Int Ed Eng* 27: 966–68, 1988.

Astbury, W. T. and H. Street. X-ray studies of the structure of hair, wool, and related fibres, I. General. *Phil Trans Roy Soc* 230: 75–101, 1931.

Astbury, W. T. and H. J. Woods. X-ray studies of the structure of hair, wool, and related fibres, II. Molecular structure and elastic properties of hair keratu. *Phil Trans Roy Soc* 232: 333–94, 1933.

Atwood, H., F. Hamada, D. K. Robinson, G. W. Orr, and R. L. Vincent. X-ray diffraction evidence for aromatic π hydrogen bonding to water. *Nature* 349: 603–84, 1991.

Atwood, J. L., S. Bott, C. Jones, and C. L. Raston. Aluminum-fused bis-p-tert-butylcalix[4]arene. A double core with two π-arene----H interactions for included methylene chloride. *J Chem Soc Chem Commun* 1349–51, 1992.

Atwood, J. L., J. E. D. Davies, and D. D. MacNicol. *Inclusion Compounds*, vols. 1 and 2. London: Academic Press, 1984.

Bacon, G. E. and N. A. Curry. A study of α-resorcinol by neutron diffraction. *Proc Roy Soc* A 235: 552–59, 1962.

Bacon, G. E. and R. S. Pease. A neutron diffraction study of potassium dihydrogen phosphate by Fourier synthesis. *Proc Roy Soc* A220: 397–421, 1953.

Bacon, G. E. and R. S. Pease. A neutron diffraction study of the ferroelectric phase transitions of potassium hydrogen phosphate. *Proc Roy Soc* A230: 359–81, 1955.

Bader, R. F. W. *Atoms in Molecules. A Quantum Theory*. Oxford: Clarendon Press, 1992, p. 304.

Bader, R. F. W. and K. E. Laidig. "Determination of Atomic and Structural Properties from Experimental Charge Distributions," In: *The Application of Charge Density Research to Chemistry and Drug Design*. Edited by G. A. Jeffrey and J. F. Piniella. pp. 23–62. New York: Plenum Press, 1991.

Badger, R. M. The relationship between the energy of the hydrogen bond and the frequencies of the O—H bonds. *J Chem Phys* 8: 288–89, 1940.

Badger, R. M. and S. H. Bauer. Spectroscopy studies of the hydrogen bond, II. The shift in O—H vibrational frequency in formation of a hydrogen bond. *J Chem Phys* 5: 839–51, 1937.

Bagley, S., M. Odelius, A. Laaksoner, and G. Widmalm. Molecular dynamics simulation of sucrose in aqueous and dimethyl oxide solution. *Acta Chem Scand* 48: 792–99, 1994.

Bailinger, W. F., P. v. R. Schleyer, T. S. S. R. Murty, and L. Robinson. Nitro groups as proton acceptors in hydrogen bonding. *Tetrahedron* 20: 1635–47, 1964.

Baker, E. L. and R. E. Hubbard. Hydrogen bonding in globular proteins. *Prog Biophys Molec Biol* 44: 97–179, 1984.

Baker, T., E. Dodson, G. Dodson, D. Hodgkin, and R. Hubbard. "The Water Structure in ZZn Insulin Crystals," In *Crystallography in Molecular Biology*. Edited by D. Moras, J. Drenth, B. Strandberg, D. Sueh, and K. Wilson. New York: Plenum Press, 1985, pp. 179–92.

Baldeshwieler, J. D. and S. S. Woodgate. Ion cyclotron resonance spectroscopy. *Accts Chem Res* 4: 114–20, 1971.

Baldwin, R. L. and S. Magusee. Helix stabilization by $Glu^-\cdots Lys^+$ salt bridges in short peptides of de novo design. *Proc Nat Acad Sci USA* 84: 8898–8902, 1987.

Baldy, A., J. Elguero, R. Faure, M. Pierrot, and E. J. Vincent. Dynamic intermolecular tautomerism of 3,5-dimethylpyrazole in the solid state by 13C C:/MAS NMR spectroscopy and X-ray crystallography. *J Am Chem Soc* 107: 5290–91, 1985.

Bamford, C. H. In general discussion of preceeding paper on dyeing of polypeptides. *Discussion Farad Soc* 16: 229, 1954.

Baran, J., H. Ratajezak, E. T. G. Lutz, N. Verhaegh, H. J. Luinge, and J. H. Van der Maas. Polarized FT-IR and Raman spectra of β-D-fructopyranose single crystals. *J Molec Struct* 326: 109–22, 1994.

Barnes, W. H. The crystal structure of Ice between 0°C and −183°C. *Proc Roy Soc (London)* 125A: 670–93, 1929.

Bartmann K. and D. Mootz. Über Bildung und Struktur Kristalliner Addukte aus einer Säure, CH_3COOH, und einer Süpersäure, CF_3SO_3H sowie HF. *Zeitschrift für anorganische und allgemeine chemie*, 601, 31–40, 1991.

Baur, W. H. and D. Kassner. The perils of Cc: Comparing the frequencies of falsely assigned space groups with their general population. *Acta Cryst.* B48: 356–69, 1992.

Bednowitz, A. L. and B. Post. Direct determination of the crystal structure of β-fumaric acid. *Acta Cryst* 21: 566–70, 1966.

Bellamy, L. J. and A. J. Owens. A simple relationship between infrared stretching frequencies and hydrogen bond distances in crystals. *Spectrochim Acta* 25A: 329–33, 1969.

Bellamy, L. J. and R. J. Pace. The significance of infrared frequency shifts in relation to hydrogen bond strengths. *Spectrochim Acta* 25A: 319–28, 1969.

Bellissent-Funel, M.-C. "Recent Structural Studies of Liquid D_2O by Neutron Diffraction," In: *Hydrogen-Bonded Liquids*. Edited by J. C. Dore and J. Teixera, NATO-ASI Series C, 329, 117–28. Netherlands: Kluwer Academic Publishers, 1991.

Benson, S. W. and E. D. Siebert. A simple two-structure model for liquid water. *J Am Chem Soc* 114: 4269–76, 1992.

Berendsen, H. J. C., J. R. Grigera, and T. P. Straatsma. The missing term in effective pair potentials. *J Phys Chem* 91: 6269–71, 1987.

Berglund, B., J. Lindgren, and J. Tegenfeldt. On the correlation between deuterium quadrupole coupling constants, O—H and O—D stretching frequencies and hydrogen bond distances. *J Molec Struct* 43: 179–91, 1978.

Berglund, G. and R. W. Vaughan. Correlations between proton chemical shift tensors, deuterium quadrupole couplings and bond distances for hydrogen bonds in solids. *J Chem Phys* 73: 2037–43, 1980.

Berkovitch-Yellin, Z. and L. Leiserowitz. Role played by C—H···O and C—H···N interactions in determining molecular packing and conformation. *Acta Cryst* B40: 159–65, 1984.

Berman, H. M. Hydration of DNA. *Current Opinion in Structural Biology* 1: 423–27, 1991.

Berman, H. M. Hydration of DNA: Take 2. *Current Opinion in Structural Biology* 4: 345–50, 1994.

Bernal, J. D. The structure of liquids. *Proc Roy Soc (Lond)* A280: 299–322, 1964.

Bernal, J. D. (1952). Referred to in M. Falk and O. Knop. "Water in Stoichiometric Hydrates." Chapt., p. 62 In: *Water. A Comprehensive Treatise*, vol. 2. Edited by F. Franks. New York, London: Plenum Press, 1973.

Bernal, J. D. and R. H. Fowler. Theory of water and ionic solution with particular reference to hydrogen and hydroxyl ions. *J Chem Phys* 1: 515–48, 1933.

Bernal, J. D. and H. D. Megaw. Function of hydrogen in intermolecular forces. *Proc Roy Soc (London)* 151A: 384–410, 1935.

Bernstein, J. Polymorphism of L-glutamic acid. Decoding the α-β phase relationships using graph-set analysis. *Acta Cryst* B47: 1004–10, 1991.

Bernstein, J., R. E. Davis, L. Shimoni, and N.-L. Chang. Patterns in hydrogen bonding: functionality and graph set analysis in crystals. *Angew Chem Int Ed Engl* 34:1555–73, 1995.

Bernstein, J., M. C. Etter, and L. Leiserowitz. "The Role of Hydrogen Bonding in Molecular Assemblies." in: *Structure Correlation*, vol. 2. Edited by H.-B. Bürgi and J. D. Dunitz. Weinhein: VCH Publishers, 1994.

Bernstein, J., M. C. Etter, and J. M. MacDonald. Decoding hydrogen bond patterns. The case of imidodiacetic acid. *J Chem Soc Perkin* II, 695–704, 1990.

Bertolasi, V., P. Gilli, V. Ferretti, and G. Gilli. Evidence for resonance-assisted hydrogen bonding. Intercorrelation between crystal structure and spectroscopic parameters in eight intramolecular hydrogen bonds. *J Am Chem Soc* 113: 4917–25, 1991.

Bertolasi, V., P. Gilli, V. Ferretti, and G. Gilli. Intermolecular N—H···O hydrogen bonds assisted by resonance heteroconjugated systems as hydrogen-bond strengthening functional groups. *Acta Cryst* B51: 1004–15, 1995.

Bethell, D. E. and N. Sheppard. The infrared spectra of H_3O^+ in acid hydrates. *J Chem Phys* 21: 142, 1953.

Betzel, C. H., W. Saenger, B. E. Hingerty, and G. M. Brown. Circular and flip-flop hydrogen bonding in β-cyclodextrin undecahydrate: a neutron diffraction study. *J Am Chem Soc* 106: 7545–57, 1984.

Beuhler, R. J., and L. Friedman. A study of the formation of high molecular weight water cluster ions (m/c < 59,000) in expansion of ionized gases. *J Chem Phys* 77: 2549–57, 1982.

Beurskens, G., G. A. Jeffrey, and R. K. McMullan. Polyhedral clathrate hydrates. VI. Lattice type and ion distribution in some new peralkyl ammonium, phosphonium and sulfonium salt hydrates. *J Chem Phys* 39: 3311–15, 1963.

Bezyszyn, M. M., J. Baran, H. Ratajczak, and A. J. Barnes. Polarized infrared study of potassium hydrogen maleate and potassium deuterium maleate single crystals. *J Molec Struct* 270: 499–515, 1992.

Bingham, C., X. Li, G. Zon, and M. Sundaralingam. Crystal and molecular structure of d(GT-GCGCAC). Investigation of the effects of base sequence on the conformation of octamer duplexes. *Biochem* 31: 12803–12, 1992.

Bingham, C. A., G. Zon, and M. Sundaralingam. Crystal and molecular structure of the A-DNA dodecamer d(CCGTACGTACGG). *J Mol Biol* 277: 738–56, 1992.

Biradha, K., R. E. Edwards, G. J. Foulds, W. T. Robinson, and G. Desiraju. (4-Dimethylamino-pyridine)$_5$(benzoic acid)$_3$(H_2O)$_{10}$, a two-dimensional clathrate hydrate. *J Chem Soc Chem Commun* 1705–7, 1995.

Bjerrum, N. Structure and properties of Ice. *Science* 115: 385–90, 1952.

Bjorkstam, J. L. Deuteron nuclear magnetic resonance study of the ferroelectric phase transition in deuterated triglycine sulfate and KD_2PO_2. *Phys Rev* 153: 599–605, 1967.

Blackwell, J., K. H. Gardner, F. J. Kolpak, R. Minke, and W. B. Claffey. Refinement of Cellulose and Chitin Structures. In: *"Fibre Diffraction Methods."* Edited by A. D. French and K. H. Gardner. Washington, DC: ACS Symposium Series 141, American Chemical Society, 1980.

Blessing, R. H. Data reduction and error analysis for accurate single crystal diffraction intensities. *Crytallogr Rev* 1: 3–58, 1987.

Blessing, R. H. On the differences between X-ray and neutron thermal vibration parameters. *Acta Cryst* B51: 816–22, 1995.

Blinc, R. *Hydrogen-bonded Ferroelectrics and Lattice Dimensionality.* In: *Molecular Interactions*, vol. 2, Edited by H. Ratajczak and W. J. Orville-Thomas. New York: John Wiley & Sons, 1980.

Blinc, R. and D. Hadzi. Deuteron coupling and hydrogen bonding in crystals. *Nature (London)* 212: 1307–9, 1966.

Blundell, T. L., D. Barlow, N. Borkakoti, and J. Thornton. Solvent-induced distortions and the curvature of α-helices. *Nature* 306: 281–83, 1983.

Boenigk, D. and D. Mootz. The system pyridine-hydrogen fluoride at low temperatures: formation and crystal structures of solid complexes with very strong NHF and FHF hydrogen bonding. *J Am Chem Soc* 110: 2135–39, 1988.

Boggs, J. E. "Quantum Mechanical Determination of Static and Dynamic Structure." In: *Accurate Molecular Structures.* Edited by A. Domenicano and I. Hargittai. UK: Oxford University Press, 1992.

Bouquiere, J. P., J. L. Finney, M. S. Lehman, P. F. Lindley, and H. F. J. Savage. High resolution neutron study of vitamin B_{12} coenzyme at 15 K. Structure analysis and comparison with the structure at 279 K. *Acta Cryst* B49: 79–89, 1993.

Boys, S. F., and P. Bernardi. The calculation of small molecular interactions by the differences of separate total energies. Some procedures with reduced errors. *Molec Phys* 19: 553–61, 1970.

Bozorth, R. M. The crystal structure of potassium hydrogen fluoride. *J Am Chem Soc* 45: 2128–32, 1923.

Brady, J. W. Molecular dynamics simulations of α-D-glucose in aqueous solution. *J Am Chem Soc* 111: 5155–63, 1989.

Brammer, L., J. M. Channock, P. L. Goggin, R. J. Goodfellow, A. G. Orpen, and T. Koetzle. The role of transition metal atoms as hydrogen bond acceptors: a neutron diffraction study of $[NPr_4^n]_2[PtC\ell_4]cis-[PtC\ell_2(NH_2Me)_2]$ at 20 K. *J Chem Soc Dalton Trans* 1789–98, 1991.

Brammer, L., M. C. McCann, R. M. Bullock, R. K. McMullan, and R. Sherwood. $Et_3NH^+Co(CO)_4^-$ hydrogen-bonded adduct or simple ion pair? Single crystal neutron diffraction study at 15 K. *Organometallics* 11: 2339–41, 1992.

Brammer, L. and R. K. McMullan. Tetragonal bromine hydrate: Structure determination by single crystal neutron diffraction at 100 K. *Abstr Am Cryst Assoc Meeting*, Albuquerque, New Mexico, Abstr. I101, p.73, 1993.

Brammer, L., D. Zhao, F. T. Ladipo, and J. Braddock-Wilking. Hydrogen bonds involving transition metal centers—A brief review. *Acta Cryst* B51: 632–40, 1995.

Bratos, S., H. Ratajczak, and P. Viot. "Properties of H-Bonding in the Infra-red Spectral Range." In: *Hydrogen-Bonded Liquids.* Edited by J. C. Dore and T. Teixeira. NATO-ASI Series C, vol. 329, pp. 221–35. Netherlands: Kluwer Academic Publishers, 1991.

Brickenkamp, C. S. and D. Panke. Polyhedral clathrate hydrates XVII. Structure of low melting hydrate of n-propylamine. A novel clathration framework. *J Chem Phys* 53: 5284–95, 1973.

Brooks, B. M., R. E. Bruccoleri, B. D. Olafson, D. J. States, S. Swaminathan, and M. Karplus. CHARMM: a program for macromolecular energy minimization and dynamics calculations. *J Comput Chem* 4: 187–217, 1983.

Brown, G. M. and H. A. Levy. Further refinement of the structure of sucrose based on neutron diffraction. *Acta Cryst* B29: 790–97, 1973.

Brown, I. D. Hydrogen bonding in perchloric hydrates. *Acta Cryst* A32: 786–92, 1976.

Brown, I. D. and R. D. Shannon. Empirical bond strength-length curves for oxides. *Acta Cryst* A29, 266–82, 1973.

Brunton, G. D. and C. G. Johnson. A neutron diffraction study of the diaquohydronium ion in yttrium oxalate trihydrate. *J Chem Phys* 62: 3797–3806, 1975.

Buckingham, A. D. and P. W. Fowler. A model for van der Waals complexes. *Can J Chem* 67: 2018–25, 1985.

Buijs, K. and G. R. Choppen. Near-infrared studies of liquid water. *J Chem Phys* 39: 2035–40, 1963.

Bürgi, H.-B. and J. D. Dunitz. Can statistical analysis of structural parameters from different crystal environments lead to quantitative energy relationships? *Acta Cryst* B44: 445–48, 1988.

Burkert, U. and N. L. Allinger. *Molecular Mechanics.* Am. Chem. Soc. Monograph 117. Washington, DC: American Chemical Society, 1982.

Burley, S. K. and G. A. Petsko. Amino-aromatic interactions in proteins. *FEBS Lett* 203: 139–43, 1985.

Burton, E. F. and W. F. Oliver. The crystal structure of Ice at low temperatures. *Proc Roy Soc* A153: 166–70, 1935.

Busing, W. R. and H. A. Levy. The effect of thermal motion on the estimation of bond lengths from diffraction methods. *Acta Cryst* 17: 142–46, 1964.

Byrn, M. P., C. J. Curtis, Y. Hsiou, S. I. Khan, P. A. Sawin, S. K. Tendick, A. Terzis, and C. E. Strouse. Porphyrin sponges: conservation of host structure in over 200 porphyrin-based lattice clathrates. *J Am Chem Soc* 115: 9480–97, 1993.

Caldwell, J., L. X. Dang, and P. A. Kollman. Implementation of non-additive intermolecular potentials by use of molecular dynamics: development of a water-water potential and water-ion cluster interactions. *J Am Chem Soc* 112: 9144–47, 1990.

Carbonnel, L. and J.-C. Rosso. Les clathrates des ethers cycliques, leur stoichiométric déduite dés diagrammes de phases eau-ethers cyclique. *Solid State Chem* 8: 302–10, 1973.

Carbonnel, L. and J.-C. Rosso. De nouveau hétérocycles azotés, générateurs d'hydrates clathrates cubiques, stoichiométric des phases nouvelles déduite des diagrammes de phase eau-hétérocycle. *Bull Soc Chim France* 7–8: 1043–50, 1976.

Castleman, H. W., Jr. *Solvated Cluster Ions. Cluster of Atoms and Molecules II.* Edited by H. Haberland. Berlin: Springer-Verlag, 1994.

Ceccarelli, C., G. A. Jeffrey, and R. Taylor. A survey of O—H····O hydrogen bond geometries determined by neutron diffraction. *J Molec Struct* 70: 255–71, 1981.

Ceccarelli, C., J. R. Ruble, and G. A. Jeffrey. The structure determination and molecular mechanics calculation of 1,6-anhydro-β-D-galactopyranose. *Acta Cryst* B36: 861–65, 1980.

Chakravorty, S. J. and E. R. Davidson. The water dimer. Correlation energy calculations. *J Phys Chem* 97: 6373–83, 1993.

Cheetham, N. W. H. and K. Lam. Molecular dynamics simulations of glycosides in aqueous solution. *Carbohydr Res* 282: 13–26, 1996.

Chenite, A. and F. Brisse. Structure and conformation of poly(ethylene oxide) PEO in the trigonal form of PEO-urea complex at 173 K. *Macromolecules* 24: 2221–25, 1991.

Chenite, A. and F. Brisse. Poly(tetrahydrofuran)urea adduct. A structural investigation. *Macromolecules* 25: 776–82, 1992.

Chenite, A. and F. Brisse. Structural investigations of urea-aliphatic polyester adducts. *Macromolecules* 26: 3055–61, 1993.

Chiari, G. and G. Ferraris. The water molecule in crystalline hydrates studied by neutron diffraction. *Acta Cryst* B38: 2331–41, 1982.

Chiba, T. Deuteron magnetic resonance study of some crystals containing an O—D····O bond. *J Chem Phys* 41: 1352–58, 1964.

Chidambaram, R. A bent hydrogen bond model for the structure of Ice I. *Acta Cryst* 14: 467–68, 1961.

Chidambaram, R., A. Sequeira, and S. K. Sikka. Neutron diffraction study of the structure of potassium oxalate monohydrate: lone-pair coordination of the hydrogen-bonded water molecules in crystals. *J Chem Phys* 41: 3616–22, 1964.

Cho, C. H., S. Singh, and G. W. Robinson. An explanation of the anomalous density maximum in water. *Phys Rev Lett* 76: 1651–54, 1996.

Choong, W., D. C. Craig, N. C. Stephenson, and J. D. Stevens. 2-Deoxy-2-fluoro-β-D-mannopyranose, $C_6H_{11}FO_5$. *Cryst Struct Commun* 4: 111–15, 1975.

Chothia, C. Hydrophobic bonding and accessible surface area in proteins. *Nature* 248: 338–39, 1974.

Chu, S. S. C. and G. A. Jeffrey. The refinement of the crystal structures of β-D-glucose and cellobiose. *Acta Cryst* B34: 830–38, 1968.

Chung, Y. J. and G. A. Jeffrey. The lyotropic liquid crystal properties of n-octyl 1-O-β-D-glucopyranoside and related n-alkyl pyranosides. *Biochem Biophys* Acta 985: 300–306, 1989.

Clark, J. R. Water molecules in hydrated organic crystals. *Rev Pure Appl Chem* 13: 50–90, 1963.

Clark, M., R. D. Cramer, III, and N. van Opdenbosch. Validation of the general purpose Tripos 5.2 force field. *J Comput Chem* 10: 982–1012, 1989.

Claussen, W. F. Suggested structures of water in inert gas hydrates. *J Chem Phys* 19: 259–66, 1951a.

Claussen, W. F. A second water structure for inert gas hydrates. *J Chem Phys* 19: 1425–26, 1951b.

Cleland, W. W. Low-barrier hydrogen bonds and low fractionation factor bases in enzymatic reactions. *Biochem* 31: 317–19, 1992.

Cleland, W. W. and M. M. Kreevoy. Low-barrier hydrogen bonds and enzymatic catalysis. *Science* 264: 1887–90, 1994.

Clementi, E., G. Corongiu, and O. G. Stradella. "Molecular Interactions and Large Molecules with KGNMOL," In: *Modern Techniques in Computational Chemistry, MOTTEC 91*. pp. 295–369. Edited by E. Clementi. Netherlands: ESCOM, 1991.

Clymer, J. W., N. Goldstein, J. L. Ragle, and E. L. Reed, Jr. Deuteron quadrupole coupling in hydrogen bonded molecules, V. Relationship of experiment and theory. *J Chem Phys* 76: 4535–73, 1982.

Clymer, J. W. and J. L. Ragle. Deuterium quadrupole coupling in methanol, salicylic acid, catechol, resorcinol and hydroquinone. *J Chem Phys* 77: 4366–73, 1982.

Cohan, N. V., M. Cotte, J. V. Iribarre, and M. Weissmann. Electrostatic energies in Ice and the formation of defects. *Trans Farad Soc* 58: 490–98, 1962.

Cohan, N. V. and M. Weissmann. Valence defects in Ice. *Nature* 201: 490, 1964.

Coulson, C. A. *Valence*. Oxford: Oxford University Press, 1952.

Coulson, C. A. In: *Hydrogen Bonding*. Edited by D. Hadzi. New York: Pergamon Press, pp. 339–60, 1959.

Coulson, C. A. *Preface to the Shape and Structure of Molecules*. Edited by R. McWeeny. Oxford: Oxford University Press, 1972.

Coulson, C. A. and M. W. Thomas. The effect of molecular vibraions on apparent bond lengths. *Acta Cryst* B27: 1354–59, 1971.

Coyle, B. A., L. W. Schroeder, and J. A. Ibers. The structure of potassium tetrahydrogen fluoride. *J Solid State Chem* I: 386–93, 1970.

Cramer, C. J. and D. G. Truhlar. PM3, SM3. A general parameterization for including aqueous solvation effects in the PM3 molecular orbital model. *J Comput Chem* 13: 1089–97, 1992.

Cramer, F. *Einschlussverbindungen*. Berlin: Springer, 1954.

Craven, B. M. Studies of hydrogen atoms in organic molecules. *Trans Am Cryst Assoc* 23: 71–81, 1987.

Craven, B. M. "Electrostatic Properties from Bragg Diffraction Data." In: *Crystallographic Computing, 4. Techniques and New Technologies*. Edited by N. N. Isaacs and M. R. Taylor, International Union of Crystalloraphy. Oxford: Oxford University Press, 1988.

Craven, B. M. and R. F. Stewart. Electrostatic properties of crystals from accurate diffraction data. *Trans Am Cryst Assoc* 25: 41–54, 1990.

Crick, F. H. C. Codon-anticodon pairing. The wobble hypothesis. *J Mol Biol* 19: 548–55, 1966.

Croasman, W. R. and R. M. K. Carson (Editors). *Two-Dimensional NMR Spectroscopy. Applications for Chemists and Biochemists*, 2nd Edition. New York: VCH Publishers, 1994.

Crofton, M. W., J. M. Price, and Y. T. Lee. "Infrared Spectroscopy of Hydrogen Bonded Charge Clusters." In: *Clusters of Atoms and Molecules, II*. Edited by H. Haberland. Berlin: Springer Verlag, 1994.

Cunningham, A. J., J. D. Payzant, and J. D. Kebarle. A kinetic study of the proton hydrate of $H^+(H_2O)_n$ equilibria in the gas phase. *J Am Chem Soc* 94: 7627–32, 1972.

Currie, M. and J. C.Speakman. The crystal structures of the acid salts of some dibasic acids. Part III. Potassium hydrogen malonate: a neutron diffraction study. *J Chem Soc A*, 1923–26, 1970.

Curtiss, L. A., D. J. Frurip, and M. Blander. Studies of molecular association in H_2O and D_2O vapors by measurement of thermal conductivity. *J Chem Phys* 71: 2703–11, 1979.

Cyvin, S. J. *Molecular Vibrations and Mean Square Amplitudes*. Amsterdam: Elsevier, 1968.

Damewood, J. R. Jr., R. A. Kumpf, W. C. F. Mühlbauer, J. J. Urban, and J. E. Eksterowitz. Parameterization of molecular mechanics calculations for the accurate description of hydrogen-bonding interactions. *J Phys Chem* 94: 6619–26, 1990.

Danford, M. D. and H. A. Levy. The structure of water at room temperature. *J Am Chem Soc* 84: 3965–66, 1962.

Dannenberg, J. J. An HMI and ab-initio molecular orbital study of the water dimer. *J Phys Chem* 92: 6869–71, 1988.

Davidson, D. W. "Clathrate Hydrates." In: *Water: A Comprehensive Treatise*. vol. 2. Edited by F. Franks. New York, London: Plenum Press, pp. 115–234, 1973.

Davidson, D. W., Y. P. Handa, C. I. Ratcliffe, J. S. Tse, and B. M. Powell. The ability of small molecules to form clathrate hydrates of structure II. *Nature* 311: 142–44, 1984.

Davidson, W. R., J. Sunner, and P. Kebarle. Hydrogen bonding of water to onium ions. Hydration of substituted pyridinium ions and related systems. *J Am Chem Soc* 101: 1675–80, 1979.

Davies, M. The physical aspects of the hydrogen bond. *Ann Reports on Progress of Chemistry* 43: 5–29. London: The Chemical Society, 1947.

Deakyne, C. A. "Ionic Hydrogen Bonds, Part II. Theoretical Calculations." In: *Molecular Structure and Energetics*, vol. 4, pp. 105–41. Edited by J. F. Liebman and A. Greenberg. New York: VCH Publishers, 1987.

Deakyne, C. A. and M. Meot-Ner. Unconventional ionic hydrogen bonds, 2. $NH^+ \rightarrow \pi$ complexes of onium ions with olefins and benzene derivatives. *J Am Chem Soc* 107: 474–79, 1985.

Deeg, A. and D. Mootz. Addukte aromat/chlorwasserstoff bei tiefen temperaturen: bieträge zu bildung und kristallstruktur. *Zeit Naturforsch* 48b: 571–76, 1993.

De Frees, D. J. and A. D. McLean. Ab-initio determination of proton affinities of small and neutral and anionic molecules. *J Comput Chem* 7: 321–33, 1986.

De La Vega, J. R. Role of symmetry in the tunnelling of the proton in double minimum potentials. *Accts Chem Res* 15: 185–91, 1982.

Del Bene, J. E. "Quantum Chemical Reaction Enthalpies." In: *Molecular Structure and Energetics*, vol. 1, pp. 319–44. Edited by J. F. Liebman and A. Greenburg. New York: VCH Publ. Inc., 1986.

Del Bene, J. E. An ab-initio study of the structures and enthalpies of the hydrogen bonded complexes of the acids, H_2O, H_2S, HCN and HCℓ with the anions OH$^-$, SH$^-$, CN$^-$ and Cℓ^-. *Struct Chem* 1: 19–27, 1988a.

Del Bene, J. E. Ab-initio molecular orbital study of the structures and energies of neutral and charged bimolecular complexes with H_2O and the hydrides AH_n (A = N, O, F, P, S and Cℓ). *J Phys Chem* 92: 2784–2880, 1988b.

Del Bene, J. E. An ab-initio molecular orbital study of the structures and energies of neutral and charged bimolecular complexes of NH_3 with the hydrides AH_n (A = N, O. F, P, S and Cℓ). *J Comput Chem* 10: 603–15, 1989.

Del Bene, J. E. Proton affinities for NH_3, H_2O and HF and their anions. A quest for the basic-set limit using the Dunning augmented correlation consistent basis sets. *J Phys Chem* 97: 107–10, 1993.

Del Bene, J. E. and H. D. Mettee. An ab-initio study of the complexes of HF with the chloromethanes. *J Phys Chem* 97: 9650–54, 1993.

Del Bene, J. E., H. D. Mettee, M. J. Frisch, B. T. Luke, and J. A. Pople. Ab-initio computation of the enthalpies of some gas-phase hydration reactions. *J Phys Chem* 87: 3279–82, 1983.

Del Bene, J. E. and J. A. Pople. Theory of molecular interactions I. Molecular orbital studies of water polymers using a minimal Slater type basis. *J Chem Phys* 52: 4858–66, 1970.

Del Bene, J. E. and I. Shavitt. Comparison of methods for determining the correlation contribution to hydrogen bond energies. *Int J Quantum Chem* 23: 445–52, 1989.

Derissen, J. L. Reinvestigation of the structure of acetic acid monomer and dimer by gas electron diffraction. *J Molec Struct* 7: 67–80, 1971.

De Santis, S. C., V. M. Coiro, F. Mazza, and G. Pochetti. The inclusion compound of deoxycholic acid with (−) camphor: a structural and energetic study. *Acta Cryst* B51: 81–89, 1995.

Desiraju, G. R. Distance dependence of C—H···O interactions in some chloroalkyl compounds. *J Chem Soc Commun*, 179–90, 1989.

Desiraju, G. R. Strength and linearity of C—H···O bonds in molecular crystals: A data base study of some terminal alkynes. *Chem Commun* 454–55, 1990.

Desiraju, G. R. The C—H···O hydrogen bond in crystals. What is it? *Accts Chem Res* 24: 290–96, 1991.

Desiraju, G. R. and B. N. Murty. Correlation between crystallographic and spectroscopic properties of C—H···O bonds in terminal acetylene. *Chem Phys Lett* 139: 360–61, 1987.

Desmeules, J. J. and L. C. Allen. Strong positive-ion hydrogen bonds: the binary complexes formed from NH_3, OH_2, FH, PH_3, SH_2 and CℓH. *J Chem Phys* 72: 4731–48, 1980.

Destro, R., R. Bianchi, and G. Morosi. Electrostatic properties of L-alanine from X-ray diffraction at 23 K and ab-initio calculations. *J Phys Chem* 93: 4447–57, 1989.

Deverell, C. "Nuclear Magnetic Resonance Studies of Electrolyte Solution." In: *Progress in Nuclear Resonance Spectroscopy*, vol. 4. Edited by J. W. Emsley, J. Feeney, and L. H. Sutcliff, London: Pergamon Press, 1969.

Dewar, M. J. S., C. Jie, and J. Yu. SAM1. The first in a new series of general purpose quantum mechanical molecular models. *Tetrahedron* 49: 5003–38, 1993.

Dickerson, R. E. and I. Geis. *The Structure and Action of Proteins*. New York, London: Harper and Row, 1969.

Dill, J. D., L. C. Allen, W. C. Topp, and J. A. Pople. Systematic study of nine hydrogen-bonded dimers involving ammonia, water and hydrofluoric acid. *J Am Chem Soc* 97: 7220–28, 1975.

Doch-Bregon, A. C., B. Chevrier, A. Podjarny, J. Johnson, J. S. de Bear, G. R. Gough, P. T. Gilham, and D. Moras. Crystallographic structure of an RNA helix [U(U-A)$_6$A]$_2$. *J Mol Biol* 209: 459–74, 1989.

Donohue, J. "Selected Topics in Hydrogen Bonding." In: *Structural Chemistry and Molecular Biology*. Edited by A. Rich and N. Davidson. San Francisco: Freeman, pp. 443–65, 1968.

Dore, J. C. "Structural Studies of Water by Neutron Diffraction." In: *Water Science Reviews, 1*: 3–92. Edited by F. Franks, London: Cambridge University Press, 1985.

Dorsey, N. E. *Properties of Ordinary Water Substance*. Am. Chem Soc. Monograph, 1968 (facsimile of 1940 publication by Reinhold Publ. Co.).

Dougall, M. W. and G. A. Jeffrey. The crystal structure of dimethyl oxalate. *Acta Cryst* 6: 831–37, 1953.

Dovesi, R., M. Causa, R. Orlando, C. Roetti, and V. R. Saunders. Ab-initio approach to molecular crystals: a periodic Hartree-Fock study of crystalline urea. *J Chem Phys* 92: 7402–16, 1990.

Dovesi, R., V. R. Saunders, and C. Roetti. CRYSTAL 92. *User Documentation*. University of Torino, Italy and SERC Daresbury Laboratory, Warrington, England, 1992.

Drago, R. S. and B. B. Wayland. A double-scale equation for correlating enthalpies of Lewis acid-base interactions. *J Am Chem Soc* 87: 3571–77, 1965.

Drago, R. S., G. C. Vogel, and T. E. Needham. A four parameter equation for predicting enthalpies of adduct formation. *J Am Chem Soc* 93: 6014–26, 1971.

Drew, H. R. and R. E. Dickerson. Structure of β-DNA dodecamer III. Geometry of hydration. *J Mol Biol* 151: 535–56, 1981.

Duer, W. C. and G. L. Bertrand. Calorimetric determination of heats of formation of hydrogen bonds. *J Am Chem Soc* 92: 2587–88, 1970.

Dunfield, L. G., A. N. Burgess, and H. A. Scheraga. Energy parameters in polypeptides, 8. Empirical potential energy algorithm for the conformational analysis of large molecules. *J Phys Chem* 82: 2609–10, 1978.

Dunitz, J. D. Nature of orientational defects in Ice. *Nature* 197: 860–62, 1963.

Dunning, T. H. and P. J. Hay. *Modern Theoretical Chemistry*. New York: Plenum Press, 1976.

Dunning, T. H. and P. J. Hay. *Methods of Electronic Structure Theory*. pp. 1–26. Edited by H. F. Shaefer. New York: Plenum Press, 1977.

Dyke, T. R., K. M. Mack, and J. S. Meunter. The structure of water dimer from molecular beam electron resonance spectroscopy. *J Chem Phys* 66: 498–510, 1977.

Dyke, T. R. and J. Muenter. Microwave spectrum and structure of hydrogen bonded dimers. *J Chem Phys* 60: 2929–30, 1974.

Dykstra, C. E. Intermolecular electrical interaction: a key ingredient in hydrogen bonding. *Accts Chem Res* 21: 355–61, 1988.

Eckert, M. and G. Zundel. Energy surfaces and proton polarizability of hydrogen-bonded chains: an ab-initio treatment with respect to the charge conduction in biological systems. *J Phys Chem* 92: 7016–23, 1988a.

Eckert, M. and G. Zundel. Motion of one excess proton between various acceptors: theoretical treatment of the proton polarizability of such systems. *J Molec Struct* 181: 141–48, 1988b.

Eigen, M. and I. De Maeyer. Self-diffusion and protonic charge transport in water and ice. *Proc Roy Soc* A247: 505–33, 1958.

Einspar, H., J.-B. Robert, R. E. Marsh, and J. D. Roberts. *Peri* interactions: an X-ray crystallographic study of the structure of 1,8-bis(dimethylamino)-napthalene. *Acta Cryst* B29: 1611–17, 1973.

Eisenberg, D. and C. A. Coulson. Energy of formation of D-defects in ice. *Nature* 199: 368–69, 1963.

Eisenberg, D. and N. Kauzman. *Structure and Properties of Water.* Oxford: Oxford University Press, 1969.

Eisenberg, D. and A. McLachlan. Solvation energy in protein folding and binding. *Nature* 319: 199–202, 1986.

Ellison, R. D. and H. A. Levy. A centered hydrogen bond in potassium hydrogen chloromaleate. A neutron diffraction structure determination. *Acta Cryst* 19: 260–68, 1965.

Emsley, J. Very strong hydrogen bonds. *Chem Soc Rev* 9: 91–124, 1980.

Emsley, J. L., Y. Y. Ma, P. A. Bates, M. Motevalli, and M. B. Hursthouse. β-Diketone interactions, Part 8. The hydrogen bonding of the enol tautomers of some 3-substituted pentane-2-4 diones. *J Chem Soc* Perkin Trans. II, 527–33, 1989.

Espinosa, E. C., Lecomte, N. E. Ghermani, J. Devémy, M. M. Rohmer, M. Bénard, and E. Molins. Hydrogen bonds: first quantitative agreement between electrostatic potential calculations from experimental X-(X+N) and theoretical ab-initio SCF models. *J Am Chem Soc* 118: 2501–2, 1996.

Etter, M. C. Decoding hydrogen-bond patterns. *Accts Chem Res* 23: 120–26, 1990.

Etter, M. C. Hydrogen bonds as design elements in organic chemistry. *J Phys Chem* 95: 4601–10, 1991.

Etter, M. C. and G. M. Frankenbach. Hydrogen-bond directed co-crystallization as a tool for designing acentric organic solids. *Materials* 1: 10–12, 1989.

Etter, M. C., R. Hoye, and G. M. Vojta. Solid-state NMR and X-ray crystallography: complementary tools for structure determination. *Cryst Rev* 1: 281–338, 1988.

Etter, M. C., J. C. MacDonald, and J. Bernstein. Graph-set analysis of hydrogen-bond patterns in organic crystals. *Acta Cryst* B46: 256–62, 1990.

Etter, M. C., S. M. Reutzel, and G. M. Vojta. Anlaysis of isotropic chemical shift data from high resolution solid-state NMR studies of hydrogen-bonded organic compounds. *J Molec Struct* 237: 165–85, 1990.

Evans, R. C. *An Introduction to Crystal Chemistry*, 2nd edition. Cambridge, UK: Cambridge University Press, 1946.

Falk, M. Infrared spectrum and hydrogen bonding in solid and liquid trimethylamine hydrate $(CH_3)N \cdot 10^1/_4H_2O$. *Can J Chem* 49: 1137–39, 1971.

Falk, M. and O. Knop. "Water in Stoichiometric Hydrates." In: *Water. A Comprehensive Treatise.* Edited by F. Franks, New York: Plenum Press, pp. 55–113, 1973.

Faraday, M. On the hydrate of chlorine. *J Sci Liter Arts* 15: 71–90, 1823.

Favier, R., J.-C. Rosso, and L. Carbonnel. Etude des systèmes binaires eau-monoamines aliphatique. Éstablissement de onze diagrammes de phases. Mise en évidence d'hydrates nouveau. *Bull Soc Chim France* Part 1: 225–35, 1981.

Feil, D. and G. A. Jeffrey. The polyhedral clathrate hydrates. Part II. Structure of the hydrate of tetra-iso-amyl ammonium fluoride. *J Chem Phys* 35: 1863–73, 1961.

Ferguson, G., J. F. Gallacher, C. Glidewell, and C. Zakaria. O—H⋯π(arene) intermolecular hydrogen bonding in the structure of 1,1,2–triphenyl ethanol. *Acta Cryst* C50: 70–73, 1994.

Ferguson, G. and J. Tyrrell. C—H⋯O hydrogen bonding. *Chem Commun* 195–97, 1965.

Ferraris, G. and M. Franchini-Angela. Survey of the geometry and environment of water molecules in crystal hydrates studied by neutron diffraction. *Acta Cryst* B28: 3572–83, 1972.

Ferraris, G., H. Fuess, and W. Joswig. Neutron diffraction study of $MgNH_4PO_4 \cdot 6H_2O$ (struvite) and survey of water molecules donating short hydrogen bonds. *Acta Cryst* B42: 253–58, 1986.

Ferraris, G. and G. Ivaldi. X—OH and O—H⋯O bond lengths in protonated oxoanions. *Acta Cryst* B40: 1–6, 1984.

Ferretti, V., V. Bertolasi, P. Gillii, and G. Gilli. "A Novel Approach to Hydrogen Bonding Theory," vol. 2. In: *Advances in Molecular Structure Research.* Edited by I. Hargittai and M Hargittai. Greenwich, CT: JAI Press, 1995.

Fersht, A. Conformational equilibria in α and δ chymotrysin. The energetics and importance of the salt bridge. *J Mol Biol* 64: 497–509, 1972.

Finney, J. L. The complementary use of X-ray and neutron diffraction in the study of crystals. *Acta Cryst* B51, 447–67, 1995.

Flensburg, C., S. Larsen, and R. F. Stewart. Experimental charge density study of methyl ammonium hydrogen succinate monohydrate. A salt with a very short O—H—O hydrogen bond. *J Phys Chem* 99: 10130–41, 1995.

Fletcher, N. H. *The Chemical Physics of Ice.* Cambridge, UK: Cambridge University Press, 1970.

Folzer, C., R. W. Hendricks, and A. H. Narten. Diffraction pattern and structure of liquid trimethylamine decahydrate at 5°C. *J Chem Phys* 54: 799–805, 1971.

Forrester, J. D., M. E. Senko, A. Zalkin, and D. H. Templeton. Crystal structure of KH_2F_3 and geometry of the $H_2F_3^-$ ion. *Acta Cryst* 16: 58–62, 1963.

Forst, R., J. Jagodzinski, and F. Frey. The disordered crystal structure of urea inclusion compound $OC(NH_2) + C_nH_{n+2}$. *Acta Cryst* B46, 70–78, 1990.

Foster, R. and C. A. Fyfe. "Nuclear Magnetic Resonance of Organic Charge-Transfer Complexes," vol. 4. In: *Progress in Nuclear Resonance Spectroscopy,* Edited by J. W. Emsley, J. Feeney, and L. H. Sutcliffe. London: Pergamon Press, 1969.

Fowler, D.L., W. V. Loebenstein, D. M. Pall, and C. A. Kraus. Some unusual hydrates of quaternary ammonium salts. *J Am Chem Soc* 62: 1140–44, 1940.

Frank, H. S. Covalency in the hydrogen bond and the properties of water and ice. *Proc Roy Soc* A247: 481–92, 1958.

Frank, H. S. and A. S. Quist. Pauling's model and the thermodynamic properties of water. *J Chem Phys* 34: 605–11, 1961.

Frank, H. S. and W.-Y. Wen. Structural aspects of ion-solvent interactions in aqueous solutions: a suggested picture of water structure. *Discussion Faraday Soc.* 24: 133–40, 1957.

Franklin, R. E. and R. G. Gosling. The structure of sodium thymonucleate fibres, I. The influence of water content. *Acta Cryst* 6: 673–77, 1953.

Franks, F. (ed.). *Water: A Comprehensive Treatise,* vols. 1–7. New York: Plenum Press, 1972–80.

Franks, F. *Polywater.* Cambridge, MIT Press, 1981.

Frazer, B. C., M. McKeown, and R. Pepinsky. Neutron diffraction study of Rochelle salt single crystals. *Phys Rev* 94: 1435–40, 1954.

Frazer, B. C. and R. Pepinsky. X-ray analysis of the ferroelectric transition in KH_2PO_4. *Acta Cryst* 6: 273–85, 1953.

French, A. D. and D. P. Miller. Comparisons of hydrogen bonding in small carbohydrate molecules by diffraction and MM3(92). *Am Chem Soc Symp Ser* 569: 235–51, 1994.

French, A. D., D. P. Miller, and A. Aabloo. Miniature crystal models of cellulose polymorphs and other carbohydrates. *Int J Biol Macromol* 15: 30–35, 1993.

Frey, M. N., T. F. Koetzle, M. S. Lehman, and W. C. Hamilton. Precision neutron diffraction structure determination of protein and nucleic acid components. XII. A study of the hydrogen bonding in the purine-pyrimidine base pair 9-methyladenine and 1-methylthymine. *J Chem Phys* 59: 915–24, 1973.

Frey, P. A., S. A. Whitt, and J. B. Tobin. A low-barrier hydrogen bond in the catalytic triad of serine proteases. *Science* 264: 1927–30, 1994.

Frisch, M.J., J. E. Del Bene, J. S. Binkley, and H. F. Schaefer. Extensive theoretical studies of the hydrogen bonded dimers $(H_2O)_2$, $(H_2O)_2H^+$, $(HF_2)_2$, $(HF_2)H^+$, F_2H^- and $(NH_3)_2$. *J Chem Phys* 84: 2279–89, 1986.

Frisch, M. J., J. A. Pople, and J. E. Del Bene. Hydrogen bonds between first-row hydrides and acetylene. *J Chem Phys* 78: 4063–65, 1983.

Frisch, M. J., J. A. Pople, and J. E. Del Bene. Molecular orbital study of the dimers $(AH_n)_2$ formed from NH_2, OH_2, FH, PH_3, SH_2 and $C\ell H$. *J Phys Chem* 89: 3664–68, 1985.

Fyfe, C. A. *Solid-State NMR for Chemists.* Guelph, Ontario: CFC Press, 1983.

Garcia-Tellado, F., S. J. Geib, S. Goswani, and A. D. Hamilton. Molecular recognition in the solid state controlled assembly of hydrogen-bonded molecular sheets. *J Am Chem Soc* 113: 9265–69, 1991.

Garneau, I., S. Raymond, and F. Brisse. A thiourea-1,5-cyclooctadiene clathrate at 173 K. *Acta Cryst* C51: 538–41, 1995.

Gaultier, J. and C. Hauw. La liaison hydrogène bifide. *Acta Cryst* B25: 546–48, 1969.

George, P., C. W. Boch, and M. Trachtman. An ab-initio study of planar hydrogen maleate with full geometry optimization. *J Phys Chem* 87: 1839–41, 1983.

Gerlt, J. A. and P. G. Gassman. An explanation for rapid enzyme-catalysed proton abstraction for carbon acids. Importance of late transition states in concerted mechanisms. *J Am Chem Soc* 115: 11552–60, 1993.

Gerstein, B. C., R. G. Pamberton, R. C. Wilson, and L. M. Ryan. High resolution NMR in randomly oriented solid with homonuclear dipolar broadening. Combined multiple pulse NMR and magic angle spinning. *J Chem Phys* 66: 361–62, 1977.

Gessler K., N. Krauss, T. Steiner, C. Betzel, A. Sarko, and W. Saenger. β-Cellotetraose as a structural model for cellulose II. An X-ray diffraction study. *J Am Chem Soc* 117:11398–406, 1995.

Gessler, K., N. Krauss, T. Steiner, C. Betzel, C. Sandman, and W. Saenger. Crystal structure of β-D-cellotetraose hemihydrate with implications for the structure of cellulose II. *Science* 266: 1027–29, 1994.

Ghermani, N., C. Lecomte, and N. Bouhmaida. Electrostatic potential from high resolution X-ray diffraction. Application to a pseudo-peptide molecule. *Zeit Naturforsch Teil N* 48: 91–98, 1993.

Giermanska, J. and M. M. Szostak. Polarized Raman and infrared spectra of OH stretching vibrations in the sucrose crystal. *J Raman Spectros* 22: 107–9, 1991.

Giguère, P. A. Bifurcated hydrogen bonds in water. *J Raman Spectros* 15: 354–59, 1984.

Giguère, P. A. The bifurcated hydrogen bond model of water and amorphous ice. *J Chem Phys* 87: 4835–39, 1987.

Giguère, P. A. and M. Pigeon-Gosselen. The nature of the free OH groups in water. *J Raman Spectros* 17: 341–44, 1986.

Gilli, G., F. Belluci, V. Ferretti, and V. Bertolasi. Evidence for resonance-assisted hydrogen bonding from crystal structure correlations on the enol form of the β-diketone fragment. *J Am Chem Soc* 111:1023–28, 1989.

Gilli, G. and V. Bertolasi. "Structural Chemistry," In: *The Chemistry of Enols*. Edited by Z. Rappoport. New York: John Wiley and Sons, pp. 714–64, 1990.

Gilli, G., V. Bertolasi, V. Ferretti, and P. Gilli. Resonance-assisted hydrogen bond. III. Formation of intermolecular hydrogen-bonded chains in crystals of β-diketones and its relevance to molecular association. *Acta Cryst* B49: 564–76, 1993.

Gilli, P., V. Bertolasi, V. Ferretti, and G. Gilli. Covalent nature of the strong homonuclear hydrogen bond. Study of the O—H····O system by crystal structure correlation methods. *J Am Chem Soc* 116: 909–15, 1994.

Gilli, P., V. Ferretti, V. Bertolasi, and G. Gilli. A novel approach to hydrogen bond theory. In *Advances in Molecular Structure Research*, vol. 2, pp. 67–102. Edited by M. Hargittai and I. Hargittai. Greenwich, CT: JAI Press, 1995.

Glennon, T. M., Y. J. Zheng, S. M. Le Grand, B. A. Shutzberg, and K. M. Merz. A force-field for monosaccharides and (1 → 4) linked polysaccharides. *J Comput Chem* 15: 1019–40, 1994.

Glusker, J. P., M. Lewis, and M. Rossi. *Crystal Structure Analysis for Chemists and Biologists.* New York: VCH Publishers, 1994.

Goertzel, R. and B. Goertzel. Linus Pauling. *A Life in Science and Politics.* New York: Basic Books, 1995.

Gould, I. R. and P. A. Kollman. Theoretical investigation of the hydrogen bond strengths in guanine-cytosine and adenine-thymine base pairs. *J Am Chem Soc* 116: 2493–99, 1994.

Gränicher, H. Gitterfehlordnung und physikalische Eigenschaftgen hexagonaler und kubischer Eiskristalle. *Zeit fur Krist* 110: 432–71, 1958.

Gray, G. W. and J. W. Goodby. *Smectic Liquid Crystals. Textures and Structures.* UK: Leonard Hill; USA, Heyden & Son, 1984.

Green, R. D. *Hydrogen-Bonding by C—H Groups.* New York: Wiley Interscience, 1974.

Grey, G. W. *Molecular Structure and Liquid Crystals.* London: Academic Press, 1967, p. 161.

Grootenhaus, P. D. J. and C. A. G. Haasnoot. A CHARM$_m$ based force field for carbohydrates using the CHEAT approach: carbohydrate hydroxy groups represented by extended atoms. *Molecular Simulations* 10: 75–95, 1993.

Guieu R., J-C. Rosso, and L. Carbonnel. Étude des systèmes binaires eau-monoamines aliphatiques. Établissement de onze diagrammes de phases. Mise en évidence d'hydrates nouyeaux. *Bull Soc Chim France* Part 1: 469–77, 1980.

Gu, Z. and A. McDermott. Chemical shielding anisotropy of protonated and deprotonated carboxylates in amino acids. *J Am Chem Soc* 115: 4282–85, 1993.

Gunnarsson, G., H. Wennerström, W. Egan, and S. Forsén. Proton and deuterium NMR of hydrogen bonds: relationship between isotope effects and the hydrogen potential. *Chem Phys Lett* 38: 96–99, 1976.

Guthrie, J. P. and R. Kluger. Electrostatic stabilization can explain the unexpected acidity of carbon atoms in enzyme-catalyzed reactions. *J Am Chem Soc* 115: 11569–72, 1993.

Ha, S. N., M. Guammona, M. Field, and J. W. Brady. A revised potential energy surface for the molecular mechanics studies of carbohydrates. *Carbohydr Res* 180: 207–21, 1988.

Haas, C. On diffusion, relaxation and defects in ice. *Phys Lett* 3: 126–29, 1962.

Hadzi, D. (ed.) *The Hydrogen Bond.* New York and London: Pergamon Press, 1957.

Hadzi, D. and S. Bratos. "Vibrational Spectroscopy of the Hydrogen Bond." In: *The Hydrogen Bond. II.* Edited by P. Schuster, G. Zundel, and C. Sandorfy. Amsterdam: North Holland Publ., 1976.

Hadzi, D., B. Orel, and A. Novak. Infrared and Raman spectra of some salts containing crystallographically symmetrical hydrogen bonds. *Spectrochim Acta* 29A: 1745–52, 1973.

Hagan, Sister M. *Clathrate Inclusion Compounds.* New York: Reinhold Publ. Co., 1962.

Hagler, A. T., E. Huler, and S. Lifson. Energy functions for peptides and proteins, I. Derivation of consistent force field including the hydrogen bond form amide crystals. *J Am Chem Soc* 96: 5319–27, 1974.

Hagler, A. T., H. A. Scheraga, and C. Nemethy. Structure of liquid water. Statistical thermodynamic theory. *J Phys Chem* 76: 3229–43, 1972.

Ham, J. T. and D. G. Williams. The crystal and molecular structure of methyl β-cellobioside methanol. *Acta Cryst* B26: 1373–83, 1970.

Hamilton, W. C. and J. A. Ibers. *Hydrogen Bonding in Solids.* New York: Benjamin, 1968.

Hankins, D., J. N. Moskowitz, and F. H. Stillinger. Water molecule interactions. *J Chem Phys* 53: 4544–54, 1970.

Hanlon, S., S. Brudno, T. T. Wu, and B. Wolf. Structural transitions of deoxyribonucleic acid in aqueous electrolyte solutions, I. Reference spectra of conformational limits. *Biochemistry* 14: 1648–60, 1975.

Hansen, N. K. and P. Coppens. Testing aspherical atom refinements on small molecule data sets. *Acta Cryst* A34: 909–21, 1978.

Hantzsch, A. Uber die isomerie-gleichgewichte des acetessigesters und die sogenannte isor-rhopesis seiner saltz. *Berichte* 43: 3049–76, 1910.

Hardy, A. D. U. and D. D. MacNicol. Crystal and molecular structure of an O—H····π hydro-gen-bonded system: 2,2-bis(2-hydroxy-5-methyl-3-t-butylphenyl) propane. *J Chem Soc Perkin Trans* 2: 1140–42, 1976.

Harlow, R. L. The structure of water as organized in an RGD peptide crystal at −80°C. *J Am Chem Soc* 115: 9838–39, 1993.

Harmon, K. M. and A. C. Akin. Hydrogen bonding, Part 38. IR and thermodynamic study of phosphorylcholine chloride calcium salt tetrahydrate and monohydrate. *J Molec Struct* 249: 173–79, 1991.

Harmon, K. M., G. F. Avei, J. Harmon, and A. C. Thiel. Hydrogen bonding, Part 23. Further studies on stoichiometry, stability and structure of the lower hydrates of tetramethylam-monium fluoride. *J Molec Struct* 160: 57–66, 1987.

Harmon, K. M. and D. M. Brooks. Hydrogen bonding, Part 49. IR and thermodynamic study of the halide hydrates of N,N′-dimethyltriethylene diammonium and N,N,N′,N′-tetra-methyl piperazinium dications. *J Molec Struct* 299: 73–89, 1993.

Harmon, K. M., D. M. Brooks, and P. K. Keefer. Hydrogen bonding, Part 50. IR and thermo-dynamic study of the hydrates of hexamethonium iodide, bromide, chloride and fluoride. *J Molec Struct* 317: 17–31, 1994.

Harmon, K. M. and N. M. Budrys. Hydrogen bonding, Part 36. NMR study of ^{14}N to C—H cou-pling as a measure of hydration of tetramethylammonium and tetrapropylammonium halides. *J Molec Struct* 249: 149–59, 1991.

Harmon, K. M., J. M. Gabriele, and J. Harmon. Hydrogen bonding, Part 30. New IR spectra-structure correlations for tetraethylammonium, tetramethylammonium and N,N-dimethyl pyrrolidinium fluoride monohydrates, tetramethyl ammonium chloride monohydrate and tetramethyl ammonium hydroxide dihydrate; evidence for a planar $(H_2OF^-)_2$ cluster. *J Molec Struct* 216: 53–62, 1989.

Harmon, K. M. and F. A. Günsel. Hydrogen bonding, Part 17. IR and NMR study of the lower hydrates of choline chloride. *J Molec Struct* 118: 267–75, 1984.

Harmon, K. M. and P. K. Keefer. Hydrogen bonding, Part 44. 1–Methyl-1,3,5,7–tetra-aza-adamantan-1-ium (N-methylhexamethylene tetramine cation) halide hydrates. IR and thermodynamic study of planar C_{2h} $(H_2OBr^-)_2$ and $(H_2OC\ell^-)_2$ clusters, a possible $C_{2h}(H_2OI^-)_2$ cluster and two higher $C\ell^-$ hydrates. *J Molec Struct* 270: 19–31, 1992.

Harmon, K. M., P. A. Mounts, and K. E. Wilson. Hydrogen bonding, Part 37. N,N,N-trimethyl-1-adamantylammonium fluoride tetrahydrate and hemihydrate; evidence for covalent bonding in the hemihydrate. *J Molec Struct* 249: 161–72, 1991.

Harmon, K. M. and B. A. Southworth. Hydrogen bonding, Part 48. IR and thermodynamic study of the lower hydrates of N-methylquinuclidinium iodide, bromide, chloride, fluoride and hydroxide; evidence for a temperature-dependent rearrangement of hydrogen bonds in the chloride monohydrate. *J Molec Struct* 298: 23–36, 1993.

Harmon, K. M., B. A. Southworth, and P. A. Mounts. Hydrogen bonding, Part 47. Stoichiometry, stability and IR spectra of N,N,N-trimethyl-1-adamantyl ammonium hydroxide hydrates:

IR evidence for a covalent HOHOH⁻ species in the monohydrate and hemihydrate. *J Molec Struct* 296: 69–78, 1993.

Harmon, K. M., P. L. Toccalino, and M. S. Janos. Hydrogen bonding, Part 29. Thermodynamics of dissociation and stoichiometric study of the lower hydrates of tetraethylammonium and tetrapropyl ammonium chloride. *J Molec Struct* 213: 193–200, 1989.

Harmony, M. D., V. W Laurie, R. K. Kuczkowski, R. H. Schwendtman, D. A. Ramsay, F.J. Lovas, W. J. Lafferty, and A. G. Maki. Molecular structures of gas-phase polyatomic molecules determined by spectroscopic methods. *J Phys Chem Ref Data* 8: 619–721, 1979.

Harris, R. K., P. Jackson, L. H. Merwin, B. J. Say, and G. Hagele. Perspectives in high resolution solid-state nuclear magnetic resonance with emphasis on combined rotation and multiple-pulse spectroscopy. *J Chem Soc Farad Trans* 84: 3649–72, 1988.

Harrison, R. W., A. Wlodawer, and L. Sjölin. Analysis of solvent structure and hydrogen exchange in proteins on the basis of neutron diffraction data from deuterated and hydrogenous crystals. *Acta Cryst* A44: 309–20, 1988.

Hehre, W. J., L. Radom, P. v. R. Schleyer, and J. Pople. *Ab Initio Molecular Orbital Theory.* New York: John Wiley & Sons, 1986.

Heinekey, D. M. and W. J. Oldham. Coordinate chemistry of dihydrogen. *Chem Rev.* 93: 913–16, 1993.

Hendricks, S. B., O. R. Wulf, G. E. Hilbert, and V. Liddell. Hydrogen bond formation between hydroxyl groups and nitrogen atoms in some organic compounds. *J Am Chem Soc* 58: 1991–96, 1936.

Hibbert, F. and J. Emsley. Hydrogen bonding and chemical reactivity. *Adv Phys Org Chem* 26: 255–379, 1990.

Hilbert, G. E., O. R. Wulf, S. B. Hendricks, and U. Liddell. The hydrogen bond between oxygen atoms in some organic compounds. *J Am Chem Soc* 58: 548–55, 1936.

Hindricks, W. and W. Saenger. Crystal and molecular structure of the hexasaccharide complex (p-nitrophenyl-maltohexaoside)Ba(I$_3$)$_2$27H$_2$O. *J Am Chem Soc* 112: 2789–96, 1990.

Hirshfeld, F. L. Difference densities by least-squares refinement: Fumaric acid. *Acta Cryst* B27: 769–81, 1971.

Hirshfeld, F. L. "The Role of Electron Density in X-ray Crystallography," In: *Accurate Molecular Structures. Their Determination and Importance.* Edited by A. Domenicano and I Hargittai, International Union of Crystallography. Oxford: Oxford University Press, pp. 237–69, 1992.

Holder, A. J. and E. M. Evieth. "SAM1: General Description and Performance Evaluation for Hydrogen Bonds." In: *Modeling the Hydrogen Bond.* ACS Symposium Series 569. Edited by D. A. Smith. Washington, DC: American Chemical Society, pp. 113–24, 1994.

Holder, G. D. and D. Manganiello. Hydrate dissociation pressure minima in multicomponent systems. *Chem Eng Sci* 37: 9–16, 1982.

Hollander, F. and G. A. Jeffrey. Neutron diffraction study of the crystal structure of ethylene oxide deuterohydrate at 80 K. *J Chem Phys* 66: 4699–4705, 1977.

Homans, S. W. A molecular mechanical force field for the conformational analysis of oligosaccharides. Comparison of theoretical and crystal structure of Manα1–3 Manβ1–4 GlcNAc. *Biochemistry* 29: 9110–18, 1990.

Hoogsteen, K. The crystal and molecular structure of a hydrogen-bonded complex between 1-methylthymine and 9-methyladenine. *Acta Cryst* 16: 907–16, 1963.

Hopfinger, A. J. *Conformation Properties of Macromolecules.* New York: Academic Press, 1973.

Hoshino, S., T. Mitsui, F. Jona, and R. Pepinsky. Dielectric and thermal study of triglycine sulfate and tryglycine fluonberyllate. *Phys Rev* 107: 1255–58, 1957.

Hoshino, S., Y. Okaya, and R. Pepinsky. Crystal structure of the ferroelectric phase of (glycine)$_3$H$_2$SO$_4$. *Phys Rev* 115: 323–30, 1959.

Hsu, B. and E. O. Schlemper. X—N deformation density studies of the hydrogen maleate ion and the imidazolium ion. *Acta Cryst* B36: 3017–23, 1980.

Huber, H. Deuterium quadrupole coupling constants. A theoretical investigation. *J Chem Phys* 83: 4591–98, 1985.

Huggins, C. M. and G. C. Pimental. Proton magnetic resonance studies of chloroform in solution. Evidence for hydrogen bonding. *J Chem Phys* 23: 1244–47, 1955.

Huggins, C. M., G. C. Pimental, and J. N. Schoolery. Proton magnetic resonance studies of the hydrogen bonding in phenol, substituted phenols and acetic acid. *J Phys Chem* 60: 1311–14, 1956.

Huggins, M. L. Electronic structure of atoms. *J Phys Chem* 26: 601–25, 1922.

Huggins, M. L. The role of hydrogen bonds in conduction by hydrogen and hydroxyl ions. *J Am Chem Soc* 53: 3190–91, 1931.

Huggins, M. L. Hydrogen bridges in ice and liquid water. *J Phys Chem* 40: 723–31, 1936a.

Huggins, M. L. Hydrogen bridges in organic compounds. *J Org Chem* 1: 405–56, 1936b.

Huggins, M. L. Structure of fibrous proteins. *Chem Rev* 32: 195–218, 1943.

Huggins, M. L. 50 Years of hydrogen bond theory. *Angew Chem Int Ed Engl* 10: 147–52, 1971.

Hughes, E. W. The crystal structure of melamine. *J Am Chem Soc* 63: 1737–52, 1941.

Huisken, F., M. Kaloudis, and A. Kulcke. Infrared spectroscopy of small size-selected water clusters. *J Chem Phys* 104: 17–25, 1995.

Hunt, M. J. and A. L. Mackay. Deuterium and nitrogen pure quadrupole resonance in deuterated amino acids. *J Mag Res* 15: 402–14, 1974.

Hunt, R. D. and L. Andrews. Infrared spectra of HF complexes with $CC\ell_4$, $CHC\ell_3$, $CH_2C\ell_2$ in solid argon. *J Phys Chem* 96: 6945–47, 1992.

Hunter, L. "The Hydrogen Bond." In: *Annual Reports on Progress of Chemistry*, vol. 43, pp. 141–54. London: The Chemical Society, 1947.

Hussain, M. S. and E. O. Schlemper. Neutron diffraction study of deuterated imidazolium hydrogen maleate: An evaluation of the isotope effect on hydrogen-bond lengths. *Acta Cryst* B36: 1104–8, 1980.

Hwang, W. J., T. P. Stockfisch, and A. T. Hagler. Derivation of class II force field, 2. Derivation and characterization of a class II force field CFF93 for alkyl functional group and alkane molecules. *J Am Chem Soc* 116: 2515–24, 1994.

Ibers, J. A. Refinement of Peterson and Levy's neutron diffraction data on KHF_2. *J Chem Phys* 40: 402–4, 1964a.

Ibers, J. A. Potential function for the stretching region in potassium acid fluoride. *J Chem Phys* 41: 25–28, 1964b.

Ichikawa, M. The O—H vs O----O distance correlation, the geometric isotope effect in OHO bonds and its application to symmetric bonds. *Acta Cryst* B34: 2074–80, 1978.

Ichikawa, M. Geometric and quantum aspects of phase transition and isotope effect in hydrogen-bonded ferroelectrics and related materials. *Ferroelectrics* 168: 177–92, 1995.

Itoh, K. and T. Mitsui. *Ferroelectrics* 5: 235–51, 1974.

Imashiro, F., S. Maeda, K. Takegoshi, T. Teyao, and A. Salka. Hydrogen bonding and conformational effects on ^{13}CNMR chemical shifts of hydroxybenzaldehydes in the solid state. *Chem Phys Letts* 99: 189–92, 1983.

Jaccard, C. *Transport Properties of Ice in Water and Aqueous Solutions.* Edited by R.H. Horne. New York: John Wiley & Sons, 1972.

James, M. N. G. and M. Matsushima. Accurate dimensions of the maleate mono-anion in a symmetrical environment not dictated by crystallographic symmetry. Imidazolium maleate. *Acta Cryst* B32: 1708–13, 1976.

Jamvóz, M. H. and Jan Cz Dobrowolski. IR study of CH_2X_2 double hydrogen bonding. *J Molec Struct* 293: 143–46, 1993.

Jeffrey, G. A. Water structure in organic hydrates. *Accts Chem Res* 2: 344–52, 1969.

Jeffrey, G. A. "Hydrate Inclusion Compounds." In: *Inclusion Compounds*, vol. 1. Edited by J. L. Atwood, J. E. D.Davies, and D. D. MacNicol. New York: Academic Press, pp. 135–90, 1984.

Jeffrey, G. A. "Hydrogen-Bonding in Crystal Structures of Nucleic Acid Components: Purines, Pyrimidines, Nucleosides and Nucleotides." In: *Numerical Data and Functional Relationships in Science and Technology*. Edited by W. Saenger. Landolt-Bornstein Series VII:1b. Berlin, Heidelberg: Springer-Verlag, pp. 277–348, 1989.

Jeffrey, G. A. "Accurate Crystal Structure Analysis by Neutron Diffraction," In: *Accurate Moleuclar Structures*. Edited by A. Domenicano and I. Hargittai. Oxford: Oxford University Press, 1992a.

Jeffrey, G. A. "Hydrogen Bonding in Carbohydrates and Hydrate Inclusion Compounds," In: *Advances in Enzymology*, vol. 65, Edited by A. Meister. New York: John Wiley & Sons, pp. 217–54, 1992b.

Jeffrey, G. A. The role of the hydrogen bond and water in biological processes. *J Molec Struct* 322: 21–25, 1994.

Jeffrey, G. A. Hydrogen-bonding: an update. *Crystallogr Rev.* 3: 213–60, 1995.

Jeffrey, G. A. and A. D. French. Mono-, oligo-, and polysaccharole crystal structures. In: Molecular Structures by Diffraction Methods. Edited by M. R. Tryter. *The Chemical Society Specialist Periodical Report*. Vol. 6, Chap. 8, 183–223, 1978.

Jeffrey, G. A., M. E. Gress, and S. Takagi. Some experimental observations on the H----O hydrogen bond lengths in carbohydrate crystal structures. *J Am Chem Soc* 99: 609–11, 1977.

Jeffrey, G. A. and L. Lewis. Cooperative aspects of hydrogen bonding in carbohydrates. *Carbohydr Res* 60: 179–82, 1978.

Jeffrey, G. A. and T. C. W. Mak. Hexamethylene tetramine hexahydrate. A new type of clathrate hydrate. *Science* 149: 178–79, 1965.

Jeffrey, G. A. and H. Maluszynska. A survey of hydrogen bond geometries in the crystal structures of amino acids. *Int J Biol Macromol* 4: 173–85, 1982.

Jeffrey, G. A. and H. Maluszynska. A survey of the geometry of hydrogen bonds in the crystal structures of barbiturates, purines and pyrimidines. *J Molec Struct* 147: 127–42, 1986.

Jeffrey, G. A. and H. Maluszynska. The stereochemistry of water molecules in the hydrates of small biological molecules. *Acta Cryst* B48: 546–49, 1990a.

Jeffrey, G. A. and H. Maluszynska. The crystal structure and thermotropic liquid crystal properties of N-n-undecyl-D-gluconamide. *Carbohydr Res* 207: 211–19, 1990b.

Jeffrey, G. A., H. Maluszynska, and J. Mitra. Hydrogen bonding in nucleosides and nucleotides. *Int J Biol Macromol* 7: 336–45, 1985.

Jeffrey, G. A. and R. K. McMullan. "The Clathrate Hydrates." In: *Progress in Inorganic Chemistry*, vol. 8. New York, London: Interscience Publishers, pp. 43–108, 1967.

Jeffrey, G. A. and J. Mitra. The hydrogen bonding patterns in the pyranose and pyranoside crystal structures. *Acta Cryst* B39: 469–80, 1983.

Jeffrey, G. A. and J. Mitra. Three-center (bifurcated) hydrogen bonding in the crystal structures of amino acids. *J Am Chem Soc* 106: 5546–53, 1984.

Jeffrey, G. A., J. R. Ruble, R. K. McMullan, D. J. DeFrees, J. S. Binkley, and J. A. Pople. Neutron diffraction at 23 K and ab-initio molecular orbital studies of the molecular structure of acetamide. *Acta Cryst* B36: 2292–99, 1980.

Jeffrey, G. A., J. R. Ruble, R. D. McMullan, D. J. DeFrees, and J. A. Pople. Neutron diffraction at 20 K and ab-initio molecular orbital studies of monofluoroacetamide. *Acta Cryst* B37: 1885–90, 1981.

Jeffrey, G. A., J. R. Ruble, and J. H. Yates. Neutron diffraction at 15 K and 120 K and ab-initio molecular orbital studies of the molecular structure of 1,2,4–triazole. *Acta Cryst* B39: 388–94, 1983.

Jeffrey, G. A., J. R. Ruble, L. M. Wingert, J. H. Yates, and R. K. McMullan. π-Bond anisotropy and C—D····O hydrogen bonding in the crystal structure of deuteronitromethane. *J Am Chem Soc* 107: 6227–30, 1985.

Jeffrey, G. A. and W. Saenger. *Hydrogen Bonding in Biological Structures*. Berlin, New York, Heidelberg: Springer-Verlag, 1991.

Jeffrey, G. A. and M. S. Shen. Crystal structure of 2,5-dimethyl-2,5-hexanediol tetrahydrate. A water hydrocarbon layer structure. *J Chem Phys* 57: 56–61, 1972.

Jeffrey, G. A. and S. Takagi. Hydrogen-bond structure in carbohydrate crystals. *Accts Chem Res* 11: 264–70, 1978.

Jeffrey, G. A. and R. Taylor. The application of molecular mechanics to the structures of carbohydrates. *J Comput Chem* 1: 99–109, 1980.

Jeffrey, G. A. and L. M. Wingert. Carbohydrate liquid crystals. *Liq Cryst* 12: 179–202, 1992.

Jeffrey, G. A. and R. A. Wood. The crystal structure of galactaric acid (mucic acid) at $-147°C$: An unusually dense hydrogen-bonded structure. *Carbohydr Res* 108: 205–11, 1982.

Jeffrey, G. A. and Y. Yeon. The correlation between hydrogen bond lengths and proton chemical shifts in crystals. *Acta Cryst* B42: 410–13, 1986.

Jeng, M.-L. H. and B. S. Ault. Infra-red matrix isolation studies of hydrogen bonds involving C—H bonds. CF_3H, $(CH_2H)_2O$, CF_3OCF_2H with selected bases. *J Molec Struct* 246: 33–44, 1991.

Joesten, M. D. and L. J. Schaad. *Hydrogen Bonding*. New York: Dekker, 1974.

Johnson, C. K. "An Introduction to Thermal-Motion Analysis." In: *Crystallographic Computing*. Edited by F. R. Ahmed. Netherlands: Munksgaard, pp. 207–19, 1970a.

Johnson, C. K. "Generalized Treatments for Thermal Motion." In: *Thermal Neutron Diffraction*. Edited by B. T. M. Willis. Oxford: Oxford University Press, 1970b.

Jona, F. and G. Shirane. *Ferroelectric Crystals*. New York: MacMillan, 1962.

Jones, R. D. G. The crystal and molecular structure of the enol tautomer of 1,3-diphenyl 1:3-propanedione (dibenzoylmethane) by neutron diffraction. *Acta Cryst* B32: 1807–11, 1976a.

Jones, R. D. G. The crystal and molecular structure of the enol form of 1-phenyl-1,3-butanedione (benzoylacetone) by neutron diffraction. *Acta Cryst* B32: 2133–36, 1976b.

Jönsson, P. G. Hydrogen-bond studies. XLIV. Neuron diffraction study of acetic acid. *Acta Cryst* B25: 2437–41, 1971.

Jönsson, P. G. and Å. Kvick. Precision neutron diffraction structure determination of protein and nucleic acid components, III. Crystal and molecular structure of the amino acid, α-glycine. *Acta Cryst* B28: 1827–33, 1972.

Jorgensen, J. D., R. A. Behlein, N. Watenabe, and T. G. Worlton. Structure of D_2O Ice VIII from in-situ powder neutron diffraction. *J Chem Phys* 81: 3211–14, 1984.

Jorgensen, W. L. Transferable intermolecular potential functions for water, alcohols and ethers. Application to liquid water. *J Am Chem Soc* 103: 335–41, 1981.

Jorgensen, W.L., J. Chandrasekhar, J. D. Madura, R. W. Imprey, and M. L. Klein. Comparison of simple potential functions for simulating liquid water. *J Chem Phys* 79: 926–35, 1983.

Jorgensen, W. L. and J. Tirado-Rives. The OPLS potential function for proteins. Energy minimizations for crystals of cyclic peptides and crambin. *J Am Chem Soc* 110: 1657–66, 1988.

Kalsbeck, N., K. Schaumburg, and S. Larsen. Short hydrogen bonds in salts of dicarboxylic acids: structural correlations from solid-state ^{13}C and ^2H NMR spectroscopy. *J Molec Struct* 299: 155–70, 1993.

Kamb, B. "Ice Polymorphism and the Structure of Water." In: *Structure Chemistry and Molecular Biology*. Edited by A. Rich and N. Davidson. San Francisco: Freeman, pp. 507–44, 1968.

Karplus, P. A. and G. E. Schulz. Refined structure of glutathione reductase at 1.54 Å resolution. *J Mol Biol* 195: 701–9, 1987.

Kassner, J. L. Jr. and D. E. Hagen. Comment on "Clustering of water on hydrated protons in a supersonic jet expansion." *J Chem Phys* 64: 1860–61, 1976.

Kato, T. and J. M. J. Fréchet. New approach to mesophase stabilization through hydrogen-bonding molecular interactions in binary mixtures. *J Am Chem Soc* 111: 8533–34, 1989.

Kavenau, J. L. *Water and Water Solute Interactions.* San Francisco: Holden-Day, 1964.

Kay, M. I. and R. Kleinberg. The crystal structure of triglycine sulfate. *Ferroelectrics* 5: 45–52, 1973.

Kazarian, S. G., P. A. Hamley, and M. Poliakoff. Is intermolecular hydrogen-bonding to the uncharged metal centers of organometallic compounds widespread in solution? A spectroscopic investigation in hydrocarbon, noble gas and supercritical fluid solutions of the interaction between fluor alcohols and $(\eta^5-C_5R_5)ML_2$ (R = H, Me; M = Co, Rh, Ir; L = Co, C_2H_4, N_2, PMe_3) and its relevance to protonation. *J Am Chem Soc* 115: 9069–79, 1993.

Kearley, G. J., F. Fillaux, M.-H. Baron, S. Bennington, and J.Tomkinson. A new look at proton transfer dynamics along the hydrogen bonds in amides and peptides. *Science* 204: 1285–89, 1994.

Kebarle, P. Ion thermochemistry and solvation from gas phase ion equilibria. *Ann Rev Phys Chem* 28: 445–76, 1977.

Keith, T. A. and M. J. Frisch. "Inclusion of Explicit Solvent Molecules in a Self-consistent Reaction Field Model of Solvation." In: *Modeling the Hydrogen Bond.* Edited by D. A. Smith. ACS Symp. Ser. 569, Washington DC: American Chemical Society, pp. 22–35, 1994.

Kendrew, J. C. Myoglobin and the structure of proteins. *Science* 139: 1259–66, 1963.

Kennard, O., W. B. T. Cruse, J. Nachman, T. Prangé, Z. Shakked, and D. Rabinovich. Ordered water structure in an A-DNA octamer at 1.7 Å resolution. *J Biomol Struct Dynam* 5: 623–48, 1986.

Ketelaar, J. A. A. Investigations in the infrared, I. The absorption and reflection spectra of KHF_2, KDF_2 and $RbHF_2$ in relation to the constitution of the bifluoride ion. *Rec Trav Chim* 60: 523–66, 1941.

Kiriyama, R., H. Kiriyama, T. Wade, N. Niizeki, and H. Hirabayashi. Nuclear magnetic resonance and X-ray studies of potassium ferrocyanide trihydrate crystal. *J Phys Soc Jap* 19: 540–45, 1964.

Kirschner, K. N. and G. C. Shields. Quantum-mechanical investigation of large water clusters. *Int J Quantum Chem* 28: 349–60, 1994.

Kitagorodskij, A. I. *Organic Chemical Crystallography.* (English Edition) New York: Consultant Bureau, 1955.

Kitaura, K. and K. Morokuma. A new energy decomposition scheme for molecular interactions within the Hartree-Fock approximation. *Int J Quant Chem* 10: 325–40, 1976.

Klein, C. L. and E. D. Stevens. "Experimental Measurements of Electron Density Distributions and Electrostatic Potentials." In: *Structure and Reactivity*, vol. 1, Edited by J. F. Liebman and A. Greenburg. New York: VCH Publ. Co., pp. 25–64, 1988.

Koehler, J. E. H., W. Saenger, and W. F. Van Gunsteren. Molecular dynamics simulation of crystalline β-cyclodextrin dodecahydrate hexahydrate. *Eur Biophys J* 15: 197–210, 1987.

Koenig, J. L. Fourier transform infrared spectroscopy of chemical systems. *Accts Chem Res* 14: 171–78, 1981.

Kollman, P. A general analysis of noncovalent intermolecular interactions. *J Am Chem Soc* 99: 4875–93, 1977.

Kollman, P. Grand challenges in computational science. Simulations on biological macromolecules. *Future Generation Computer Systems* 5: 207–11, 1989.

Kollman, P. A. and L. C. Allen. The theory of the hydrogen bond. *Chem Rev* 72: 283–303, 1972.

Kolpak, F. J. and J. Blackwell. Determination of the structure of Cellulose II. *Macromolecules* 9: 273–78, 1976.

Kopka, M. L., A. V. Fratini, H. R. Drew, and R. E. Dickerson. Ordered water structure around a B-DNA dodecamer. A quantitative study. *J Molec Struct* 163: 129–46, 1983.

Krijn, M. P. C. M. and D. Feil. Electron density distributions in hydrogen bonds. A local density-functional study of α-oxalic acid dihydrate and comparison with experiment. *J Chem Phys* 89: 4199–4208, 1988.

Kroon, J., J. A. Kanters, J. G. C. M. van-Duijneveldt-van der Rydt, F. B. van-Duijneveldt, and J. A. Vliegenhart. O—H····O hydrogen bonds in molecular crystals. A statistical and quantum-chemical analysis. *J Molec Struct* 24: 109–29, 1975.

Krumm, S., K. Kurowa, and T. Rebare. "Infra-red Studies of C—H····O=C Hydrogen Bonding in Polyglycine II." In: *Conformation in Biopolymers*. Edited by G. N. Ramachandran. London: Academic Press, pp. 439–47, 1967.

Kuchitsu, K. "The Potential Energy Surface and the Meaning of Internuclear Distances." In: *Accurate Molecular Structures*. Edited by A. Domenicano and I. Hargittai. Oxford: Oxford University Press, pp. 14–43, 1992.

Kuchitsu, K. and Y. Morino. Estimation of anharmonic potential constants, I and II. *Bull Chem Soc Jap* 38: 805–24, 1965.

Kuhs, W. F., J. L. Finney, C. Vettier, and D. V. Bliss. Structure and hydrogen bonding in Ices VI, VII and VIII by neutron powder diffraction. *J Chem Phys* 81: 3612–23, 1984.

Kuhs, W. F. and M. S. Lehman. The structure of ice Ih by neutron diffraction. *J Phys Chem* 87: 4312–13, 1983.

Kuhs, W. F. and M. S. Lehman. The structure of ice Ih. *Water Science Rev* 2: 1–66, 1985.

Kulinska, K. and A. Laaksonen. Hydration of cytidine, 2'-deoxycytidene and their phosphate salts in aqueous solution. A molecular dynamics computer simulation study. *J Biomol Struct Dynamics* 11: 1307–12, 1994.

Kumaraswamy, V. S., P. F. Lindley, C. Slingsby, and I. D. Glover. An eye lens protein-water structure: 1.2 Å resolution of $\gamma\beta$-crystallin at 150 K. *Acta Cryst* D52:611–22, 1996.

Küppers, H., Å. Kvick, and J. Olovsson. Hydrogen bond studies, CXLII. Neutron diffraction study of the two very short hydrogen bonds in lithium hydrogen phthalate-methanol. *Acta Cryst* B37: 1203–6, 1981.

Küppers, H. and S. M. Jessen. Geometric conditions for the formation of short intramolecular hydrogen bonds in dicarboxylic acids and their acid salts. Crystal structures of two cyclopropane derivatives containing such bonds: ammonium hydrogen caronate and potassium hydrogen caronate hydrate. *Zeit fur Krist* 203: 167–82, 1993.

Kurikara, K., Y. Ohtu, Y. Tanaka, Y. Aayama, and T. Kunitake. Molecular recognition of sugars by monolayers of resorcinol-dodecanal cyclotetramer. *J Am Chem Soc* 113: 444–50, 1991.

Kuroki, S., A. Takahashi, I. Ando, A. Shoji, and T. Osaki. Hydrogen bonding structural study of solid peptides-polypeptides containing α-glycine residue by ^{17}O NMR spectroscopy. *J Molec Struct* 323: 197–208, 1991.

Kvick, Å, T. F. Koetzle, R. Thomas, and F. Takusagawa. Hydrogen bond studies, 85. A very short asymmetrical intramolecular hydrogen bond. A neutron diffraction study of pyridine-2,3-dicarboxylic acid ($C_7H_5NO_4$). *J Chem Phys* 60, 3866–74, 1974.

Larson, J. W. and T. B. McMahon. Gas-phase bifluoride ion. An ion-cyclotron resonance determination of the hydrogen bond energy in FHF^- from gas-phase fluoride transfer equilibrium measurements. *J Am Chem Soc* 104: 5848–49, 1982.

Larson, J. W. and T. B. McMahon. Strong hydrogen bonding in gas-phase anions. An ion cyclotron resonance determination of fluoride binding energetics to Brønsted acids from gas-phase fluoride exchange equilibria measurements. *J Am Chem Soc* 105: 2944–50, 1983.

Larson, J. W. and T. B. McMahon. Chelation of gas-phase anions. An ion cyclotron resonance study of cooperative binding to fluoride and chloride ions. *J Phys Chem* 88: 1083–86, 1984.

Larson, J. W. and T. B. McMahon. Hydrogen bonding in gas-phase anions. The energetics of interactions between the cyanide ion and Brønsted acids determined from ion cyclotron cyanide exchange equilibria. *J Am Chem Soc* 107: 6230–56, 1987.

Larsson, K., J. Lindgren, and J. Tegenfeldt. Reorientation of water molecules in solid hydrates. *J Chem Soc Farad Trans* 87: 1193–1200, 1991.

Latajko, Z. and S. Scheiner. Basis-sets for molecular interactions, 2. Application to H_3N—HF, H_3N—HOH, H_2O—HF, $(NH_3)_2$ and H_3CH—OH_2. *J Comput Chem* 8: 674–81, 1987.

Latimer, W. M. and W. H. Rodebush. Polarity and ionization from the standpoint of the Lewis theory of valence. *J Am Chem Soc* 42: 1419–33, 1920.

Lecomte, C. "Experimental Electron Densities of Molecular Crystals and Calculation of Electrostatic Properties from High Resolution X-ray Diffraction." In: *Advances in Molecular Structure Research*, vol. 1, Edited by M. Hargittai and I. Hargittai. London; Greenwich CT: JAI Press, pp. 261–302, 1995.

Lee, J. C., Jr., E. Peris, A. L.Rheingold, and R.H. Crabtree. An unusual type of H----H interaction. Ir—H----H—O and Ir—H----H—N hydrogen bonding and its involvement in σ-bond metathesis. *J Am Chem Soc* 116: 11014–19, 1994.

Leeflang, B. R., J. F. G. Vliegenhart, L. M. J. Kroon-Batenburg, B. P. van Eijck, and J. Kroon. A ^{1}H NMR and MD study of itnramolecular hydrogen bonds in methyl β-cellobioside. *Carbohydr Res* 270: 41–61, 1992.

Legon, A. C. Pulsed-nozzle Fourier transform microwave spectroscopy of weakly bound dimers. *Ann Rev Phys Chem* 34: 275–300, 1983.

Legon, A. C. The nature of ammonium and methylammonium halides in the vapor phase: hydrogen bonding versus proton transfer. *Chem Soc Rev*, 153–63, 1993.

Legon, A. C. and D. J. Millen. Determination of properties of hydrogen-bonded dimers by rotational spectroscopy and a classification of dimer geometries. *Faraday Disc Chem Soc* 73: 71–87, 1982.

Legon, A. C. and D. J. Millen. Gas phase spectroscopy and the properties of hydrogen-bonded dimers: HCN—HF as the spectroscopic prototype. *Chem Rev* 86: 635–57, 1986.

Legon, A. C. and D. J. Millen. Hydrogen bonding as a probe of electron densities. Limiting gas-phase nucleophilicities and electrophilicities of B and HX. *J Am Chem Soc* 109: 356–58, 1987a.

Legon, A. C. and D. J. Millen. Directional character strength and nature of the hydrogen bond in gas-phase dimers. *Accts Chem Res* 20: 39–46, 1987b.

Legon, A. C. and D. J. Millen. The nature of the hydrogen bond to water in the gas phase. *Chem Soc Revs* 21: 71–78, 1992.

Legon, A. C., D. J. Millen, P. I. Mjöberg, and S. C. Roger. A method for the determination of the dissociation energies D_o and D_e for hydrogen bonded dimers from the intensities of rotational transitions and its application to HCN^- HF. *Chem Phys Lett* 55: 157–59, 1987.

Legon, A. C., D. J. Millen, and S. C. Rogers. Spectroscopic investigations of hydrogen bonding interactions in the gas phase, I. The determination of the geometry, dissociation energy, potential constants and electric dipole moment of the hydrogen-bonded hetero-dimer HCN—HF from its microwave rotational spectrum. *Proc Roy Soc Lond* A370: 213–37, 1980.

Lehman, T. A. and M. M. Bursey. *Ion Cyclotron Resonance Spectroscopy*. New York: Wiley-Interscience, 1976.

Leiserowtiz, L. Molecular packing modes. Carboxylic acids. *Acta Cryst* B32: 775–802, 1976.

Leiserowtiz, L. and M. Tuval. Molecular packing modes. N-methylamides. *Acta Cryst* B34: 1230–47, 1978.

Lemieux, R. U. How water provides the impetus for molecular recognition in aqueous solution. *Accts Chem Res* 29: 373–80, 1996.

Leonard, G. A., K. McAuley-Hecht, T. Brown, and W. N. Hunter. Do C—H----O hydrogen bonds contribute to the stability of nucleic acid base pairs? *Acta Cryst* D51: 136–37, 1995.

Lesk, A. M. *Protein Architecture. A Practical Approach.* New York: Oxford University Press, 1991.

Levitt, M. and M. F. Perutz. Aromatic rings act as hydrogen-bond acceptors. *J Mol Biol* 201: 751–54, 1988.

Libnau, F. O., J. Toft, A. A. Christy, and O. M. K. Valheim. Structure of liquid water determined from infrared temperature profiling and evolutionary curve resolution. *J Am Chem Soc* 116: 8311–16, 1994.

Lichtenhalter, F. W. 100 Years "Schlüssel-Schloss-Prinzip": What made Emil Fischer use this analogy. *Angew Chem Int Ed Engl* 33: 2364–74, 1994.

Liddell, U. and N. F. Ramsey. Temperature dependent magnetic shielding in ethyl alcohol. *J Chem Phys* 19: 1608, 1951.

Liddell, U. and O. R. Wulf. The character of the absorption of some amines in the near infrared. *J Am Chem Soc* 55: 3574–88, 1933.

Liebman, J. F. *Are the Numerical Magnetudes of Proton Affinities Intuitively "Plausible"?* In: "Molecular Structure and Energetics," vol. 4, pp. 40–70. Edited by J. F. Liebman and H. Greenberg. New York: VCH Publishers, 1987.

Lifson, S., A. T. Hagler, and P. Dauber. Consistent force field studies of intermolecular forces in hyrogen-bonded crystals, 1. Carboxylic acids and the C=O----H hydrogen bonds. *J Am Chem Soc* 101: 5111–21, 1979.

Lii, J.-H. and N. L. Allinger. Directional hydrogen bonding in the MM3 force field, I. *J Phys Org Chem* 7: 591–609, 1994.

Lines, M. E. and A. M. Glass. *Principles and Applications of Ferroelectrics and Related Materials.* Oxford: Clarenden Press, 1977.

Lipkowski, J., R. Luboradzhi, K. Udechin, and Y. Dyadin. A layer clathrate hydrate structure of tetrapropyl ammonium fluoride. *J Incl Phenom* 13: 295, 1992.

Lipkowski, J., K. Suwinska, K. Udachin, T. Rodionava, and Y. Dyadin. A novel clathrate hydrate structure of tetraisoamyl ammonium fluoride. *J Incl Phenom* 9: 275–76, 1990.

Lipson, H. S. *Crystals and X-rays.* New York: Springer-Verlag, 1970.

Llamas-Saiz, A. L. and C. Foces-Foces. N—H----N(sp^2) hydrogen bond interactions in crystals. *J Molec Struct* 238: 367–82, 1990.

Llamas-Saiz A. L., C. Foces-Foces, and J. Elguero. Proton sponges. *J Molec Struct* 328: 297–323, 1994.

Llamas-Saiz, A. L., C. Foces-Foces, O. Mo, M. Yañez, and J. Elguero. Nature of the hydrogen bond: crystallographic versus theoretical description of O—H----N(sp^2) hydrogen-bond. *Acta Cryst* B48: 700–713, 1992.

Llewellyn, F. J., E. G. Cox, and T. H. Goodwin. Crystalline structure of sugars, Part IV. Pentaerythritol. *J Chem Soc* 883–94, 1937.

Loehlin, J. H. and Å. Kvick. Tetramethyl ammonium monohydrate. *Acta Cryst* B34, 3488–90, 1978.

Londono, J. D., J. L. Finney, and W. F.Kuhs. Formation, stability and structure of helium hydrate at high pressure. *J Chem Phys* 97: 547–52, 1992.

Londono, J. D., W. F. Kuhs, and J. L. Finney. Neutron diffraction studies of ices III and IX on under-pressure and recovered samples. *J Chem Phys* 98: 4878–88, 1993.

Lonsdale, K. The structure of the Benzene Ring in C$_6$ (CH$_3$)$_6$. *Proc Roy Soc Lond* A.123: 494–515, 1929.

Lord, R. C. and R. E. Merrifield. Strong hydrogen bonds in crystals. *J Chem Phys* 21: 166–67, 1953.

Löwig, C. Ueber bromhydrate und fester bromkohlenstoff. *Ann Phys Chem (Poggendorf)* 16: 376–80, 1829.

Lundgren, J. O. and I. Olovsson. "The Hydrated Proton in Solids," In: *The Hydrogen Bond—Recent Developments in Theory and Experiments.* Edited by P. Schuster. Amsterdam: North Holland Publishing Co., 1976.

Lundgren, J. O. and R. Tellgren. Hydrogen bond studies LXXXVL. An asymmetric non-centered $H_5O_2^+$ ion: neutron diffraction study of phenylsulphonic acid tetrahydrate $(H_5O_2)^-[C_6H_2(NO_3)]^-2.H_2O$. *Acta Cryst* B30: 1937–47, 1974.

Lutz, B., J. van der Maas, and J. A. Kanters. Spectroscopic evidence for $\equiv$C—H$\cdots$O interaction in crystalline steroids and reference compounds. *J Molec Struct* 325: 203–14, 1994.

Lutz, E. T. G., Y. S. J. Veldhuizen, J. A. Kanters, J. H. van der Maas, J. Baran, and H. Ratajczan. A variable low-temperature FTIR study of crystalline β-fructopyranose and deuterated analogues. *J Molec Struct* 270: 381–93, 1992.

Lyusternik, L. A. *Convex Figures and Polyhedra.* New York: Dover Publishing Co., 1963.

MacNicol, D. D., F. Toda, and R. Bishop. *Comprehensive Supramolecular Chemistry*, Vol. 6. New York: Pergamon, 1996.

Maidique, M. A., A. Von Hippel, and W. B. Westphal. Transfer of protons through pure ice Ih. Single crystals. III. Extrinsic versus Intrinsic polarization. Surface versus volume conditions. *J Chem Phys* 54: 150–60, 1971.

Mak, T. C. W. Hexamethylenetetramine hexahydrate: A new type of clathrate hydrate. *J Chem Phys* 43: 2799–2805, 1965.

Mak, T. C. W. Orientation of methanol enclathrated in the B-hydroquinone lattice: an x-ray crystallographic study. *J Chem Soc Perkin Trans II*: 1435–37, 1982.

Mak, T. C. W. Crystal structure of tetraethyl ammonium fluoride-water (4/11) $4(C_2H_5)_4N^+F^-\cdot11H_2O$, a clathrate hydrate containing linear chains of edge-sharing $(H_2O)_4F$ tetrahedra and bridging water molecules. *J Incl Phenom* 3: 347–54, 1985.

Mak, T. C. W., H. J. Brunslot, and P. T. Beurskens. Tetraethylammonium chloride tetrahydrate, a double channel host lattice constructed from $(H_2O)_4C\ell^-$ tetrahedra linked between vertices. *J Incl Phenom* 4: 295–302, 1986.

Mak, T. C. W. and Zhou, Cong-Du. "Inclusion Compounds." In: *Crystallography in Modern Chemistry.* New York: John Wiley & Sons, 1992.

Mandelcorn, L. *Non-stoichiometric Compounds.* New York: Academic Press, 1959.

Maple, J. R., J. J. Hwang, T. P. Stockfusel, U. Dinur, M. Waldman, C. S. Ewig, and A. T. Hagler. Derivation of class II force fields, I. Methodology and quantum force field for the alkyl functional group and alkane molecules. *J Comput Chem* 15: 162–82, 1994.

Marchi, R. P. and H. Eyring. Application of significant structure theory to water. *J Phys Chem* 68: 221–28, 1964.

Marsh, R. E. Centrosymmetric or noncentrosymmetric? *Acta Cryst* B42: 193–98, 1986.

Marsh, R. E. and I. Bernal. More space group changes. *Acta Cryst* B51: 300–307, 1995.

Marsh, R. E. and F. H. Herbstein. More space group changes. *Acta Cryst* B44: 77–88, 1988.

Marsh, R. E. and V. Schomaker. Some incorrect space groups in Inorganic Chemistry, vol. 16. *Inorg Chem* 18: 2331–36, 1979.

Mathiouthi, M. and M. O. Portman. Hydrogen-bonding and the sweet taste mechanism. *J Molec Struct* 237: 327–38, 1990.

Maurin, J. K., A. Les, and M. Winnicka-Maurin. Resonance-assisted hydrogen bonds between oxime and carboxyl groups. Comparison of tetrameric structures of 4–methyl-2–oxopentanoic acid oxime and levalinic acid oxime. *Acta Cryst* B51: 232–40, 1995.

Mayo, S. L., B. D. Olafson, and W. A. Goddard, III. DREIDING. A generic force field for molecular simulations. *J Phys Chem* 94: 8897–8909, 1990.

Mazzi, F., F. Jona, and R. Pepinsky. Preliminary X-ray study of the non-ferroelectric phases of Rochelle salt. *Zeit Krist* 108: 359–74, 1957.

McGaw, B. J. and J. A. Ibers. Nature of the hydrogen bond in sodium hydrogen fluoride. *J Chem Phys* 39: 2677–84, 1963.

McLachlan, D. and J. P. Glusker (eds). *Crystallography in North America*. Buffalo, NY: American Crystallographic Association, 1983.

McLaughlin, S. and R. E. Margolskee. The sense of taste. *American Scientist* 82: 538–45, 1994.

McMahon, T. B. and J. L. Beauchamp. A versatile trapped ion cell for ion cyclotron resonance. *Rev Sci Instr* 43: 509–12, 1972.

McMahon, T. B. and J. W. Larson. Gas-phase bifluoride ion. An ion cyclotron resonance determination of the hydrogen bond energy in FHF^- from gas-phase fluoride transfer equilibrium measurements. *J Am Chem Soc* 104: 5848–49, 1982.

McMullan, R. K. and G. A. Jeffrey. Hydrates of tetra-n-butyl and tetra-iso-amyl quaternary ammonium salts. *J Chem Phys* 31: 1231–34, 1959.

McMullan, R. K., T. Jordan, and G. A. Jeffrey. Polyhedral clathrate hydrates. XII. The crystallographic data on hydrates of ethylamine, dimethylamine, trimethylamine, *n*-propylamine (two forms), *iso*-propylamine, diethylamine (two forms), and *tert*-butylamine. *J Chem Phys* 47: 1218–22, 1967.

McMullan, R. K. and Å. Kvick. Neutron diffraction study of the structure II clathrate hydrate: $3.5Xe \cdot 8CC\ell_4 \cdot 136D_2O$ at 13 and 100 K. *Acta Cryst* B46: 390–99, 1990.

McMullan, R. K., Å. Kvick, and P. Popelier. Structures of cubic and orthorhombic phases of acetylene by single-crystal neutron diffraction. *Acta Cryst* B48: 726–31, 1992.

McMullan, R. K., T.C. W. Mak, and G. A. Jeffrey. Polyhedral clathrate hydrates XI. Structure of tetramethylammonium hydroxide pentahydrate. *J Chem Phys* 44: 2338–45, 1966.

McMullan, R. K., R. Thomas, and J. F. Nagle. Structures of the paraelectric and ferroelectric phases of $NaD_3(SO_3)_2$ by neutron diffraction: A vertex model for the ordered ferroelectric state. *J Chem Phys* 77: 637–47, 1982.

Meot-Ner (Mautner), M. The ionic hydrogen bond and ionic solvation, 1. $NH^+ \cdots O$, $NH^+ \cdots N$ and $O—H^+ \cdots O$ bonds. Correlations with proton affinity. Deviations due to structural effects. *J Am Chem Soc* 106: 1257–63, 1984a.

Meot-Ner (Mautner), M. The ionic hydrogen bond and ionic solvation, 2. Solvation of onium ions by one to seven H_2O molecules. Relations between monomolecular, specific and bulk hydration. *J Am Chem Soc* 106: 1265–73, 1984b.

Meot-Ner (Mautner), M. "Ionic Hydrogen Bonds. Part I. Thermochemistry, Structural Implications and Role in Ion Solvation," In: *Molecular Structure and Energetics*. vol. 4, Edited by J. F. Liebman and A. Greenberg. New York: VCH Publishing, pp. 71–104, 1987.

Meot-Ner (Mautner), M. Ionic hydrogen bond and ion solvation, 6. Interaction energies of the acetate ion with organic molecules. Comparison of CH_3COO^- with Cl^-, CN^- and SH^-. *J Am Chem Soc* 110: 3854–58, 1988.

Meot-Ner (Mautner), M. and C. A. Deakyne. Unconventional ionic hydrogen bonds, 1. $CH^{\delta+} \cdots X$ complexes of quaternary ions with n^- and π^- donors. *J Am Chem Soc* 107: 469–73, 1985.

Meot-Ner (Mautner), M. and I. W. Sieck. The ionic hydrogen bond, 1. Sterically hindered bonds. Solvation and clustering of protonated amines and pyridines. *J Am Chem Soc* 105: 2956–61, 1984.

Meot-Ner (Mautner), M. and L. W. Sieck. The ionic hydrogen bond and ion solvation, 5. $O—H \cdots O$ bonds. Gas-phase solvation and clustering of alkoxide and carboxylate ions. *J Am Chem Soc* 108: 7525–26, 1986.

Mettee, H. D. Vapor-phase dissociation energy of $(HCN)_2$. *J Phys Chem* 77: 1762–68, 1973.

Mezei, M., K. L. Beveridge, H. M. Berman, J. M. Goodfellow, J. L. Finney, and S. Neidle.

Monte Carlo studies on water in dCpG/proflavin crystal hydrate. *J Biomol Struct Dynam* 1: 287–97, 1983.

Mhin, B. J., S. S. Kim, H. S.Kim, C. W. Yoon, and K. S.Kim. Ab-initio studies of the water hexamer near degenerate structures. *Chem Phys Lett* 176: 41–45, 1991.

Mikenda, W. and S. Steinbock. Stretching frequency versus bond distance correlation of O—D(H)⋯F hydrogen bonds in solid hydrates. *J Molec Struct* 326: 123–30, 1994.

Millen, D. J. Vibrational spectra and vibrational states of simple gas-phase hydrogen-bonded dimers. *J Molec Struct* 100: 351–77, 1983.

Mitra, J. and C. Ramakrishnan. Analysis of OH⋯O hydrogen bonds. *Int J Pept Protein Res* 9: 27–48, 1977.

Mó, O., M. Yañez, and J. Elguero. Cooperative effects in the cyclic timer of methanol. An ab-initio molecular orbital study. *J Molec Struct (Theochem)* 314: 73–81, 1994.

Mó, M. Yañez, and J. Elguero. Cooperative (nonpairwise) effects in water timers: an ab-initio molecular orbital study. *J Chem Phys* 97: 6628–38, 1992.

Moore, T. S. and T. F. Winmill. The state of amines in aqueous solution. *J Chem Soc* 101: 1635–76, 1912.

Mootz, D. and K. Bartman. Hydrate der fluorsulfonsäure das schmelzdiagram des systems $FSO_3H \cdot H_2O$ und die kristallstruktur des monohydrats $(H_2O)FSO_3$. *Z Anorg Alleg Chem* 592: 171–78, 1991a.

Mootz, D. and K. Bartman. Zur kristallchemie von supersauren: Bildung und struktur die tiefschmelzenden addukte $SbF_6 \cdot CF_3SO_3H$ und $SbF_5 \cdot 7HF$. *Zeit Naturforsch* 46b: 1659–63, 1991b.

Mootz, D. and D. Boenigk. Fluorides and fluoro acids. 12. Complex-anion homology and isomerism in the crystal structures of two potassium poly(hydrogen fluorides) $KF \cdot 2.5HF$ and $KF \cdot 3HF$. *J Am Chem Soc* 108: 6634–36, 1986.

Mootz, D. and D. Boenigk. Poly(hydrogen fluorides) with tetramethylammonium cation: Preparation, stability ranges, crystal structures. $(H_nF_{n+1})^-$ anion homology, hydrogen bonding F—H⋯F. *Zeit Anorg Alleg Chem* 544: 159–66, 1987.

Mootz, D. and A. Deeg. 2–Butyne and hydrogen chloride cocrystallized: Solid-state geometry of $C\ell$—H⋯π hydrogen bonding to carbon-carbon triple bond. *J Am Chem Soc* 114: 5887–88, 1992.

Mootz, D. and A. Merscherz-Quack. Zur kenntris der höchshen hydrate der schwelefsaüre. Bilding und struktur von $H_2SO_4 \cdot 6.5H_2O$ and $H_2SO_4 \cdot 8H_2O$. *Zeit Naturforsch* 42b: 1231–36, 1987.

Mootz, D., E.-J. Oellers, and M. Wiebeke. First examples of type I clathrate hydrates of strong acids. Polyhydrates of hexafluorophosphoric, tetrafluoroboric and perchloric acid. *J Am Chem Soc* 109: 1200–1202, 1987.

Mootz, D., U. Ohms, and W. Poll. Scheltzdiagramm H_2O—HF und strukturen der 1:1 und einer 1:2 phase. *Inorg Alleg Chem* 479: 75–83, 1981.

Mootz, D. and W. Poll. Kristallstruktur der 1:4 phase im system Wasser-fluorowasserstoff und eine neue untersuchung einer der 1:2 phasen. *Z Anorg Alleg Chem* 484: 158–64, 1982.

Mootz, D. and W. Poll. Fluoride und fluorosaüren VI. Verbindungsbildung und kristallstructuren im system ammoniak-fluor wasserstoff. *Zeit Naturforsch* 39b: 290–97, 1984a.

Mootz, D. and W. Poll. Fluoride und fluorosaüren IX. Verbindungsbildung im system NOF-HF sowie. Kristallstructuren von NOF.3HF und NOF.4HF. *Zeit Naturforsch* 39b, 1300–1305, 1984b.

Mootz, D. and M. Schilling. Trifluoroacetic acid tetrahydrate. A unique change from an ionic to a molecular crystal structure on deuteration. *J Am Chem Soc* 114: 7435–39, 1992.

Mootz, D. and R. Seidel. Zum kristallstruktur der metastabilen phase β.NaOH.4H$_2$O. *Zeit Anorg Allg Chem* 582: 162–68, 1990a.

Mootz, D. and R. Seidel. Polyhedral clathrate hydrates of a strong base. Phase relations and crystal structures in the system tetramethylammonium hydroxide water. *J Incl Phenom* 8: 139–57, 1990b.

Mootz, D. and D. Stäben. Clathratehydrate von tetramethylammonium hydroxide: Neue phases und kristallstrukturen. *Zeit Naturforsch* 47b: 263–74, 1992.

Mootz, D. and D. Stäben. Die hydrate von *tert*-butanol: kristallstructur von $Me_3COH \cdot 2H_2O$ und $Me_3COH \cdot 7H_2O$. *Zeit Naturforsch* 48b: 1325–30, 1993.

Mootz, D. and H.-G. Wussow. Crystal structures of pyridine and pyridine trihydrate. *J Chem Phys* 74: 1517–22, 1981.

Morokuma, K., Molecular orbital studies of hydrogen bonds, III. C=O—H····O hydrogen bond in H_2CO—H_2O and H_2CO—$2H_2O$. *J Chem Phys* 55: 1236–44, 1971.

Morokuma, K. Why do molecules interact? The origin of electron donor acceptor complexes, hydrogen bonding and proton affinity. *Accts Chem Res* 10: 294–300, 1977.

Mortimer, M., E. A. Moore, A. Healy, and W. F Pearson. A comparison of ab-initio, NMR and results for a strong OHF hydrogen bond. *J Molec Struct* 271: 149–54, 1992.

Mulliken, R. S. Molecular compounds and their spectra, II. *J Am Chem Soc* 64: 811–24, 1952.

Murphy, V. J., D. Rabinovich, and G. Parkin. False minima and the perils of a polar axis in X-ray structure solutions: Molecular structures of $W(PMe_3)_4H_2X_2$ (X=F, Cℓ, Br) and $W(PMe_3)_4H_2F_2(H_2O)$. *J Am Chem Soc* 117: 9762–63, 1995.

Murray-Rust, P. and J. P. Glusker. Directional hydrogen-bonding to sp^2 and sp^3 hybridized oxygen atoms and its relevance to ligand-macromolecular interactions. *J Am Chem Soc* 106: 1018–25, 1984.

Murray-Rust, P., W. C. Stallings, C. T. Monti, R. K. Preston, and J. P. Glusker. Intermolecular interactions of the C—F bond: the crystallographic environment of fluorinated carboxylic acids and related structures. *J Am Chem Soc* 105: 5761–66, 1983.

Murthy, A. S. M. and C. N. R. Rao. Recent theoretical studies of the hydrogen bond. *J Molec Struct* 6: 253–82, 1970.

Nagle, J. F. Lattice statistics of hydrogen-bonded crystals I. The residual entropy of ice. *J Math Phys* 7: 1484–91, 1986.

Nagle, J. F. "Proton Transport in Condensed Matter," In: *Proton Transfer in Hydrogen Bonded Systems*. Edited by T. Bountis. New York: Plenum Press, pp. 17–27, 1992.

Naito, A., S. Ganapathy, P. Raghunathan, and C. A. McDowell. Determination of the ^{14}N quadrupole coupling tensor and the ^{13}C chemical shielding tensors in a single crystal of L-serine monohydrate. *J Chem Phys* 79: 4173–79, 1983.

Nakamoto, K., M. Margolis, and R. E. Rundel. Stretching frequencies as a function of distances in hydrogen bonds. *J Am Chem Soc* 77: 6480–86, 1955.

Narten, A. H., M. D. Danford, and H. A. Levy. X-ray radial distribution functions for water from 0°C to 200°C. *Oak Ridge National Laboratory Report ORNL 3397*, Oak Ridge, TN: Oak Ridge National Laboratory, 1966.

Neidle, S., H. M. Berman, and H. S. Shieh. Highly structured water network in crystals of a deoxydinucleoside-drug complex. *Nature* 288: 129–33, 1980.

Nemethy, G., M. S. Pottle, and H. A. Scheraga. Energy parameters in polypeptides, 9. Updating the geometrical parameters, non-bonded interactions and hydrogen bond interactions for the naturally occurring amino acids. *J Phys Chem* 87: 1883–87, 1983.

Nemethy, G. and H. A. Scheraga. A model for the thermodynamic properties of liquid water. *J Chem Phys* 36: 3382–3400, 1962.

Newton, M. D. Small water clusters as theoretical models for structural and kinetic properties of ice. *J Phys Chem* 87: 4288–92, 1983.

Newton, M. D., G. A. Jeffrey, and S. Takagi. Application of ab-initio molecular orbital calculations to the structural moieties of carbohydrates. 5. *J Am Chem Soc* 101: 1997–2002, 1979.

Novak, A. "Hydrogen Bonding in Solids. Correlation of Spectroscopic and Crystallographic Data." In: *Structure and Bonding*. 18: 177–216. New York, Berlin, Heidelberg: Springer-Verlag, 1974.

Novoa, J. J., B. Tarron, W. H. Whango, and J. M. Williams. Interaction energies associated with short intermolecular contacts of C—H bonds. Ab-initio computational study of the C—H⋯O contact interaction in CH_4—OH_2. *J Chem Phys* 95: 5179–86, 1991.

Nyholm, P.-G., I. Pascher, and S. Sundell. The effect of hydrogen bonds on the conformation of glycosphingolipids. Methylated and unmethylated cerbroside studied by X-ray single crystal analysis and model calculations. *Chem Phys Lipids* 52: 1–10, 1990.

Ojamäe, L. and K. Hermansson. Water molecules in different crystal surroundings: vibrational O—H frequencies from ab-initio calculations. *J Chem Phys* 96: 9035–44, 1992.

Ojamäe, L., K. Hermansson, C. Pisane, M. Causa, and C. Roetti. Structural, vibrational and electronic properties of a crystalline hydrate from ab-initio periodic Hartree-Fock calculations. *Acta Cryst* B50: 268–78, 1994.

Ojamäe, L., I. Shavitt, and S. J. Singer. Potential energy surfaces and vibrational spectra of $H_5O_2^+$ and large hydrated proton complexes. *Int J Quantum Chem* 29: 657–68, 1995.

Oki, M. and H. Iwamura. Steric effects on the O—H⋯π interaction in 2–hydroxybiphenyl. *J Am Chem Soc* 89: 576–79, 1967.

Onsager, L. and L. K. Runnels. Mechanism for self-diffusion in ice. *Proc Natl Acad Sci USA*, 50: 208–10, 1963.

Onsager, L. and L. K. Runnels. Diffusion and relaxation phenomena in ice. *J Chem Phys* 50: 1089–1103, 1969.

O'Reilly, D. E. and T. Tsang. Deuteron magnetic resonance and proton relaxation times in ferroelectric ammonium sulfate. *J Chem Phys* 46: 1291–1300, 1967.

Orentlicher, M. and P. O. Vogelhut. Structure and properties of liquid water. *J Chem Phys* 45: 4719–24, 1966.

Osawa, E. and K. B. Lipkowitz. *Published Force Field Parameters. Appendix I. Reviews in Computational Chemistry*, Vol. VI. Edited by K. B. Lipkowitz and D. B. Boyd. New York: VCH Publishers, 1995.

Ott, K.-H. and B. Meyer. Molecular dynamics simulations of maltose in water. *Carbohydr Res* 281: 11–34, 1996.

Ottersen, T. On the structure of the peptide linkage. The structures of formamide and acetamide at −165°C and an ab-initio study of formamide, acetamide and N-methylformamide. *Acta Chem Scand* A29: 939–44, 1975.

Paleos, C. M. and D. Tsiourvas. Thermotropic liquid crystals formed by intermolecular hydrogen bonding interactions. *Angew Chem Int Ed Engl* 34: 1696–1711, 1995.

Palin, D. E. and H. M. Powell. The structure of molecular compounds, Part III. Crystal structure of addition compounds of quinol with certain volatile compounds. *J Chem Soc* 208–21, 1947.

Panunto, T. W., Z. Urbánczyk-Lipkowska, R. B. Johnson, and M. C. Etter. Hydrogen bond formation in nitroanilines: the first step in designing acentric materials. *J Am Chem Soc* 109: 7786–97, 1987.

Parry, G. A. The crystal structure of uracil. *Acta Cryst* 7: 313–20, 1954.

Pascher, I., M. Lundmark, P.-G. Nyholm, and S. Sundell. Crystal structures of membrane proteins. *Biochem Biophys Acta* 1113: 339–73, 1992.

Pauling, L. The theoretical prediction of the physical properties of many electron atoms and ions. *Proc Roy Soc London* A114: 181–211, 1927.

Pauling, L. The principles determining the structure of complex ionic crystals. *J Am Chem Soc* 51: 1010–26, 1929.

Pauling, L. The nature of the chemical bond. Application of results obtained from the quantum mechanics and from a theory of paramagnetic susceptibility to the structure of molecules. *J Am Chem Soc* 53: 1367–1400, 1931.

Pauling, L. The structure and entropy of ice and of other crystals with some randomness of atomic arrangement. *J Am Chem Soc* 57: 2680–84, 1935.

Pauling, L. *The Nature of the Chemical Bond*. Ithaca, NY: Cornell University Press, 1939.

Pauling, L. "The Structure of Water." In: *Hydrogen Bonding*. Edited by D. Hadzi and H. W. Thompson. New York: Pergamon Press, pp. 1–6, 1959.

Pauling, L. and R. B. Corey. Configurations of polypeptide chains with favored orientations around single bonds: two new pleated sheets. *Proc Natl Acad Sci USA* 37: 729–40, 1951.

Pauling, L., R. B. Corey, and H. R. Branson. The structure of proteins. Two hydrogen-bonded helical configurations of the polypeptide chain. *Proc Natl Acad Sci USA* 37: 205–11, 1951.

Pauling, L. and R. E. Marsh. The structure of chlorine hydrate. *Proc Natl Acad Sci USA* 38: 112–18, 1952.

Pawelka, Z. and Zeegers-Huyskens. The strange behavior of the hydrogen bond. Complexes of 1.8-bis (dimethylamino)napthalene in solution. *J Molec Struct* 200: 565–72, 1989.

Payzant, J. D., A. J. Cunningham, and P. Kebarle. Gas phase solvation of the ammonium ion by NH_3 and H_2O and stabilities of mixed clusters $NH_4^+(NH_3)_n(H_2O)_n$. *Can J Chem* 51: 3242–49, 1973.

Pedireddi, V. R. and G. R. Desiraju. A crystallographic scale of carbon acidity. *J Chem Soc Chem Commun*, 988–90, 1992.

Pertsin, A. J. and A. I. Kitaigorodskij. *The Atom-Atom Potential Method. Application to Organic Molecular Solids*. Berlin, Heidelberg: Springer-Verlag, 1987.

Perutz, M. F. X-ray analysis of hemoglobin. *Science* 146: 863–69, 1963.

Peterson, K. I. and W. Klemperer. Water-hydrocarbon interactions. Rotational spectroscopy and structure of the water-acetylene complex. *J Chem Phys* 81: 3842–45, 1984.

Peterson, S. W. and H. A. Levy. A single crystal neutron diffraction determination of the hydrogen position in potassium bifluoride. *J Chem Phys* 20: 704–10, 1952.

Peterson, S. W. and H. A. Levy. A single crystal neutron diffraction study of heavy ice. *Acta Cryst* 10: 70–76, 1957.

Pfeiffer, P. Zur kenntnis der sauren salze der carbonsäuren. *Berichte* 47: 1580–95, 1914.

Philp, D. and J. Fraser Stoddart. Self-assembly in natural and unnatural systems. *Angew Chem Int Ed Engl* 35: 1154–96, 1996.

Pickering, S. U. The hydrate theory of solutions. Some compounds of alkylamines and ammonium with water. *Trans Chem Soc I* 63: 141–95, 1893.

Pimental, G. C. and A. L. McClellan. *The Hydrogen Bond*. San Francisco: Freeman, 1960.

Pimental, G. C. and C. H. Sederholm. Correlation of infrared stretching frequencies and hydrogen bond distances in crystals. *J Chem Phys* 24: 639–41, 1956.

Piraud, B., G. Baudoux, and F. Durant. A data base study of intermolecular HN----O hydrogen bonds for carboxylates, sulfonates and monohydrogen phosphates. *Acta Cryst* B51: 103–7, 1995.

Pisani, C., R. Dovesi, and C. Roetti. *Hartree-Fock ab-initio Treatment of Crystalline Systems. Lecture Notes in Chemistry, 48*. Heidelberg: Springer-Verlag, 1988.

Podlahová, J. and J. Loub. Structure of 4–deoxy-4–fluoro-β-D-fructopyranose, $C_6H_{11}FO_5$. *Acta Cryst* C40: 1284–86, 1984.

Polian, A. and M. Grimsditch. New high-pressure phase of H_2O: Ice X. *Phys Rev Lett* 52: 1312–14, 1984.

Pople, J. A. Molecular association in liquids II. A theory of the structure of water. *Proc Roy Soc London* A,205: 163–78, 1951.

Pople, J. A., W. G. Schneider, and H. J. Bernstein. *High Resolution Nuclear Magnetic Resonance*. New York: McGraw Hill, 1959.

Powell, H. M. "Clathrates." In: *Non-stoichiometric Compounds*. Edited by L. Mandelcorn. New York, London: Academic Press, 1964.

Preissner, R., V. Enger, and W. Saenger. Occurrence of bifurcated three-center hydrogen bonds in proteins. *Fed Eur Biochem Soc* 288: 192–96, 1991.

Purcell, K. F. and R. S. Drago. Theoretical aspects of the linear enthalpy wave number shift relation for hydrogen-bonded phenols. *J Am Chem Soc* 89: 2874–79, 1967.

Ragle, J. L., G. Minott, and M. Mokarram. Deuterium quadrupole coupling in solid complexes of diethyl ether, acetone and mesitylene with chloroform-d. *J Chem Phys* 60: 3184–88, 1974.

Rahman, A. and F. H. Stillinger. Hydrogen-bond patterns in liquid water. *J Am Chem Soc* 95: 7943–48, 1973.

Ramanadham, M. and R. Chidambaram. "Amino Acids: Systematics of Molecular Structure Conformation and Hydrogen Bonding." In: *Advances in Crystallography*. Edited by R. Srinivasan. New Delhi: Oxford and IBH Publishing Co., 1978.

Rashin, A. A. Buried surface area. Conformational entropy and protein stability. *Biopolymers* 23: 1605–19, 1984.

Rasmussen, K. Conformation and anomeric ratio of α-D-glucopyranose in different energy functions. *Acta Chem Scand* A36: 323–27, 1982.

Ratajczak, H. and W. J. Orville-Thomas. Hydrogen-bond studies. Part 1. The relation between vibrational frequencies and bond lengths in O—H---O hydrogen bonded systems. *J Molec Struct* 1: 449–61, 1967–68.

Ratajczak, H. and W. J.Orville-Thomas. *Molecular Interactions*, vol. 1. New York: John Wiley & Sons, 1980. Edited by H. Ratajczak and W. J.Orville-Thomas.

Ratcliffe, E. I. and D. E. Irish. "The Nature of the Hydrated Proton, Part One. The Solid and Gaseous States." In: *Water Science Reviews*, vol. 2. Edited by F. Franks. UK: Cambridge University Press, pp. 149–214, 1985.

Ravishanker, G., S. Vijayakumar, and D. I. Beveridge. "STRIPS. An Algorithm for Generating Two-Dimensional Hydrogen-Bond Topology Diagrams for Proteins." In: *Modeling the Hydrogen Bond*. Edited by D. A. Smith. *Am Chem Soc Symp Ser 569*. Washington DC: American Chemical Society, 1994.

Raymond, S., B. Henrissat, D. T. Qui, Å. Kvick, and H. Chanzy. The crystal structure of methyl β-cellotrioside monohydrate 0.25 ethanolate and its relationship to cellulose II. *Carbohydr Res* 277: 209–29, 1995a.

Raymond, S., D. Heyraud, D. T. Qui, Å. Kvick, and H. Chanzy. Crystal and molecular structure of β-D-cellotetraose hemihydrate as a model of cellulose II. *Macromolecules* 28: 2096–2100, 1995b.

Rebek, J. Molecular recognition and biophysical organic chemistry. *Accts Chem Res* 23: 399–404, 1990.

Reuben, J. Isotopic multiplets in the carbon-13 NMR spectra of polyols with partially deuterated hydroxyls. Fingerprints of molecular structure and hydrogen bonding effects in the ^{13}C NMR spectra of monosaccharides with partially deuterated hydroxyls. *J Am Chem Soc* 106: 6180–86, 1984.

Reuben, J. Intramolecular hydrogen bonding as reflected in the deuterium isotope effects on carbon-13 chemical shifts. Correlation with hydrogen bond energies. *J Am Chem Soc* 108: 1735–38, 1986.

Richards, F. M. Areas, volumes, packing and protein structure. *Ann Rev Biophys Bioeng* 6: 151–76, 1977.

Ripmeester, J. A. and C. I. Ratcliffe. ^{129}Xe NMR studies of clathrate hydrates: New guests for structure II and structure H. *J Phys Chem* 94: 8773–76, 1990.

Robertson, J. M. and A. R. Ubbelohde. Structure and thermal properties associated with some hydrogen bonds in crystals, I. The isotope effect. *Proc Roy Soc* A170, 222–40, 1939.

Robiette, A. G. and J. L. Duncan. High resolution vibration-rotation spectroscopy. *Ann Rev Phys Chem* 34: 245–72, 1983.

Rodriguez, J. Semi-empirical study of compounds with intramolecular O—H---O hydrogen bonds, II. Further verification of a modified MNDO method. *J Comput Chem* 15: 183–89, 1994.

Rohlfing, C. M., L. C. Allen, and R. Ditchfield. Proton chemical shift tensors in hydrogen-bonded dimers of RCOOH and ROH. *J Chem Phys* 79: 4958–66, 1983.

Rosenberg, J. M., N. C. Seeman, R. O. Day, and A. Rich. RNA double-helical fragment at atomic resolution II. The crystal structure of sodium quanylyl-3′,5′-cytidine monohydrate. *J Mol Biol* 104: 145–67, 1976.

Rosso, J.-C. and L. Carbonnel. Clathrates et autres hydrates des aldéhydes et des cétones. Stoichiométric des combinaisons déduite des diagrammes de phase eau-aldéhydes et eau-cétones. *Bull Soc Chim France* 9–10: I-381–87, 1978.

Rundle, R. E. and M. Parasol. O—H stretching frequencies in very short and possibly symmetrical hydrogen bonds. *J Chem Phys* 20: 1487–88, 1952.

Rush, J.J., P. Leung, and T. I. Taylor. Motions of water molecules in potassium ferrocyanide trihydrate, water and ice. A neutron scattering study. *J Chem Phys* 45: 1312–17, 1966.

Ryan, L. M., R. E. Taylor, A. J. Paff, and B. C. Gerstein. An experimental study of resolution of proton chemical shifts in solids: combined multiple pulse NMR and magic angle spinning. *J Chem Phys* 72: 508–15, 1980.

Ryan, M. D. "Effect of Hydrogen Bonding on Molecular Electrostatic Potentials." In: *Modeling the Hydrogen Bond.* Edited by D. A. Smith. American Chemical Society Symposium Series 569, Washington, DC, 1994.

Rybak, S., B. Zeriorski, and K. Szalewicz. Many-body symmetry-adapted perturbation theory of intermolecular interactions H2O and HF dimers. *J Chem Phys* 95: 6577–6601, 1991.

Saenger, W. Circular hydrogen bonds. *Nature (London)* 279: 343–44, 1979.

Saenger, W. "Structural Aspects of Cyclodextrins and Their Inclusion Compounds." In: *Inclusion Compounds,* vol. II. Edited by J. E. D. Davies and J. L. Atwood. New York: Academic Press, pp. 231–59, 1984a.

Saenger, W. *Principles of Nucleic Acid Structure.* Berlin: Springer, 1984b.

Saenger, W., C. H. Betzel, B. E. Hingerty, and G. M. Brown. Flip-flop hydrogen bonding in a partially disordered System. *Nature (London)* 296: 581–83, 1982.

Saenger, W. and K. Lindner. OH clusters with homodromic circular arrangement of hydrogen bonds. *Angew Chem Int Ed Engl* 19: 398–99, 1980.

Sakya, P., J. M. Seddon, and R. H. Templer. Lyotropic phase behaviour of n-octyl-1-O-β-D-glucopyranoside and its thio derivative n-octyl-1-S-β-D-glucopyranoside. *J Phys II, France,* 1311–31, 1994.

Salem L. Theoretical interpretation of force constants. *J Chem Phys* 38: 1227–36, 1963.

Sastry, D. L., K. Takegoshi, and C. A. McDowell. Determination of the ^{13}C chemical-shift tensors in a single crystal of methyl α-D-glucopynonosute. *Carbohyd Res* 165: 161–71, 1987.

Saupe, T., C. Krüger, and H. A. Staub. 4,5-Bis(dimethylamino)phenanthrene and 4,5-bis(dimethylamino)-9,10-dihydrophenanthrene: synthesis and 'proton sponge' properties. *Angew Chem Int Ed Engl* 25: 451–53, 1986.

Savage, H. Water structure in vitamin B_{12} coenzyme crystals I and II. *Biophys J* 50: 947–80, 1986a.

Savage, H. F. J. "Water Structure in Crystalline Solids, Ices and Proteins." In: *Water Science Review*, vol. 2. Edited by F. Franks. Cambridge: Cambridge University Press, pp. 67–147, 1986b.

Savage, H. F. J., C. J. Elliot, C. M. Freeman, and J. L. Finney. Lost hydrogen bonds and buried surface area: rationalizing stability in globular proteins. *J Chem Soc Farad Trans* 89(15): 2609–17, 1993.

Savage, H. F. J. and J. Finney. Repulsive regularities of water structure in ices and crystalline hydrates. *Nature (London)* 322: 717–20, 1986.

Schaefer, T. A relationship between hydroxyl proton chemical shifts and torsional frequencies in some ortho-substituted phenol derivatives. *J Phys Chem* 79: 1888–90, 1975.

Scheiner, S. Ab-initio studies of hydrogen bonds: the water dimer paradgm. *Ann Rev Phys Chem* 45: 23–56, 1994.

Scheiner, S. and M. Čuma. Relative stability of hydrogen and deuterium bonds. *J Am Chem Soc* 118: 1511–21, 1996.

Scheiner, S. and T. Kar. The non-existence of specially stabilized hydrogen bonds in enqymes. *J Am Chem Soc* 117: 6970–75, 1995.

Schlemper, E. O. and W. C. Hamilton. Neutron diffraction study of the structures of ferroelectric and paraelectric ammonium sulfate. *J Chem Phys* 44: 4498–4509, 1966.

Schlenk, W. Jr. Die harnstoff-addition der aliphatischen verbindungen. *Ann Chem* 565: 204–40, 1949.

Schlenk, W. Jr. Die thioharnstoff-addition der organishe verbindungen. *Ann Chem* 573: 141–62, 1951.

Schneider, B., D. Cohen, and H.M. Berman. Hydration of DNA bases: analysis of crystallographic data. *Biopolymers* 32: 725–50, 1992.

Schneider, B., D. M. Cohen, L. Schleiffer, A. R. Srinvasan, W. K. Olson, and H. M. Berman. A systematic method for studying the spatial distribution of water molecules around nucleic acid bases. *Biophys J* 65: 2291–2303, 1993.

Schroeder, W. Die geschehichte der gas hydrate Ahren's sammlung chemischer und chemischer-technisches. *Vortrage* 29: 1–98, 1927.

Schuster, P., G. Zundel, and C. Sandorfy. *The Hydrogen Bond. Recent Developments in Theory and Experiment.* Vols. I–III. Amsterdam: North Holland, 1976.

Schwartz, B. and D. G. Drueckhammer. A simple method for determining the relative strengths of normal and low-barrier hydrogen bonds in solution. Implications for enzyme catalysis. *J Am Chem Soc* 117: 11902–5, 1995.

Sciortino, F., A. Geiger, and H. E. Stanley. Effects of defects in molecular mobility in liquid water. *Nature* 354: 218–21, 1991.

Seeman, N. C., J. M. Rosenberg, F. L. Suddath, J. J. P. Kim, and A. Rich. RNA double helical fragment at atomic resolution I. The crystal and molecular structure of sodium adenylyl-3′,5′-uridine. *J Mol Biol* 104: 109–44, 1976.

Seiler, P. "Measurement of Accurate Bragg Intensities." In: *Accurate Molecular Structures.* Edited by A. Domenicano and I. Hargittai. Oxford: Oxford University Press, 1992.

Seiler, P., G. R. Weisman, E. D. Glendening, F. Weinhold, V. B. Johnson, and J. D.Dunitz. Observation of an eclipsed C_{sp3}—CH_3 bond in a tricyclic orthoamide. Experimental and theoretical evidence for C—H····O hydrogen bonds. *Angew Chem Int Ed Eng* 26: 1175–77, 1987.

Senderowitz, H., C. Parish, and W. C. Still. Carbohydrates: United atom AMBER. Parameterization of pyranoses and simulations yielding anomeric free energies. *J Am Chem Soc* 118: 2078–84, 1996.

Shallcross, F. V. and G. B. Carpenter. The crystal structure of cyanoacetylene. *Acta Cryst* 11: 490–96, 1958.

Shallenberger, H. S. and T. E. Acree. Molecular theory of sweet taste. *Nature* 216: 480–82, 1967.

Shibuya, I. and T. Mitsui. The ferroelectric phase transitions in (glycine)$_3$ H$_2$S$_0$$_4$ and critical x-ray scattering. *J Phys Soc Japan* 16: 479–89, 1961.

Sidgwick, N. V. and R. K. Callow. Abnormal benzene derivatives. *J Chem Soc* 125: 527–38, 1924.

Siegle, C. and M. Weithase. Spin-gitter relaxation der protonen in hexagonalein eis. *Zeit Phys* 219: 364–80, 1969.

Singh, U. C. and P. A. Kollman. A water dimer potential based on ab-initio calculations using Morokuma component analysis. *J Chem Phys* 83: 4033–40, 1985.

Smith, A. E. The crystal structure of urea-hydrocarbon and thiourea-hydrocarbon complexes. *J Chem Phys* 18: 150–51, 1950.

Smith, A. E. The crystal structure of the urea-hydrocarbon complexes. *Acta Cryst* 5: 224–35, 1952.

Soda, G. and T. Chiba. Deuteron magnetic resonance study of cupric sulphate pentahydrate. *J Chem Phys* 50: 439–55, 1969.

Solodovinikov, S. F., T. M. Polyanskaya, V. I. Alekseev, L. S. Aladka, Y. A. Dyadin, and V. V. Balakin. A new type of clathrate hydrate. The crystal structure of (*iso*-C$_5$H$_{11}$)$_4$P.Br.32H$_2$O. *Kristallografiya* 27: 247–54, 1982; *Sov Phys Crystall* 27: 151–58, 1982.

Souhassou, M., C. Lecomte, N.-E. Ghermani, M. M. Rohmer, R. Wiest, M. Benard, and R. H. Blessing. Electron distributions in peptides and related molecules, 2. An experimental and theoretical study of (Z)-N-acetyl-α,β-dehydrophenylalanine methylamide. *J Am Chem Soc* 114: 2371–82, 1992.

Spackman, M. A. A simple qualitative model of hydrogen bonding. *J Chem Phys* 85: 6587–6601, 1986.

Spackman, M. A. A simple qualitative model of hydrogen bonding. Application to more complex systems. *J Phys Chem* 91: 3179–86, 1987.

Spackman, M. A. and A. S. Brown. Charge densities from X-ray diffraction data. *Ann Rpts Prog Chem for 1994C*. 173–212, Roy. Soc. Chem., UK. 1995.

Spackman, M. A. and R. F. Stewart. "Electrostatic Properties From Accurate Diffraction Data." In: *Methods and Applications in Crystallographic Computing*. Edited by S. R. Hall and T. Ashida. Oxford: Clarenden Press, 1984.

Speakman, J. C. "Acid Salts of Carboxylic Acids. Crystals with Some 'Very Short' Hydrogen Bonds." In: *Structure and Bonding*, vol. 12. Berlin, Heidelberg, New York: Springer-Verlag, pp. 141–99, 1972.

Stäben, D. and D. Mootz. Die kristallinen hydrate von tetramethyl ammonium fluoride. Bilding, struktur, wasserstoff brüchen bindung. *Zeit Naturforsch* 48b: 1057–64, 1993.

Stäben, D. and D. Mootz. The 7.25 hydrate of tert-butylamine. A semi-clathrate and complex variant of the cubic 12A structure type. *J Incl Phen and Molec Recog in Chem* 22: 145–54, 1995.

Starikov, E. B. Three-dimensional crystal orbital calculations on mononucleotide crystallohydrates. II. Diprotonated mononucleotides. *Int J Quantum Chem* 22: 145–58, 1995.

Staub, H. A. and T. Saupe. 'Protonenschwamm' verbindungen and die geometric von wasserstoff bruchen: aromatic stickstoffbasen mit undewöhnlicher bazisitat. *Angew Chem* 100: 895–909, 1988.

Staub, H. A., T. Saupe, and G. Krüger. 4,5–Bis(dimethylamino) fluorene. A new proton sponge. *Angew Chem Int Ed Eng* 22: 731–32, 1983.

Steiner, T., S. A. Mason, and W. Saenger. Cooperative O—H···O hydrogen bonds in β-cyclodextrin-ethanol-octahydrate at 15 K. A neutron diffraction study. *J Am Chem Soc* 112: 6184–90, 1990.

Steiner, T., S. A. Mason, and W. Saenger. Disordered guest and water molecules. Three-center and flip-flop O—H----O hydrogen bonds in crystalline β-cycloextrin ethanol octahydrate at T = 295 K. A neutron and X-ray diffraction study. *J Am Chem Soc* 113: 5676–87, 1991.

Steiner, T. and W. Saenger. Geometric analysis of non-ionic O—H----O hydrogen bonds and non-bonding arrangements in neutron diffraction studies of carbohydrates. *Acta Cryst* B48: 819–27, 1992a.

Steiner, T. and W. Saenger. Covalent bond lengthening in hydroxyl groups involved in three-center and in cooperative hydrogen bonds. Analysis of low-temperature neutron diffraction data. *J Am Chem Soc* 114: 7123–26, 1992b.

Steiner, T. and W. Saenger. Geometry of C—H----O hydrogen bonds in carbohydrate crystal structures. Analysis of neutron diffraction data. *J Am Chem Soc* 114: 10146–54, 1992c.

Steiner, T. and W. Saenger. Role of C—H----O hydrogen bonds in the coordination of water molecules. Analysis of neutron data. *J Am Chem Soc* 115: 4540–47, 1993a.

Steiner, T. and W. Saenger. The ordered water cluster in vitamin B_{12} coenzyme at 15 K is stabilized by C—H----O hydrogen bonds. *Acta Cryst* D49: 592–93, 1993b.

Steiner, T. and W. Saenger. Reliability of assigning O—H----O hydrogen bonds to short intermolecular O----O separations in cyclodextrin and oligosaccharide crystal structures. *Carbohydr Res* 259: 1–12, 1994.

Steiner, T., W. Saenger, and R. E. Lechner. Dynamics of orientationally disordered hydrogen bonds and of water molecule in a molecular cage: a quasi-elastic neutron scattering study of β-cyclodextrin 11H_2O. *Molec Phys* 72: 1211–32, 1991.

Steiner, T., E. B. Starikov, A. M. Amado, and J. J. C. Teixeira-Dias. Weak hydrogen bonding, part 2. The hydrogen bonding nature of short C—H----π contacts: crystallographic, spectroscopic and quantum mechanical studies of some terminal alkanes. *J Chem Soc Perkin Trans I*, 1321–26, 1995.

Steinwender, E., E. T. G. Lutz, J. H. van der Maas, and J. A.Kanters. 2–Ethynladamantan-2–ol: a model compound with distinct OH----π and CH----O hydrogen bonds. *Vibrational Spect* 4: 217–29, 1993.

Stevens, E. D. Low temperature experimental electron density distribution in formamide. *Acta Cryst* B34: 544–51, 1978.

Stevens, E. D. and P. Coppens. Experimental electron density distributions of hydrogen bonds. High resolution study of α-oxalic acid dihydrate at 100 K. *Acta Cryst* B36: 1864–76, 1980.

Stevens, E. D., M. S. Lehman, and P. Coppens. Experimental electron density distribution of sodium hydrogen diacetate. Evidence for covalency in a short hydrogen bond. *J Am Chem Soc* 99: 2829–31, 1977.

Stevens, R. D., R. Bau, D. Milstein, O. Blum, and T. F. Koetzle. Concept of the H($\delta+$)----H($\delta-$) interaction. A low temperature neutron diffraction study of *cis*[IrH(OH)(PMe$_3$)]PF$_6$. *J Chem Soc Dalton Trans*, 1429–32, 1990.

Stevenson, D. P. On the monomer concentration in liquid water. *J Phys Chem* 69: 2145–52, 1965.

Stewart, J. J. P. Optimization of parameters for semi-empirical methods, II. *J Comput Chem* 10: 209–64, 1989.

Stewart, R. F. Electron population analysis with rigid pseudoatoms. *Acta Cryst* A32: 565–74, 1976.

Stewart, R. F. Mapping electrostatic potentials from diffraction data. *God Jugosl Cent Kristalogr* 17: 1–24, 1982.

Stewart, R. F. "Electrostatic Properties of Molecules from Diffraction Data." In: *The Application of Charge Density Research to Chemistry and Drug Design.* Edited by G. A. Jeffrey and J. Piniella, NATO-ASI Series B, 250, New York: Plenum Press, 63–101, 1991.

Stewart, R. F. and B. M. Craven. Molecular electrostatic potentials from crystal diffraction. The neurotransmitter γ-aminobutyric acid. *Biophys J* 65: 998–1005, 1993.

Stillinger, F. H. Water revisited. *Science* 209: 451–57, 1980.

Stillinger, F. H. and A. Rahman. Improved simulation of liquid water by molecular dynamics. *J Chem Phys* 60: 1545–54, 1974.

Stipanovic, A. J. and A. Sarko. Packing analysis of carbohydrates and polysaccharides, 6. Molecular and crystal structure of cellulose II. *Macromolecules* 9, 851–57, 1976.

Su, Z. and P. Coppens. On the mapping of electrostatic propeties from the multipole description of charge density. *Acta Cryst* A48: 188–97, 1992.

Sundaralingam, M. and Y. C. Sikhardu. Water inserted α-helical segments implicate reverse turns as folding intermediates. *Science* 244: 1333–37, 1989.

Sutor, D. J. The C—H⋯O hydrogen bond in crystals. *Nature (London)* 195: 68–69, 1962.

Sutor, D. J. Evidence for the existence of C—H⋯O hydrogen bonds in crystals. *J Chem Soc* 1105–10, 1963.

Symons, M. C. R. The structure of liquid water. *Nature* 239: 257–59, 1972.

Symons, M. C. R. Water structure and reactivity. *Accts Chem Res* 14: 179–87, 1981.

Symons, M. C. R. Liquid water—the story unfolds. *Chemistry in Britain*, 491–94, 1989.

Szezesniak, M. M., G. Chalasínski, S. M. Cybulski, and P. Cieplak. Ab-initio study of the potential energy surface of CH_4–H_2O. *J Chem Phys* 98: 3078–85, 1993.

Takusagawa, F. and T. F. Koetzle. Neutron diffraction study of quinolinic acid recrystallized from D_2O. Evaluation of temperature and isotope effects in the structure. *Acta Cryst* B35: 2126–35, 1979.

Taylor, J. C., M. H. Mueller, and R. L. Hitterman. A neutron diffraction study of ferroelectric KFCT ($KFe(CN)_6$·$3D_2O$) above the Curie temperature. Abstr. J6, *Meeting of the Am Crystallogr Assoc* Atlanta, GA, 1967.

Taylor, R. An empirical potential for the O—H⋯O hydrogen bond. *J Molec Struct* 71: 311–25, 1981.

Taylor, R. and O. Kennard. Crystallographic evidence of the existence of C—H⋯O, C—H⋯N and C—H⋯Cℓ hydrogen bonds. *J Am Chem Soc* 104: 5063–70, 1982.

Taylor, R., O. Kennard, and W. Versichel. Geometry of the NH⋯O=C hydrogen bond. 1. Lone-pair directionality. *J Am Chem Soc* 105: 5761–66, 1983.

Taylor, R., O. Kennard, and W. Versichel. Geometry of the NH⋯O=C hydrogen bond. 3. Hydrogen bond distances and angles. *Acta Cryst* B40: 280–88, 1984a.

Taylor, R., O. Kennard, and W. Versichel. Geometry of the NH⋯O=C hydrogen bond. 2. Three-center (bifurcated) and four-center (trifurcated) bonds. *J Am Chem Soc* 106: 244–48, 1984b.

Teeter, M. M. Water structure of a hydrophobic protein at atomic resolution: Pentagon rings of water molecules in crystals of Crambin. *Proc Natl Acad Sci USA* 81: 6014–18, 1984.

Teeter, M. M. Water-protein interactions. Theory and experiment. *Ann Rev Biophys Chem* 20: 577–600, 1991.

Teller, R. G. and R. Bau. *Crystallographic Studies of Transition Metal Hydride Complexes.* In: "Structure and Bonding," vol. 44, pp. 1–82. New York: Springer-Verlag, 1981.

Texter, J. Nucleic-acid-water interactions. *Prog Biophys Mol Biol* 33: 83–97, 1978.

Thomas, J. A., J. Tellgren, and I. Olovsson. Hydrogen-bond studies LXXXIV. An X-ray diffraction study of the structures of $KHCO_3$ and $KDCO_3$ at 298, 219, 95 K. *Acta Cryst* B30: 1155–66, 1974.

Tomiie, Y. The electron distribution and the location of the bonded hydrogen atoms in crystals. *J Phys Soc Jpn* 13: 1030–37, 1958.

Tomkinson, J. The vibrations of hydrogen bonds. *Spectrochem Acta* 48A: 329–48, 1992.

Toyoda, H., S. Waku, H. Shibata, and Y. Tanaka. Some ferroelectric properties of triglycine sulfate. *J Phys Soc Japan* 14: 105, 1959.

Tsang, T. and D. E. O'Reilly. Deuteron magnetic resonance of ferroelectric potassium ferrocyanide trihydrate. *J Chem Phys* 43: 4234–49, 1965.

Tse, Y. C. and M. D. Newton. Theoretical observations on the structural consequences of cooperativity in H----O hydrogen bonding. *J Am Chem Soc* 99: 611–13, 1977.

Tucker, E. E. and E. Lippert. "High Resolution Nuclear Magnetic Resonance Studies of Hydrogen Bonding." In: *The Hydrogen Bond. Recent Developments in Theory and Experiment.* Edited by P. Schuster, G. Zundel, and C. Sandorfy. Amsterdam: North Holland, 1976.

Turi, L. and J. J. Dannenberg. Molecular orbitals of C—H----O H-bonded complexes. *J Phys Chem* 97: 7899–7909, 1993a.

Turi, L. and J. J. Dannenberg. Molecular orbital study of acetic acid aggregation, I. Monomers and dimers. *J Phys Chem* 97: 12197–12204, 1993b.

Ubbelohde, A. R. and K.J. Gallagher. Acid-base effects in hydrogen bonds in crystals. *Acta Cryst* 8: 71–83, 1955.

Ubbelohde, A. R. and I. Woodward. Structure and thermal properties of crystals. VI. The role of hydrogen bonds in Rochelle salt. *Proc Roy Soc* A185: 448–64, 1945.

Umeyana, H. and K. Morokuma. Origin of alkyl substituent effect in the proton affinities of amines, alcohols and ethers. *J Am Chem Soc* 98: 4400–4404, 1976.

Umeyana, H. and K. Morokuma. The origin of hydrogen bonding: an energy decomposition study. *J Am Chem Soc* 99: 1316–32, 1977.

Uvarov E. B., D. R. Chapman, and A. Isaacs. *The Penguin Dictionary of Science.* London: Penguin Books, 1971.

Van Alsenoy, C., A. D. French, M. Cao, S. Q. Newton, and I. Schäfer. Ab-initio and molecular mechanics studies of the distorted linkage in raffinose. *J Am Chem Soc* 116: 9590–95, 1994.

van Eijck, B. P., W. T. M. Mooij, and J. Kroon. Attempted prediction of the crystal structures of six monosaccharides. *Acta Cryst* B51: 99–102, 1995.

Van Gunsteren, W. F. *GROMOS. Groningen Molecular Simulation Computer Program Package.* University of Groningen, The Netherlands, 1987.

Van Gunsteren, W. F. and H. J. C. Berendsen. Computer simulation by molecular dynamics. Methodology, applications and perspectives in chemistry. *Angew Chem Int Edit Engl* 29: 992–1023, 1990.

van Mourik, T. and F. B. van Diujneveldt. Ab-initio calculations on the C—H----O hydrogen-bonded systems CH_4H_2O, CH_3NH_2–H_2O and $CH_3NH_3^+$—H_2O. *J Molec Struct Theochem* 341: 63–73, 1995.

Vanquickenborne, L. G. *Quantum Chemistry of the Hydrogen Bond in Intermolecular Forces.* Edited by P. L. Huyskens, W. A. P. Loel, and T. Zeegens-Huysken. Berlin, New York: Springer Verlag, pp. 31–53, 1991.

Veeman, W. S. ^{13}C Chemical shift tensors in organic single crystals. *Phil Trans Roy Soc Lond* A299: 629–41, 1981.

Vinogradov, S. N. "Structural Aspects of Hydrogen Bonding in Amino Acids, Peptides, Proteins and Model Systems." In: *Molecular Interactions*, vol. 2. Edited by H. Ratajczak and W. J. Orville-Thomas. New York: John Wiley & Sons, 1980.

Vinogradov, S. N. , and R. H. Linnell. *Hydrogen Bonding.* New York: Van Nostrand Reinhold, 1971.

Voet, D. and A. Rich. The crystal structures of purines, pyrimidines and their intermolecular complexes. *Prog Nucl Acid Res & Mol Biol* 10: 183–265, 1970.

von Stackelberg, M. and H. R. Muller. On the structure of gas hydrates. *J Chem Phys* 19: 1319, 1951.

Vos, W. L., L. W. Finger, R. J. Hemley, and H. Mao. Novel H_2–H_2O clathrates at high pressure. *Phys Rev Lett* 71: 3150–53, 1993.

Wahl, M. C., S. T. Rao, and M. Sundaralingam. The structure of r(UUCGCG) has a 5-UU-over-hang exhibiting Hoogsteen-like trans U·U base pairs. *Nature Structural Biology* 3: 24–31, 1996.

Wahl, M. C. and M. Sundaralingam. New crystal structures of nucleic acids and their complexes. *Current Opinion in Structural Biology* 5: 282–95, 1995.

Waku, S., J. Hirabayashi, H. Iwasaki, and R.Kiriyama. Ferroelectrics in potassium ferrocyanide trihydrate. *J Phys Soc Jap* 14: 973, 1959.

Wall, T. T. and D. F. Hornig. Raman intensities of HDO and structure of liquid water. *J Chem Phys* 43: 2079–87, 1965.

Walrafen, E. "Raman and Infrared Spectral Investigations of Water Structure." In: *Water: A Comprehensive Treatise*, vol 1. Edited by F. Franks. New York, London: Plenum Press, 1972.

Warshel, A., A. Papazyan, and P. Kollman. On low barrier hydrogen bonds and enzyme catalysis. *Science* 269: 102–3, 1995.

Warshel, A. and S. T. Russell. Calculation of electrostatic interactions in biological systems and in solutions. *Quart Rev Biophys* 17(3): 283–422, 1984.

Watson, J. D. "The Importance of Weak Interactions." In: *Molecular Biology of the Gene*. New York: Benjamin, 1965.

Watson, J. D. *The Double Helix*. New York: The New American Library Inc., 1968.

Watson, J. D. and Crick, F. H. C. Structures of deoxyribose nucleic acid. *Nature (London)* 171: 737–38, 1953.

Waugh, J. S., F. B. Humphrey, and D. M. Yost. Magnetic resonance spectrum of a linear three-spin system. The configuration of the bifluoride ion. *J Phys Chem* 57: 486–90, 1953.

Wei, S., Z. Shi, and A. N. Castleman. Mixed cluster ions as a structure probe: Experimental evidence for clathrate structure of $(H_2O)_{20}H^+$ and $(H_2O)_{21}H^+$. *J Chem Phys* 94: 3268–70, 1991.

Wei, Y., R. Barton, and B. Robertson. Electron-density and electrostatic potential distributions in nucleoside analogs, I. An experimental study of 3′-O-acetyl-2′-deoxy-5-methoxymethyl uridine. *Acta Cryst* B50: 161–73, 1994.

Weiderman, E. G. and G. Zundel. Field-dependent mechanism of anomalous proton conductivity and the polarizability of hydrogen bonds with tunneling proton. *Zeit Naturforsch* 25: 627–34, 1970.

Weiner, J. J., P. A. Kollman, D. T. Nguyen, and D. A. Case. An all atom force field for simulation of proteins and nucleic acids. *J Comput Chem* 7: 230–52, 1986.

Weiner, P. K. and P. A. Kollman. AMBER: assisted model building with energy refinement. A general method for modeling molecules and their interactions. *J Comput Chem* 2: 287–303, 1981.

Weiner, S. J., P. A. Kollman, D. A. Case, U. C. Singh, C. Ghio, G. Alagona, S. Profita, Jr. and P. Weiner. A new force field for molecular mechanical simulation of nucleic acids and proteins. *J Am Chem Soc* 106: 765–84, 1984.

Wells, A. F. *Structural Inorganic Chemistry*. Oxford: Clarendon Press, 1962.

Wenes, D. and S. A. Rice. A new model for liquid water. *J Am Chem Soc* 94: 8983–9000, 1972.

Werner, A. Über Haupt- und nebenvalenzen und die constitution der ammoniumverbindungen. *Liebigs Ann* 322: 261–97, 1902.

Westhof, E. Hydration of oligonucleotides in crystals. *Int J Biol Macromol* 9: 186–92, 1987.

Westhof, E. Water: an integral part of nucleic acid structure. *Ann Rev Biophys Biochem* 17: 125–44, 1988.

Westhof, E. and D. L. Beveridge. "Hydration of Nucleic Acids." In: *Water Science Reviews*, vol. 4. Edited by F. Franks. Oxford: Cambridge Univ. Press, pp. 24–136, 1989.

Whalley, E. The O—H distance in ice. *Molec Phys* 28: 1105–8, 1974.

Wiebeke, M. and D. Mootz. Clathrathydrate starker säuren isostrukturelle hexahydrate der hexa-fluoroarsen (V)-säure und hexafluoroantimon (V)-säure. *Zeit Krist* 183: 1–13, 1988.

Wilcox, G. L., F. A. Quiocho, C. Leventhal, S. C. Harvey, G. M. Maggiona, and J. A. McCammon. Symposium overview. Minnesota conference on supercomputing in biology: proteins, nucleic acids and water. *J Computer-aided Molecular Design* I: 271–81, 1987.

Williams, J. M. and S.W. Peterson. A novel example of the $[H_2O_2]^+$ ion. A neutron diffraction study of $HAuC\ell_2 \cdot 4H_2O$. *J Am Chem Soc* 91: 776–77, 1969.

Williams, J. M., S. W. Peterson, and H. A. Levy. The aquated hydrogen ion $H_7O_3^+$. A high precision neutron diffraction study of sulfosalicylic acid trihydrate (SSATH). *Abstrs L.17 Am Crystallogr Assoc Meeting*, Winter, 1972.

Wolf, B. and S. Hanlon. Structural transitions of deoxyribonucleic acid in aqueous electrolytic solutions, II. The role of hydration. *Biochemistry* 14: 1661–74, 1975.

Yang, X. and A. N. Castleman. Large protonated water clusters $H^+(H_2O)_n$ ($1 \geq n > 60$). The production and reactivity of clathrate-like structures under thermal conditions. *J Am Chem Soc* 111: 6845–46, 1989.

Yu, L. J. Hydrogen bond-induced ferroelectric crystals. *Liquid Cryst* 14: 1303–9, 1993.

Zhang, B. L., C. Z. Wang, K. M. Ho, C. H. Xu, and C. T. Chan. The geometry of small fullerene cages, C_{20} to C_{70}. *J Chem Phys* 97: 5007–11, 1992.

Zheng, J. and K. M. Merz, Jr. Study of hydrogen bonding interactions relevant to biomolecular structure and function. *J Comput Chem* 13: 1151–69, 1992.

Zundel, G. "Easily Polarizable Hydrogen Bonds—Their Interactions with the Environment—IR Continuum and Anomalous Proton Conductivity." In: *The Hydrogen Bond—Recent Developments in Theory and Experiment*, vol. II. Edited by P. Schuster, G. Zundel, and C. Sandorfy. Netherlands: North Holland Publishing Co., 1976.

Zundel, G. Proton polarizability and proton transfer processes in hydrogen bonds and cation polarizabilities of other cation bonds—their importance to understanding molecular processes in electrochemistry and biology. *Trends in Phys Chem* 3: 129–56, 1992.

Zundel, G. Hydrogen-bonded chains with large proton polarizability as charge conductors in proteins. Bacteriorhodopsin and the Fo sub-unit of *E. coli*. *J Molec Struct* 322: 33–42, 1994.

Zundel, G. and B. Brzezinski. "Proton Polarizability of Hydrogen Bond Systems Due to Collective Proton Motion—With a Remark on the Proton Pathways in Bacteriorhodopsin." In: *Proton Transfer in Condensed Matter*. Edited by T. Bountis. New York: Plenum Press, 1992.

Note: The distribution of journal references is as follows. *J Am Chem Soc*, 15%; *J Chem Phys*, 11%; *Acta Cryst*, 10%; *J Mol Struct*, 6%; *Nature* and *Science*, 4%; and all others (each); <2%.

Index

Hydrogen bonds range from the very strong, comparable with covalent bonds, to the very weak, comparable with van der Waals forces. Most hydrogen bonds are weak attractions with a binding strength about one-tenth of that of a normal covalent bond. Nevertheless, they are very important. Without them, all wooden structures would collapse, cement would crumble, oceans would vaporize, and all living things would disintegrate into inanimate matter.

An easy-to-read supplement to the often brief descriptions of hydrogen bonding found in most undergraduate chemistry and molecular biology textbooks, *An Introduction to Hydrogen Bonding* describes and discusses the current ideas concerning hydrogen bonding, ranging from the very strong to the very weak, with introductions to the experimental and theoretical methods involved. Ideal for courses in chemistry and biochemistry, it will also be useful for structural biology and crystallography courses. For students and researchers interested in supramolecular chemistry, biological structure and recognition, and other sophisticated concepts and methodologies, it provides a careful selection of key references from the vast hydrogen bonding literature.

ABOUT THE AUTHOR

George A. Jeffrey is Professor Emeritus at the University of Pittsburgh. He has received the Hudson Award of the American Chemical Society, the Buerger Award of the American Crystallographic Association, an Alexander von Humboldt Senior U.S. Scientist Award from Germany, a Doctor Honoris Causa from the Universidade Technica de Lisboa, Portugal, and in 1998 will receive a Haworth Award from the Royal Society of Chemistry, Britain. He is co-author of *Hydrogen Bonding in Biological Structures* (1991).

Cover design: *Snow on a Distant Planet*
by Ed Atkeson/Berg Design

ISBN 0-19-509549-9